69.00
2I

D1233872

Algorithms and Combinatorics 9

Bernhard Korte   László Lovász
Hans Jürgen Prömel   Alexander Schrijver
Editors

Paths, Flows, and VLSI-Layout

With 164 Figures

Springer-Verlag

Berlin Heidelberg New York
London Paris Tokyo
Hong Kong Barcelona

Bernhard Korte
Hans Jürgen Prömel
Institute of Operations Research/
Research Institute for Discrete Mathematics
University of Bonn
Nassestraße 2
D-5300 Bonn
Fed. Rep. of Germany

László Lovász
Department of Computer Science
Eötvös Loránd University
Múzeum krt. 6–8
H-1088 Budapest
Hungary

Alexander Schrijver
Mathematical Centre
Kruislaan 413
NL-1099 SJ Amsterdam
The Netherlands

Mathematics Subject Classification (1980):
05 C XX, 68 E XX, 90 C XX

ISBN 3-540-52685-4 Springer-Verlag Berlin Heidelberg New York
ISBN 0-387-52685-4 Springer-Verlag New York Berlin Heidelberg

Library of Congress Cataloging-in-Publication Data
Paths, flows, and VLSI-layout / Bernhard Korte . . . [et al.]. p. cm. –
(Algorithms and combinatorics; 9)
Includes bibliographical references and indexes.
ISBN 0-387-52685-4
1. Integrated circuits – Very large scale integration – Design and construction – Data
processing. 2. Computer-aided design. I. Korte, B. H. (Bernhard H.), 1938–.
II. Series. TK7874.P397 1990 621.39'5 – dc 20 90-10307 CIP

© Springer-Verlag Berlin Heidelberg 1990
Printed in the United States of America

2141/3140-543210 – Printed on acid-free paper

Preface

VLSI (very large scale integration) is nowadays a magic phrase in science and technology. Amongst other striking aspects this field has turned out to be an example par excellence for applications of discrete mathematics and combinatorial optimization. We therefore judged it opportune and useful to organize a scientific meeting focusing on paths, flows and VLSI-layout in order to contribute to the further propagation of the results obtained, and to stimulate cooperation and exchange of ideas and problems.

The meeting was held from June 20 to July 1, 1988, at the University of Bonn in the shape of a *Summer School and Workshop*. The announcement of this summer school found a strong echo in the scientific community. From a large number of applicants we were able to select 33 particularly qualified junior scientists and doctoral students from 13 different countries. In addition, the postgraduate students of the Research Institute of Discrete Mathematics at Bonn took part in the summer school.

We were very happy that most of the intended speakers accepted our invitation, including several of the most prominent researchers in the field. Each was asked to present three one-hour lectures, giving an in-depth survey of their and related research. Stimulated by an attentive and appreciative audience and by a pleasant, informal atmosphere, the lectures turned out to be very instructive and inspiring, and altogether they constitute a rich body of fascinating information on a many-facetted field.

We have asked the speakers to formalize and extend their lectures into written surveys, giving a more complete account of the topics presented. This book presents these surveys.

The Summer School and Workshop was generously sponsored by the Volkswagen Foundation. It also benefited by the support of several programs of the Alexander von Humboldt Foundation, the Hungarian Academy of Sciences (Magyar Tudományos Akadémia), the German Research Association (Deutsche Forschungsgemeinschaft) and the Donor's Association of German Sciences (Stifterverband für die Deutsche Wissenschaft). Only through this invaluable support was this meeting possible; for this the editors wish to express their sincere thanks and appreciation.

We are very grateful to Beatrix Schaschek and Gerd Bieding for typing and to the postgraduate students of the Institute, in particular to Frank Pfeiffer for proof reading and editorial help, and to Springer-Verlag for competent cooperation.

Oberwolfach, January 1990

Bernhard Korte
László Lovász
Hans Jürgen Prömel
Alexander Schrijver

Table of Contents

Network Flow Algorithms 101
Andrew V. Goldberg, Éva Tardos and Robert E. Tarjan

Routing Problems in Grid Graphs 165
Michael Kaufmann and Kurt Mehlhorn

Steiner Trees in VLSI-Layout 185
Bernhard Korte, Hans Jürgen Prömel and Angelika Steger

Cycles Through Prescribed Elements in a Graph 215
Michael V. Lomonosov

Contributors

Francisco Barahona
University of Waterloo
Department of Combinatorics
and Optimization
Waterloo, Ontario
Canada N2L 3G1

Fan R.K. Chung
Bell Communications Research
445 South Street
Morristown, NJ 07960, USA

Paul Erdős
Hungarian Academy of Sciences
Mathematical Institute
Realtonada u. 13–15
H-1053 Budapest
Hungary

András Frank
Institute for Operations Research
Research Institute for
Discrete Mathematics
Nassestraße 2
D-5300 Bonn, Fed. Rep. of Germany
and
Department of Computer Science
Eőtvős Loránd University
Múzeum krt. 6–8
H-1088 Budapest
Hungary

Andrew V. Goldberg
Stanford University
Computer Science Dept.
Stanford, CA 94305, USA

Michael Kaufmann
Universität des Saarlandes
FB 10 Informatik
D-6600 Saarbrücken
Fed. Rep. of Germany

Bernhard Korte
Institute for Operations Research
Research Institute for
Discrete Mathematics
Nassestraße 2
D-5300 Bonn, Fed. Rep. of Germany

Michael V. Lomonosov
Department of Mathematics
and Computer Science
Ben Gurion University of the Negev
P.O. Box 653
Beer Sheva 84105, Israel

László Lovász
Department of Computer Science
Eőtvős Loránd University
Múzeum krt. 6–8
H-1088 Budapest, Hungary

Kurt Mehlhorn
Universität des Saarlandes
FB 10 Informatik
D-6600 Saarbrücken
Fed. Rep. of Germany

Hans Jürgen Prömel
Institute for Operations Research
Research Institute for
Discrete Mathematics
Nassestraße 2
D-5300 Bonn, Fed. Rep. of Germany

Neil Robertson
Department of Mathematics
Ohio State University
Columbus, OH 43210, USA

Alexander Schrijver
Mathematical Centre
Kruislaan 413
NL-1099 SJ Amsterdam
The Netherlands

Paul D. Seymour
Bellcore
445 South Street
Morristown, NJ 07960, USA

Angelika Steger
Institute for Operations Research
Research Institute for
Discrete Mathematics
Nassestraße 2
D-5300 Bonn, Fed. Rep. of Germany

Éva Tardos
School of Operations Research
Upson Hall
Cornell University
Ithaca, NY 14850, USA

Robert E. Tarjan
Department of Computer Science
Princeton University
Princeton, NJ 08544, USA
and
Bell Laboratories
600 Mountain Avenue
Murray Hill, NJ 07974, USA

Richard Vitray
Department of Mathematics
Appalachian State University
Boone, NC 28608, USA

Introduction

Finding disjoint paths in a graph belongs to the classical problems in graph theory. A fundamental characterization was found in 1927 by K. Menger: the maximum number of pairwise disjoint paths between a given 'source' and a given 'sink' in a graph is equal to the minimum size of a 'cut' separating source and sink. Menger's interest was motivated by a question in topology, namely characterizing the furcation number of points in certain topological spaces called *Kurven*. (D. Kőnig provided a proof of an essential missing link in Menger's inductive proof, viz. the nontrivial base of the induction.)

Menger's result forms the basis for research on paths in graphs, continuing until the present day. The origins of flow research can be traced back to the spring of 1955, when T.E. Harris mentioned the following problem to L.R. Ford, Jr and D.R. Fulkerson:

"Consider a rail network connecting two cities by way of a number of intermediate cities, where each link of the network has a number assigned to it representing its capacity. Assuming a steady state condition, find a maximal flow from one given city to the other."

Ford and Fulkerson observed that this problem is a special case of a linear programming problem, of a character similar to the transportation and transshipment problems studied in the 1940s by F.L. Hitchcock and Tj. C. Koopmans (which formed an important motivation for studying linear programming). Hence the problem can be solved with G.B. Dantzig's simplex method.

In 1956, Ford and Fulkerson published a direct 'labeling' method finding the maximum flow. The method implies the famous *max-flow min-cut theorem*: the maximum amount of flow is equal to the minimum capacity of any cut separating the two cities. If all capacities are integer, the maximum flow found is integer. Later it was observed that Menger's theorem and Ford and Fulkerson's theorem can be derived from each other by a simple construction, providing a link between paths and flows.

The results by Menger and Ford and Fulkerson prepared the path for an immense flow of further research. It was also Ford and Fulkerson who considered the problem of finding *multicommodity flows*, where distinct commodities (= pairs of source and sink) have to be connected by flows in such a way that the total amount of flow through any edge is bounded by a given capacity. This problem arises e.g. in transportation and communication networks, where several goods or messages have to be conveyed between distinct commodities simultaneously.

The problem can be formulated again as a linear programming problem, and hence it can be solved with the simplex method. Ford and Fulkerson gave an alternative LP formulation of the multicommodity flow problem, allowing the application of a column generation technique.

In 1963, T.C. Hu extended Ford and Fulkerson's labeling algorithm for 1-commodity flow problems to 2-commodity flow problems. The algorithm gives a half-integer flow if all capacities are integer. Moreover, Hu described a *max-biflow min-cut theorem*, extending the max-flow min-cut theorem.

With respect to extending his method to more than two commodities, Hu remarked in his 1963 paper:

"Although the algorithm for constructing maximum bi-flow is very simple, it is unlikely that similar techniques can be developed for constructing multicommodity flows. The linear programming approach used by Ford and Fulkerson to construct maximum multicommodity flows in a network is the only tool now available."

and this last statement still applies today.

No efficient algorithm was found for finding an optimum *integral* two-commodity flow (although B.L. Rothschild and A. Whinston showed in 1966 that if the capacities satisfy a certain parity condition, the optimum flow is integer).

New light on path and flow research was thrown in the 1970s by the breakthrough of 'complexity theory', due to the pioneering new insights of J. Edmonds, S.A. Cook and R.M. Karp. The introduction of polynomial time as a complexity criterion and of the complexity classes P, NP and co-NP led to a re-evaluation of the problems and results encountered so far. Thus Menger's theorem and the max-flow min-cut theorem were identified as *good characterizations* (i.e., as showing membership of $NP \cap$ co-NP). E.A. Dinits, and independently Edmonds and Karp, showed in 1970 that a variant of Ford and Fulkerson's labeling algorithm terminates in polynomial time.

Edmonds and Karp also gave a polynomial-time algorithm for the *min-cost flow problem*. The algorithm is based on scaling the capacities with the result that the number of iterations depends on the number of bits required to describe the capacities. As an open problem, Edmonds and Karp stated:

"A challenging open problem is giving a method for the minimum-cost flow problem having a bound on computation which is a polynomial in the number of nodes, and is independent of both costs and capacities."

On the negative side, in 1974 D.E. Knuth showed that the integer multicommodity flow problem is NP-complete, even if we take all capacities equal to one. In addition, J.F. Lynch proved in 1975 that the problem: "given a planar graph and pairs $r_1, s_1, \ldots, r_k, s_k$ of vertices, find k pairwise vertex-disjoint paths connecting r_i and s_i for $i = 1, \ldots, k$ respectively" is NP-complete. Moreover, S. Even, A. Itai and A. Shamir showed in 1976 that the integer two-commodity flow problem is NP-complete, and S. Fortune, J.E. Hopcroft and J. Wyllie showed in 1980 that finding two vertex-disjoint paths connecting two given pairs of vertices in a directed graph is NP-complete.

In 1979, L.G. Khachiyan proved with the ellipsoid method that linear programming is solvable in polynomial time, thereby showing that in particular the (fractional) multicommodity flow problem is polynomially solvable. The complexity status of the undirected integer k-commodity flow problem with all capacities equal to one and k fixed, remained unsettled in the 1970s.

Our brief historical sketch now arrives at the 1980s, which have shown further achievements on paths and flows.

As a first breakthrough we mention the solution of Edmonds and Karp's problem cited above, by É. Tardos in 1985: the min-cost flow problem is solvable in *strongly polynomial time,* i.e., the number of iterations is bounded by a polynomial in the number of vertices. The algorithm is based on a scaling technique sharpening the one given by Edmonds and Karp.

In a subsequent paper, Tardos showed that linear programs with a $\{0, \pm 1\}$-constraint matrix are solvable in strongly polynomial time, implying that min-cost multicommodity flow problems are solvable in strongly polynomial time.

Another breakthrough is the new algorithm of N. Robertson and P.D. Seymour (1986) for finding k vertex-disjoint paths in an undirected graph between k specified pairs of terminals. For any fixed k, the running time of the algorithm is polynomially bounded. In fact, the running time has order $O(|V|^2|E|)$, with a constant heavily depending on k. Even for $k = 3$, the constant is very high, and the challenge remains to find a practical algorithm for the three disjoint paths problem.

The algorithm of Robertson and Seymour is in fact one of the spin-offs of their impressive *Graph Minors* project, encompassing over 20 difficult papers and implying several new and deep results in graph theory.

The 1980s have also exhibited the increasing interest of disjoint path and tree methods for the design of *very large-scale integrated (VLSI) circuits* (chips). A basic technical problem here is to interconnect certain sets of 'pins' by 'wires'. Constraints are that the wires should follow certain prescribed 'channels' on the chip, and that wires belonging to different sets of pins should not touch each other. In its simplest form, the problem mathematically amounts to finding disjoint trees in a graph, each spanning a given set of vertices. If each such set consists of just two vertices, we have a disjoint paths problem.

In practice, however, technology adds several constraints and objectives to the problem, like minimizing the total length of wires, or minimizing the area covered by the wires, or minimizing the number of transitions of the wires from one 'layer' of the chip to another *(via minimization).* Moreover, the *VLSI-routing problem* can be integrated with the *VLSI-placement problem,* deciding the positions of the pins (or of modules containing the pins) on the chip. A typical mathematical problem related to this is embedding a given graph in the rectangular grid, in such a way that vertices are lattice points, and edges follow edges of the grid. Objectives are minimizing the total area covered, or minimizing the total number of crossing edges.

The general VLSI-layout problem is NP-complete and in practice very difficult, in particular due to the generally huge and complicated input and the often inadequate modelling of technological constraints and objectives. To solve the problem one must be content with man-machine interactive procedures, involving various heuristic stages and generally not leading to an optimal solution. A challenge is to mechanize as many subroutines in the procedure as possible by efficient computer algorithms, and this has inspired a stream of research also interesting from the mathematical point of view.

Fundamental questions concerning area and time required by a chip have given further impetus to the field called *communication complexity*. Moreover, an important tool in designing large layouts (including placement) are the *graph separator theorems*, which decompose large graphs into smaller, about equal-sized graphs. These two fields form an illustration of the fact that besides graph theory also extremal combinatorics is of increasing importance as a tool in computer science, especially in complexity theory. The very large numbers involved in VLSI seem to tend to infinity and provoke asymptotic analysis of VLSI-problems. This has led to a fruitful interaction between VLSI-design and extremal combinatorics.

The above sketches a lively field of research, having a variety of applications and bringing together several disciplines: discrete mathematics (graph theory, extremal combinatorics), operations research (combinatorial optimization, linear programming), computer science, complexity theory, VLSI-design.

We hope that this volume with 12 contributions covering a wide range of the above mentioned area will serve as a compendium demonstrating the liveliness of this area, as well as the state of the art as directions for future work in this field of research.

For the convenience of the reader we will give in the following a guided tour through the chapters of the book by sketching briefly their contents.

The recent progress on (one-commodity) *flow* techniques is discussed in the comprehensive survey "**Network Flow Algorithms**" by A.V. Goldberg, É. Tardos and R.E. Tarjan. Besides the maximum flow problem, it covers minimum cost flows and 'generalized flows'. In recent years, for each of these problems algorithms of improved efficiency have been designed. The paper gives a full account of these methods, including all "fastest currently known algorithms for network flow problems", highlighting the new ideas and linking them to results from the past.

For the maximum flow problem, among the methods described are the new push/relabel method of Goldberg and Tarjan, and a variant based on excess scaling due to Ahuja, Orlin and Tarjan.

For the minimum-cost flow problem, the research reported is to a large extent motivated by Tardos' strongly polynomial algorithm found in 1985, which is based on a new type of cost-scaling. Besides this algorithm, and the classical Edmonds-Karp algorithm, the paper discusses recent algorithms by Ahuja, Bertsekas, Fujishige, Gabow, Galil, Goldberg, Orlin, Tarjan, and others, using (generalized) cost-scaling or capacity-scaling, or a mixture of these.

The paper concludes with two polynomial-time algorithms for the generalized flow problem, a 'combinatorial' algorithm due to Goldberg, Plotkin and Tardos (1988) and the currently fastest method, based on linear programming, due to Vaidya (1989).

Various aspects of multicommodity flows and disjoint paths and trees are considered in the surveys by Frank, Robertson and Seymour, Korte, Prömel and Steger, Schrijver, and Kaufmann and Mehlhorn.

The article "**Packing Paths, Circuits, and Cuts – A Survey**" by A. Frank forms a general overview of the field, giving several theorems, ideas of proofs and algorithms, and references. Although the general disjoint paths problem is NP-complete, there are several useful special cases that are tractable and lead to interesting theory. An example of such a case occurs when the underlying graph is planar and all terminals are on the boundary of the infinite face. Similarly, adding 'parity' conditions appears to be helpful. Frank reviews research in this direction by Okamura, Robertson, Seymour, Schrijver, Lomonosov, Hu, Rothschild, Whinston, Papernov, Frank, and Karzanov.

In particular, Frank considers *rectangular grid graphs*, i.e., graphs that are subgraphs of the rectangular grid in the euclidean plane. These graphs are of particular interest in the context of VLSI-design, since the permitted routing channels on a chip often form a rectangular grid. Results and methods due to Frank and to Preparata and Lipski are discussed.

The paper also reviews results on the complexity of the disjoint paths problem when we fix the number of pairs of terminals to be connected, on the maximization problem (with proofs of theorems by Rothschild and Whinston, Lomonosov, and Lovász), and on the related problems of packing cuts and circuits (with proofs of theorems by Seymour and Frank, Sebő and Tardos).

The paper "**An Outline of a Disjoint Paths Algorithm**" by N. Robertson and P. Seymour focuses on their disjoint paths algorithm. For any fixed number k of pairs of terminals it terminates in polynomial time. This deep result was established in their paper "Graph Minors XIII: The disjoint paths problem", part of the great Graph Minors project, yielding a proof of Wagner's conjecture: for every infinite collection of graphs, one is a minor of another. Proving this formed the original goal of the project.

Robertson and Seymour's survey sketches the algorithm and the (much more complicated) proof of the algorithm. The central idea of the algorithm is a result which says roughly that either the input graph has (more or less) bounded genus, or it has an 'irrelevant' vertex whose deletion does not change the problem.

A basic concept in the algorithm (as well as in the whole Graph Minors project) is the *tree-width* of a graph. It turns out that irrelevant vertices can be deleted from the graph until the graph has low tree-width, in which case the graph can be cut up by low-order separations. The reduction process terminates with graphs being embeddable on a surface of bounded genus.

The algorithm extends to the disjoint trees problem: given disjoint subsets $X_1, \ldots, X_k$ of the vertex set of a graph, find pairwise vertex-disjoint trees $T_1, \ldots, T_k$ where T_i covers $X_i (i = 1, \ldots, k)$. Such a tree T_i is called a *Steiner tree* for X_i. For fixed $\sum |X_i|$ the algorithm is polynomial-time, in fact again $O(|V|^2|E|)$.

Combined with Robertson and Seymour's proof of Wagner's conjecture this implies that each graph property that is closed under taking minors, can be tested in time $O(|V|^3)$.

Although finding disjoint trees is important for VLSI-design, it must be admitted that Robertson and Seymour's algorithm is as yet mainly of theoretical importance because of the very high coefficient of the running time polynomial.

The disjoint trees problem is one of the problems also discussed in the survey **"Steiner Trees in VLSI-Layout"** by B. Korte, H.J. Prömel and A. Steger. The first part of the paper reviews methods for finding a minimum-length Steiner tree (for a given set of vertices) and for packing Steiner trees. Both problems are NP-complete, even for rectangular grid graphs.

For the minimum-length Steiner tree problems the methods discussed are exact (exponential-time) algorithms due to Dreyfuss and Wagner (1972) based on dynamic programming and to Lawler (1976) based on implicit enumeration, and approximative algorithms due to Kou, Markowsky and Berman (1987) and to Wu, Widmayer and Wong (1986).

The second part of the paper describes a hierarchical approach for solving the routing problem in VLSI-layout, illustrated by application to the design of the central processing unit chip and the memory managing unit chip of the new IBM ES/370 chip set. They lead to graphs with 18,679,520 vertices each, with 10,196 and 5,204 nets (= sets of pins to be connected), respectively.

Fundamental in the approach is the strategy of searching for short Steiner trees connecting the nets. Intuitively speaking, such a strategy may not only lead successfully to a packing of wires as required, but would also cut down the length of the wires. The latter reduces both the fault-susceptibility and the cycle-time of the chip. The paper describes how this strategy can be implemented by growing short Steiner trees with the help of modifications of the above-mentioned Steiner tree algorithms.

Other VLSI-design systems adopt a routing strategy using the topology of the chip. This strategy was initiated by Pinter (1983). Suppose we are at a stage that certain connections have to be made on one layer of the chip (the assignment of connections to layers and vias has been performed in a previous stage). Usually pins are grouped together on 'modules', which take up some area on the layer. Now the strategy is first to select, for each connection to be made, a *global routing* with respect to the modules. That is, we select the *homotopy* of each connection in the topological space obtained from the layer by deleting the areas covered by the modules. After having made these selections, we try to find a *local routing*, that is, disjoint connections, following the channels, of the required homotopy. If we fail, we modify the global routing and try again.

For determining a global routing, one usually applies heuristics. It turns out that in several cases there exist efficient computer algorithms for finding a local routing given a global routing.

Mathematically, the local routing problem amounts to the following. Let a planar graph G be given, together with some paths $P_1, \ldots, P_k$ in G and some faces $F_1, \ldots, F_p$ of G, in such a way that each terminal of each P_i is incident to at least one F_j. Find pairwise disjoint paths $\tilde{P}_1, \ldots, \tilde{P}_k$ in G where $\tilde{P}_i$ is homotopic to P_i in the surface obtained by deleting $F_1, \ldots, F_p$ from the plane ($i = 1, \ldots, k$). That is, $\tilde{P}_i$ arises by shifting P_i over the plane but not over any of the 'holes' $F_1, \ldots, F_p$

(the end points of P_i being fixed). The problem has a direct extension if we wish to pack trees instead of just paths.

The vertex-disjoint mode of this problem is polynomially solvable, whereas the edge-disjoint mode is NP-complete, although there are some interesting special cases that are polynomially solvable.

The papers by Schrijver and by Kaufmann and Mehlhorn survey this homotopic approach.

After an introduction (without proofs) to disjoint path and tree problems in general, the paper "**Homotopic Routing Methods**" by A. Schrijver concentrates on theorems and algorithms for the homotopic routing problem and for related packing and cut problems for graphs embedded on surfaces. It gives a sketch of a polynomial-time algorithm for the above-mentioned homotopic routing problem in the vertex-disjoint mode. (Earlier algorithms are due to Leiserson and Maley for grid graphs and to Robertson and Seymour for $p \leq 2$.) The algorithm is based on solving a certain system of linear inequalities in integer variables, and shifting the paths P_i accordingly.

The same method is shown to imply a conjecture raised by the Graph Minors project concerning disjoint circuits of prescribed homotopies in a graph embedded on a compact surface. An extension of the algorithm finds disjoint *trees* of prescribed homotopies. It implies that the (nonhomotopic) disjoint trees problem in planar graphs is polynomially solvable, provided that all terminals to be connected can be covered by a fixed number of faces.

The paper also discusses fractional solutions to the edge-disjoint version of the homotopic routing problem, and it describes a link between Okamura's theorem on disjoint paths in planar graphs and a circuit packing theorem for graphs embedded on the Klein bottle.

The article "**Routing Problems in Grid Graphs**" by M. Kaufmann and K. Mehlhorn focuses on the edge-disjoint version of the homotopic routing problem. This in general being NP-complete, they survey algorithms for VLSI-design.

Clearly, if all terminals lie on the boundary of the infinite face of a planar graph and there are no holes, the homotopic problem reduces to the nonhomotopic problem. This relates to a theorem of Okamura and Seymour, which states that a cut condition is sufficient for a solution, assuming a parity condition. The paper starts discussing a polynomial-time algorithm for this case. Next it is shown how this algorithm can be extended to solve the edge-disjoint homotopic routing problem for grid graphs (again assuming a parity condition). It turns out that the cuts of interest can be restricted to cuts with at most one 'bend' (they consist of at most one horizontal and one vertical piece).

The paper moreover reviews several other algorithms for routing in grid graphs, including 'generalized switchboxes' and rectangles with one rectangular hole. To a large extent these algorithms are derived from those for the homotopic case.

The surveys by Lovász and by Chung relate to the problem of minimizing the layout area of a chip. This is important in practice, as a small area increases the reliability and reduces the cycle-time. Moreover, small chips are cheaper and easier to test.

The paper "**Communication Complexity: A Survey**" by L. Lovász concentrates on the fundamental question: what does communication cost? For instance: how much time and space does it require? In many devices, communication is significantly slower and costlier than local computation, and it is the real bottleneck in solving certain problems. VLSI-technology is one of the fields where problems of this type arise. Central is the problem of finding out how two strings of information (generally far apart) relate to each other (e.g., are they equal? do they correspond to disjoint sets?).

One of the results proved is a *time-area trade-off theorem* of Thompson (1979), a lower bound on the product of time and area required for testing the equality of two strings. Other bounds discussed are lower bounds due to Mehlhorn and Schmidt (1982) and to Hajnal, Masao and Turán (1988) (using Szemerédi's deep 'regularity lemma'), and upper bounds due to Aho, Ullman and Yannakakis (1983), Lovász and Saks (1988) and Hajnal.

A basic concept in communication complexity, analogous to 'algorithm' in ordinary complexity theory, is that of *protocol*. There are deterministic, non-deterministic, and randomized protocols. The paper displays a communication analogue to NP-theory (it turns out that $P_{comm} = NP_{comm} \cap co-NP_{comm} \neq NP_{comm}$), and describes links with classical complexity theory (including lower bounds on depths of Boolean circuits) and with polyhedral combinatorics.

The paper "**Separator Theorems and Their Applications**" by F.R.K. Chung considers the problem of splitting a large graph into two about equal-sized parts, by deleting a relatively small number of vertices. Thus a subset S of vertices in a graph is called a *separator* if by removing S, the remaining graph can be partitioned into two parts, A and B say, so that no edge connects A and B and so that $|A| \leq |B| \leq 2|A|$. For instance, Lipton and Tarjan proved in 1977 the important result that each planar graph with N nodes has a separator of size $2\sqrt{2}\sqrt{N}$.

Such results can be applied to the VLSI-design problem: a given graph has to be embedded into a rectangular grid (so that vertices of the graph are vertices of the grid, and edges of the graph form paths along edges of the grid). If the graph to be embedded on the grid is large but has small separators (hereditarily), the problem can be attacked with a divide-and-conquer strategy. Moreover, if the routing region itself is the large graph, we can again apply separation techniques to obtain a divide-and-conquer routing algorithm.

Leiserson showed in 1980 how small layout areas can be obtained for classes of graphs for which good separator theorems are available. Thus the Lipton-Tarjan theorem yields a chip of size $O(N \cdot \log^2 N)$ for any planar graph with N nodes.

The paper first reviews several separator theorems for trees (a tree always has a separator of size 1), and colored trees. Next the planar separator theorem of Lipton and Tarjan is discussed, together with sharpenings due to Chung and to Miller, and extensions to graphs of higher genus by Djidjev (1981) and Gilbert, Hutchinson and Tarjan (1984).

Finally, the paper describes how separator theorems can be used to construct small 'universal graphs' – graphs which contain all graphs in a prescribed class as subgraphs. Such graphs are also important in VLSI-design as they can model chips that should be able to perform a prescribed class of tasks.

The survey "**On Some Applications of the Chinese Postman Problem**" by F. Barahona is method-oriented. Starting point is the method of Edmonds and Johnson (1973) for handling the *Chinese postman problem*: find a shortest cycle in an undirected graph covering each edge at least once. Barahona modified the algorithm to an $O(n^{3/2} \log n)$ algorithm for planar graphs, based on Lipton and Tarjan's planar separator theorem.

The paper gives a number of applications to various problems. The first application is in physics, viz. to finding minimum energy configurations (*ground states*) in Ising spin glasses. Secondly, an application to via minimization in VLSI-design is described. This is the problem of minimizing the number of layer switchings in a routing on a multi-layered chip. The paper describes a new derivation of an $O(n^{3/2} \log n)$ time algorithm for via minimization in a two-layer chip.

Thirdly, it is shown how planar multicommodity flow problems can be handled with the help of the Chinese postman algorithm. As a last application the author shows how the problem of finding a maximum cut can be solved in polynomial time if the input graph is not contractible to the graph K_5. (In general, this problem is NP-complete.)

The research on paths, flows and VLSI-layout exhibits a strong and fruitful interplay between operations research and computer science on one side, and the purer parts of discrete mathematics like graph theory and extremal combinatorics, on the other. The last three surveys to be mentioned review various aspects of pure discrete mathematics. These interesting surveys also illuminate methods and problems typical in graph theory and extremal combinatorics, and they possibly will challenge the reader to improve his skill in handling these fields, either as a tool or as a topic of interest in itself.

The paper "**Cycles Through Prescribed Elements in a Graph**" by M.V. Lomonosov discusses some approaches to the problem of deciding whether a graph contains a cycle through a specified set of vertices and edges. The author in particular reviews results obtained jointly by him and Kelmans.

The paper "**Representativity of Surface Embeddings**" by N. Robertson and R. Vitray is an expository paper on the *representativity* of a graph embedded on a compact surface. By definition, it is the minimum number of vertices of the graph through which.a nontrivial closed curve on the surface may pass. It forms a measure of how closely the embedded graph mimics the surface. The concept was developed by Robertson and Seymour in connection with the Graph Minors project.

The paper develops an elementary theory of representativity, outlines recent developments by Archdeacon, Roundby, Thomassen and the authors themselves, and discusses connections to Graph Minors.

Finally, in the paper "**Some of My Old and New Combinatorial Problems**", P. Erdős surveys the state of affairs regarding his favourite problems. They cover a wide variety of important and challenging open problems in graph theory and extremal combinatorics. The old problems mentioned are due to Erdős-Ko-Rado,

Erdős-Rado, Erdős-Faber-Lovász, Erdős-Lovász, and Erdős alone. The more recent problems cover Ramsey theory, extremal graph theory, infinite graphs, random structures, and hypergraphs. It is again fascinating to see how these problems all have stimulated new and original developments in combinatorics, and sometimes have opened new interesting fields of research.

On Some Applications
of the Chinese Postman Problem

Francisco Barahona

1. Introduction

The Chinese Postman Problem was introduced by Edmonds and Johnson (1973). Given a graph $G = (V, E)$, $T \subseteq V$, with $|T|$ even, and a set of integer weights $w(e) \geq 0$, for $e \in E$ the problem can be formulated as

$$\text{minimize } \sum w(e)\, x(e)$$

subject to

(1.1)
$$\sum_{e \in \delta(v)} x(e) \equiv \begin{cases} 1\,(\text{mod } 2), & \text{if } v \in T, \\ 0\,(\text{mod } 2), & \text{if } v \in V \setminus T. \end{cases}$$

$$x(e) \in \{0, 1\}, \text{ for } e \in E.$$

We use $\delta(S)$ to denote the set of edges with exactly one endnode in S, for $S \subseteq V$. We denote the number of nodes in G by n, and the number of edges by m.

One way to solve problem (1.1) is to reduce it to a perfect matching problem in a complete graph K_p, where $p = |T|$. One could compute shortest paths between all pairs of nodes in T, use these values as edge-weights in K_p and find a minimum weighted perfect matching. The drawback of this approach is that one might lose the sparsity of the original graph. This is important, for instance in planar graphs, if a computationally efficient algorithm is desired. The second approach that Edmonds and Johnson proposed is a direct algorithm reminiscent of Edmonds' algorithm for matching. For complete graphs, the complexity of this procedure is $O(n^3)$ provided that one use the data structures of Lawler (1976). For sparse graphs, one can use the data structures of Ball and Derigs (1983) and Galil et al (1986) to obtain an $O(nm \log n)$ procedure.

Planar graphs have small separators as shown by Lipton and Tarjan (1979). This suggests that one should decompose the graph, solve the problem in the pieces, and use a primal version of the algorithm to put the pieces together. This leads to an $O(n^{3/2} \log n)$ procedure, see Barahona (1987).

Using this algorithm, Edmonds and Johnson proved that (1.1) is equivalent to the linear program

$$\text{minimize } \sum w(e)\, x(e)$$

subject to

(1.2)
$$\sum_{e\in\delta(S)} x(e) \geq 1,$$

for every set $S \subseteq V$, with $|S \cap T|$ odd,

$$x \geq 0.$$

We have made the assumption that the edge-weights are integer, however, problem (1.2) does not necessarily have an integer optimal dual solution. Seymour (1981a) proved that if the sum of the weights of the edges of every circuit is even then the dual variables can be chosen to be integer. In Barahona (1987) we showed that the algorithm of Edmonds and Johnson can be slightly modified to produce this dual integer solution.

Given a solution $\bar{x}$ of (1.1), the set $\{e \in E \mid \bar{x}(e) = 1\}$ is called a *T-join*. Given $S \subseteq V$, with $|S \cap T|$ odd, the set $\delta(S)$ is called a *T-cut*. Seymour's Theorem says that if G is bipartite then the cardinality of a minimum T-join is equal to the maximum number of edge-disjoint T-cuts.

If there is a node u that covers all the cycles of odd weight then the dual solution can also be chosen to be integer, the problem can be reduced to the former case by splitting u and putting in an edge of weight one. We shall use this in Section 2. In fact, Gerards (1988) proved that if the graph does not have an *odd K_4* then the same min-max relation holds.

In this paper we survey some of the applications of the Chinese Postman Problem, namely, finding ground states of Ising spin glasses, via minimization, planar multicommodity flows, and max cut in graphs not contractible to K_5. In the last Section we show how to formulate all these problems as linear programs of polynomial size.

2. Spin Glasses

The problem of spin glasses is of great interest both in solid state physics and in statistical physics. The materials studied are magnetic alloys such as 1% of magnetic impurities embedded in a non-magnetic material. A property of these systems is the cusp in the magnetic susceptibility at a well defined temperature, indicating a phase transition, but the question whether there is another phase transition is not yet solved.

Between two impurities there is an energy interaction

$$-J(r_{12})S_1 \cdot S_2,$$

where S_i is the magnetic moment (spin) of the impurity i. The interaction $J(r)$ varies as $\frac{\cos(Kr)}{r^3}$, where r is the distance between the impurities, and K a physical constant.

The energy of a spin configuration is given by the Hamiltonian

$$H = \sum -J(r_{ij})S_i \cdot S_j.$$

In order to produce a mathematical model, several simplifications have been introduced. The first step is to substitute the random distribution of impurities by a disposition at the vertices of a regular lattice. We study two-dimensional lattices here.

The second step is to associate to each edge (i, j) of the lattice an interaction J_{ij} chosen randomly from $\{-1, 1\}$. Finally the three-dimensional vectors S_i are replaced by one-dimensional vectors whose values can be ± 1. It is believed that despite its strong simplifications, this model retains the relevant features of real spin glasses.

One of the problems of interest in this model is the study of the configurations of minimum energy called *ground states*. As the temperature approaches absolute zero, the system will move to such a configuration.

Finding a ground state is the following discrete quadratic program

$$\text{minimize} \sum -J_{ij} S_i S_j$$

(2.1)   subject to

$$S_i \in \{-1, 1\}, \text{ for all } i.$$

Given a spin configuration, i.e., an assignment of values to the variables $\{S_i\}$, we say that the interaction ij is violated if

$$J_{ij} > 0 \text{ and } S_i \neq S_j,$$

or

$$J_{ij} < 0 \text{ and } S_i = S_j.$$

It is easy to prove that problem (2.1) is equivalent to finding the spin configuration that minimizes

$$\sum \{|J_{ij}| : ij \text{ is violated}\}.$$

A cycle containing an odd number of negative interactions will have at least one violated interaction. For that reason, physicists call them *frustrated* cycles. Actually frustrated cycles have an odd number of violated interactions, whereas non-frustrated cycles have an even number of violated interactions. So the problem is equivalent to

$$\text{minimize} \sum |J_{ij}| \, x_{ij}$$

subject to

(2.2) $$\sum_{ij \in C} x_{ij} \equiv \begin{cases} 1 \, (\text{mod } 2), & \text{if } C \text{ is frustrated}, \\ 0 \, (\text{mod } 2), & \text{if } C \text{ is not frustrated}, \end{cases}$$

for every cycle C,

$$x_{ij} \in \{0, 1\}, \text{ for every interaction } ij.$$

It is enough to impose the constraints of (2.2) for a set of cycles that contains a cycle basis of the graph. Two-dimensional lattices are planar graphs, so if we choose the set of faces, problem (2.2) is a Chinese Postman Problem in the dual graph.

Under planar duality, T-cuts correspond to frustrated cycles, so it follows from the Theorem of Edmonds and Johnson that (2.2) is equivalent to the linear program

$$\text{minimize} \sum |J_{ij}| \, x_{ij}$$

subject to

(2.3)
$$\sum_{ij \in C} x_{ij} \geq 1,$$

for every frustrated cycle C,

$x_{ij} \geq 0$, for every interaction ij.

Let us recall that the interactions J_{ij} take the values ± 1, so in the dual graph the node corresponding to the exterior face covers all the cycles of odd weight. As we mentioned in the introduction, the dual problem of (2.3) has an optimal solution that is integer. So the minimum number of violated interactions is equal to the maximum number of edge-disjoint frustrated cycles. This min-max relation shows that there is a simple way to prove that a set of violated interactions defines a ground state. The energy needed to violate the interactions can be distributed into a set of disjoint frustrated cycles. See Figure 2.1.

Toulouse (1977) observed that one should find a minimum pairing of frustrated faces (i.e. a minimum T-join in the dual graph), but he did not know the existence of a polynomial algorithm. In Bieche (1979) and Bieche et al (1980) the problem was solved as a matching problem in a complete graph whose nodes are the frustrated faces. In Barahona (1980) and Barahona et al (1982b) we used a primal version of the Chinese Postman algorithm. This primal algorithm was needed to identify clusters of affine spins. These are clusters of spins that have the same relative orientation in all the ground states. Although there might exist exponentially many ground states, these clusters can be identified in polynomial time by post optimality analysis.

In order to compute the entropy of the ground states one needs to know their number. This can be also computed in polynomial time, for that we modified the procedure of Kasteleyn (1967), which counts perfect matchings in a planar graph, to take into account the complementary slackness conditions of linear programming. This is a polynomial procedure that counts optimum perfect matchings in a planar graph, see Barahona (1980).

In order to simulate an infinite grid one can use a toroidal grid, i.e., a planar grid whose opposite sides are identified. In this case the problem reduces to a Chinese Postman Problem with two additional parity constraints. In Barahona (1981) we proved that the unweighted version of the problem ($J_{ij} = \pm 1$) can be solved in polynomial time. This gives a polynomial algorithm to find a cut of maximum cardinality in a toroidal graph. We used a procedure to enumerate perfect matchings in a toroidal graph given by Kasteleyn (1961). We do not know a polynomial algorithm to solve the weighted version of the problem.

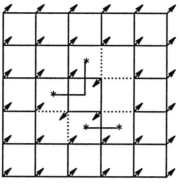

———— positive interaction   * frustrated face

············ negative interaction   ↙ ↗ spin orientation

violated interaction

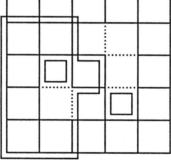

frustrated cycles

Fig. 2.1

In a more general case the problem is

$$\text{minimize} \sum -J_{ij}S_iS_j + F \sum S_i$$

subject to

$$S_i \in \{-1, 1\}, \text{ for all } i.$$

The linear term $F \sum S_i$ represents an exterior magnetic field. This reduces to a max cut problem in a graph that is planar grid plus one universal node. Let us call *quasi-planar* a graph G with the property that there is a vertex v such that $G \setminus v$ is planar. In Barahona (1982) we proved that the max cut problem for quasi-planar graphs is NP-Hard. We used a reduction from the stable set problem in planar cubic graphs.

One can use the system (2.3) as a linear programming relaxation for this case and for other non-planar grids. The first computational experience with three-dimensional grids was presented in Barahona and Maccioni (1982a). We solved several problems to optimality without having to resort to Branch and Bound. Some further computational experience with a system of inequalities equivalent to (2.3) has been published in Barahona et al (1988).

3. Via Minimization

We now present an application to the design of VLSI circuits. A layout of a two-layer chip, where the assignment of wires to layers has not yet been performed is called a *transient routing*. Given a transient routing, the via minimization problem consists of finding an assignment that minimizes the number of connections (vias) between the layers.

Pinter (1984) gave the following formulation. Given a picture like Figure 3.1, and two colours, we have to colour the wires in such a way that when two of them cross they have different colours. We should minimize the total number of times that wires change colours.

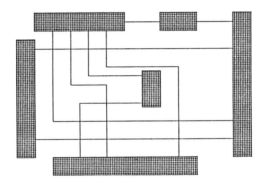

Fig. 3.1

It was shown by Pinter (1984) and Chen et al (1983) that this problem reduces to a max cut problem in a planar graph. They gave an $O(n^3)$ algorithm, where n is the number of crossings. In Barahona (1988a) we presented a simpler derivation and pointed out that the problem can be solved in $O(n^{3/2} \log n)$ time. Here we present another simple derivation. It is based on the following Lemma.

Lemma 3.1. *Given a bipartite graph embedded in the plane, one can always colour the edges with two colours in such a way that:*

(i) *if u is a node of even degree then edges incident with u and adjacent in the embedding have different colours.*

(ii) *if u is a node of odd degree then edges incident with u and adjacent in the embedding have different colours, except for one pair of them.*

Proof. First identify the nodes of odd degree. Partition this set into pairs. For each pair $\{u, v\}$ add a path with "artificial" nodes and edges between u and v. The number of nodes in this path should have the same parity as any other path between u and v. Notice that when we add those artificial edges we may loose the planarity of the graph, but this is not a problem.

Now we have an Eulerian bipartite graph. It is easy to find an Euler tour that does not cross itself at any node. Given that tour, we just give an alternating colouring to the edges in the same order as they appear.   □

This proof is due to W. Pulleyblank.

Now given the picture of the circuit we construct a graph G by inserting a node at every crossing, if a segment of a line does not cross any other line, we also insert a node at the extremity of that segment. The edges correspond to the line segments. Clearly G is a planar graph. If G is bipartite, it follows from Lemma 3.1 that no via is needed. If G is not bipartite we have to find the minimum number of edges to be broken into two pieces to make G bipartite. This will be the minimum number of vias. For every odd (resp. even) cycle we should break an odd (resp. even) number of edges. So the problem can be formulated as

$$\text{minimize} \sum x_{ij}$$

subject to

(3.2)
$$\sum_{ij \in C} x_{ij} \equiv \begin{cases} 1 \,(\text{mod } 2), & \text{if } C \text{ is odd,} \\ 0 \,(\text{mod } 2), & \text{if } C \text{ is even,} \end{cases}$$

for every cycle C,
$$x_{ij} \in \{0, 1\}, \text{ for every edge } ij.$$

It is enough to impose the constrains (3.2) for a set of cycles that contains a cycle basis. So if we pick the faces of G, we have a Chinese Postman Problem in the dual graph. The set T is the set of odd faces in G. It follows from the Theorem of Edmonds and Johnson that (3.2) is equivalent to the linear program

$$\text{minimize} \sum x_{ij}$$

subject to

(3.3)
$$\sum_{ij \in C} x_{ij} \geq 1,$$

for every odd cycle C,
$$x_{ij} \geq 0, \text{ for every edge } ij.$$

This might not have an integer optimal dual solution. It follows from Seymour's Theorem that such an integer vector exists if the degree of every node of G is even. Let us replace every edge of G by two parallel edges, and denote by G' this new graph. The graph G' satisfies this "evenness" condition. This shows that the minimum number of vias (for the original problem) is equal to one half of the maximum number of edge-disjoint odd cycles of G'.

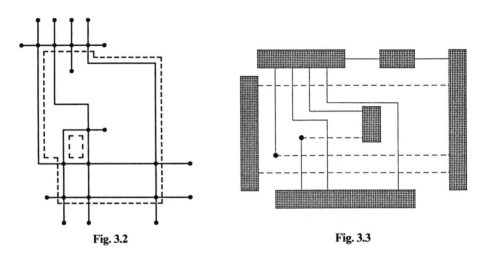

Fig. 3.2   Fig. 3.3

We could decide not to solve the Chinese Postman Problem exactly, but to apply a heuristic. In this case, a set of disjoint odd cycles gives information about the quality of our heuristic solution.

Figure 3.2 shows the graph associated with the example of Figure 3.1. We indicate a set of odd cycles that prove the optimality of the solution shown in Figure 3.3.

In a more general statement of the problem, there are wire segments that are required to be in one layer, or there is a preference for one of the layers. This can be modeled by adding an extra node to G. We join this node to all the segments that have to be fixed. We put appropriate weights in those new edges to ensure that the preassignment is satisfied and we should solve a max cut problem in this new graph. This is similar to the Spin Glass Problem within a magnetic field.

For this case two approaches can be tried. First we should find a heuristic solution, then we can use linear programming to evaluate the quality of that solution. One can also try to find an optimal solution by using a cutting plane approach. More details about this method appear in Barahona et al (1988).

4. Planar Multicommodity Flows

Let $G = (V, E)$ be a graph and $c : E \to \mathbb{R}$ a capacity function. Let $E^- = \{e : c(e) < 0\}$, and $E^+ = E \setminus E^-$. If $e \in E^+$ then $c(e)$ is the capacity of e. If $e \in E^-$

then $c(e)$ represents a demand between the endnodes of e. Let $\mathscr{C}$ be the family of cycles that contain exactly one edge in E^-. A multicommodity flow is a function $f : \mathscr{C} \to \mathbb{R}$, such that

$$\sum \{f(C) : e \in C\} \begin{cases} \leq c(e) & \text{if } e \in E^+, \\ \geq -c(e) & \text{if } e \in E^-, \end{cases}$$

$$f(C) \geq 0, \text{ for } C \in \mathscr{C}.$$

For $S \subseteq V$, a cut $\delta(S)$ is called *negative* if

$$\sum \{c(e) : e \in \delta(S)\} < 0.$$

The existence of a negative cut implies the non-existence of a flow. Seymour (1981a) proved that if G is planar then there is a flow if and only if there is no negative cut. He also proved that if the capacities are integer and the sum of the capacities of the edges in every cut is even then the flow can be chosen to be integer valued. Matsumoto et al (1986) reduced the problem to $O(n)$ matching problems and gave an $O(n^{5/2} \log n)$ algorithm. In Barahona (1987) we pointed out that this can be solved in $O(n^{3/2} \log n)$ time.

As in the two preceding sections, this also reduces to a Chinese Postman Problem in the dual graph.

We can find, if one exists, a negative cut by solving

$$\text{minimize} \sum c(e)\, x(e)$$

subject to

(4.1)
$$\sum_{e \in F} x(e) \equiv 0 \,(\text{mod } 2), \text{ for } F \in \mathbb{F},$$

$$x(e) \in \{0, 1\}, \text{ for } e \in E,$$

where $\mathbb{F}$ denotes the set of faces. Since $\mathbb{F}$ contains a cycle basis of G, any vector satisfying this system of equations is the incidence vector of a cut.

Let us define $d(e) = |c(e)|$, for $e \in E$, and

$$x'(e) = \begin{cases} x(e) & \text{if } e \in E^+, \\ 1 - x(e) & \text{if } e \in E^-. \end{cases}$$

Problem (4.1) is equivalent to

$$\text{minimize} \sum d(e)\, x'(e)$$

subject to

(4.2)
$$\sum_{e \in F} x'(e) \equiv \begin{cases} 1 \,(\text{mod } 2), & \text{if } |F \cap E^-| \text{ is odd}, \\ 0 \,(\text{mod } 2), & \text{if } |F \cap E^-| \text{ is even}, \end{cases}$$

$$\text{for } F \in \mathbb{F},$$

$$x'(e) \in \{0, 1\}, \text{ for } e \in E.$$

This is a Chinese Postman Problem in the dual graph, so it is equivalent to

$$\text{minimize } \sum d(e)\, x(e)$$

subject to

(4.3)
$$\sum_{e \in C} x(e) \geq 1,$$

for every cycle C, such that $|C \cap E^-|$ is odd,

$$x(e) \geq 0, \text{ for every edge } e.$$

This problem is similar to (2.3).

There is no negative cut if and only if the vector $\bar{x}$ defined as

$$\bar{x}(e) = \begin{cases} 0 & \text{if } e \in E^+, \\ 1 & \text{if } e \in E^-, \end{cases}$$

is an optimum of (4.2) and (4.3). In this case the Chinese Postman algorithm provides a vector y such that

(4.4)
$$y \geq 0,$$

(4.5)
$$\sum \{y_C : e \in C\} \leq c(e), \text{ for } e \in E^+,$$

(4.6)
$$\sum \{y_C : e \in C\} = -c(e), \text{ for } e \in E^-,$$

(4.7)
$$y_C > 0 \Rightarrow |C \cap E^-| = 1.$$

Conditions (4.4) and (4.5) follow from the inequalities of the dual problem of (4.3). Conditions (4.6) and (4.7) are the complementary slackness conditions that the pair $(\bar{x}, y)$ must satisfy. So the vector y defines a flow. If the capacities are integers and the sum of them is even for every cut, then a slight modification of this algorithm produces an integer vector y. See Barahona (1987) for details.

5. Max Cut in Graphs Not Contractible to K_5

Given a graph $G = (V, E)$ and a weight function $w : E \to \mathbb{Z}$, the problem of finding a cut $\delta(S)$ such that its weight $\sum \{w(e) : e \in \delta(S)\}$ is as large as possible is called the max cut problem. This problem is NP-Hard even for graphs having degree at most three, see Yannakakis (1978). Hadlock (1975) proved that it is polynomially solvable for planar graphs, in fact this is problem (4.1). Wagner (1964) gave a way to decompose graphs not contractible to K_5. Using it we proved that the max cut problem for this class of graphs reduces to a sequence of Chinese Postman problems, see Barahona (1983). We shall sketch that procedure.

Let $G = (V, E)$ be a connected graph, and let $Y \subseteq V$ be a minimal articulation set (that is, the deletion of Y produces a disconnected graph, but no proper subset of Y has this property). Choose nonempty subsets T_1, T_2 of V, such that (T_1, Y, T_2) is a partition of V, and no edge joins a node in T_1 to a node in T_2. Add a set Z of new edges joining each pair of nonadjacent nodes in Y. Let $G_1 = (V_1, E_1)$, $G_2 = (V_2, E_2)$ be subgraphs so that $V_i = T_i \cup Y$, $i = 1, 2$, $E_1 \cup E_2 = E \cup Z$, and $G_1 \cap G_2 = (Y, E_1 \cap E_2)$ is a complete graph. Then if $|Y| = k$, $1 \leq k \leq 3$, G is called a k-sum of G_1 and G_2. We give the weight zero to the edges in Z.

Let us denote by $\mathbf{W}$ the class of connected graphs not contractible to K_5. Wagner (1964,1970) has shown that any graph $G \in \mathbf{W}$ can be obtained by means of k-sums starting from planar graphs and copies of V_8, this is the graph of Figure 5.1. This decomposition can be found in polynomial time, see Truemper (1986).

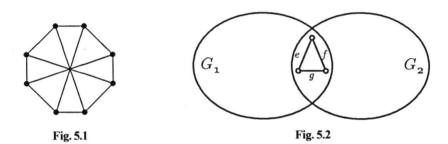

Fig. 5.1   Fig. 5.2

The algorithm we use is similar to that of Cornuéjols et al (1985) for the Traveling Salesman Problem in graphs with three-edge cutsets.

The algorithm works inductively. If the present graph is planar we solve problem (4.1). If the present graph is V_8, we solve the problem by enumeration. Otherwise G is a k-sum of G_1 and G_2, where G_2 is a planar graph or V_8. We need a way to decompose the problem. Let us denote by $\lambda(S, T, H)$ the maximum weight of a cut of the graph H, containing the edge-set S and having empty intersection with the edge-set T. We write $\lambda(H)$ instead of $\lambda(\emptyset, \emptyset, H)$.

If $k = 3$, let e, f and g be the edges in $G_1 \cap G_2$. A cut contains 0 or 2 of these edges. See Figure 5.2.

The edge-weights in G_2 are taken to be the same as for G. Then, the max cut problem is solved in G_1 where all the edge-weights are taken to be the same as for G, except for e, f, g, which are redefined as the solution of the following system of linear equations:

$$w'(e) + w'(f) = \lambda(\{e, f\}, \emptyset, G_2) - \lambda(\emptyset, \{e, f, g\}, G_2),$$

$$w'(f) + w'(g) = \lambda(\{f, g\}, \emptyset, G_2) - \lambda(\emptyset, \{e, f, g\}, G_2),$$

$$w'(e) + w'(g) = \lambda(\{e, g\}, \emptyset, G_2) - \lambda(\emptyset, \{e, f, g\}, G_2).$$

We have that $\lambda(G) = \lambda(G_1) + \lambda(\emptyset, \{e, f, g\}, G_2)$.

If $k = 2$, let e be the edge in $G_1 \cap G_2$. We take in G_2 the same weights as for G. In G_1 we only redefine the weight of e as

$$w'(e) = \lambda(\{e\}, \varnothing, G_2) - \lambda(\varnothing, \{e\}, G_2).$$

Then $\lambda(G) = \lambda(G_1) + \lambda(\varnothing, \{e\}, G_2)$.

If $k = 1$, the problem is solved independently in G_1 and G_2.

Now consider the multicommodity flow problem defined in Section 4. Seymour (1981b) proved that for graphs not contractible to K_5 there is a flow if and only if there is no negative cut. This problem can be decomposed in an analogous way, so it also reduces to a sequence of Chinese Postman problems, an algorithm for it appears in Barahona (1988b).

One can also use a polyhedral approach for the max cut problem. Given a cut C, the incidence vector of C, x^C is defined by

$$x^C(e) = \begin{cases} 1 & \text{if } e \in C, \\ 0 & \text{if } e \notin C. \end{cases}$$

The cut polytope, denoted by $P_C(G)$, is the convex hull of incidence vectors of cuts of G. Then the max cut problem can be formulated as

$$\text{maximize } wx$$

$$\text{subject to}$$

$$x \in P_C(G).$$

In order to use linear programming techniques, one needs to know a system of inequalities sufficient to define $P_C(G)$. Let us call $S(G)$ the system:

$$\sum_{e \in F} x(e) - \sum_{e \in C \setminus F} x(e) \leq |F| - 1,$$

(5.1)   for each cycle C, $F \subseteq C$, $|F|$ odd,

$$0 \leq x(e) \leq 1, \text{ for } e \in E.$$

Since the intersection of a cut and a cycle has even cardinality, it is clear that every incidence vector of a cut satisfies (5.1), thus these inequalities are valid for $P_C(G)$.

Grötschel, Lovász and Schrijver (1981) showed that one can use the ellipsoid method to optimize in polynomial time a linear function over a polyhedron P, if there is a polynomial algorithm to solve the so-called *separation* problem:

Given a vector $\bar{x}$, prove that it belongs to P or else find a hyperplane that separates $\bar{x}$ from P.

In order to solve the separation problem for $S(G)$ notice that these inequalities can be written as

$$\sum_{e \in F} (1 - x(e)) + \sum_{e \in C \setminus F} x(e) \geq 1,$$

for each cycle C, $F \subseteq C$, $|F|$ odd.

Given $\bar{x}$, we are looking for a minimum weighted cycle, where some edges have the weight $\bar{x}(\cdot)$, and an odd number of edges have weight $1 - \bar{x}(\cdot)$. From the given graph G we form a graph G' with two nodes i' and i'', for every node i of G. For every edge ij of G we put edges $i'j'$ and $i''j''$ with weight $\bar{x}(ij)$, and edges $i'j''$ and $i''j'$ with weight $1 - \bar{x}(ij)$. Now, for node i of G we find a shortest path from i' to i''. The minimum over the nodes of the lengths of the corresponding shortest path is the weight of the required cycle. As the computation of a shortest path takes $O(n^2)$ time, the entire procedure has time complexity $O(n^3)$.

Since we can solve the separation problem for the polyhedron defined by $S(G)$, one may ask what is the class of graphs such that $S(G)$ is sufficient to define the cut polytope. In Barahona and Mahjoub (1986) we proved that $P_C(G)$ is defined by $S(G)$ if and only if G is not contractible to K_5, this proof was based on the work of Seymour (1981b) on the matroids with the "sum of circuits property". In Barahona (1988b) we gave a direct proof of it.

6. Using Polynomial Size Linear Programming

The problems in the last four Sections have been formulated as linear programs with exponentially many inequalities. In what follows we shall see how to formulate them as linear programs of polynomial size. So they can be solved with any polynomial algorithm for linear programming. Such a system of inequalities is called *compact*.

Let K_n be a complete graph, $n \geq 3$, we denote by $T(K_n)$ the following system:

$$x(e) + x(f) + x(g) \leq 2,$$
$$x(e) - x(f) - x(g) \leq 0,$$
$$-x(e) + x(f) - x(g) \leq 0,$$
$$-x(e) - x(f) + x(g) \leq 0,$$

for every triangle $\{e, f, g\}$ of K_n.

Let Q be the polyhedron defined by $T(K_n)$, let G be a subgraph of K_n, and let Q' be the projection of Q along the variables associated to the edges in $K_n \setminus G$. In Barahona (1988b) we proved that this projection is exactly the polyhedron defined by $S(G)$. So if G is not contractible to K_5, then this projection is the cut polytope of G. In this case the max cut problem can be formulated as

$$\text{maximize} \sum_{e \in G} w(e)\, x(e)$$

(6.1)   subject to

$$x \text{ satisfies } T(K_n).$$

On the other hand, if G contains K_5 as a minor then there is at least one facet defining inequality of $P_C(G)$ that is not in $S(G)$. See Barahona and Mahjoub (1986) for a procedure that produces these inequalities. We can state the following.

Theorem 6.2. *Let G be a subgraph of K_n. The value of the optimum of (6.1) is the value of a max cut of G, for every weight function w, if and only if G is not contractible to K_5.*

Now we can use planar duality to derive a compact system for the Chinese Postman Problem in planar graphs. In what follows the graph may have loops and parallel edges. The sets of parallel edges will be called *parallel classes*. Given a graph G, we shall denote by $\overline{G}$ the graph obtained from G by deleting loops and keeping only one representative of each parallel class.

Given a graph G, we call $U(G)$ the system

$$x(e_0) - x(e_i) = 0, \quad i = 1, \ldots, p,$$

for each parallel class $\{e_0, \ldots, e_p\}$ of G,

$$x(e) + x(f) + x(g) \leq 2,$$

$$x(e) - x(f) - x(g) \leq 0,$$

$$-x(e) + x(f) - x(g) \leq 0,$$

$$-x(e) - x(f) + x(g) \leq 0,$$

for every triangle $\{e, f, g\}$ of $\overline{G}$,

$$0 \leq x(e) \leq 1,$$

for every edge e of $\overline{G}$ that does not belong to a triangle.

Let $H = (V, E)$ be a planar graph, and G the dual graph of H. Cuts of G correspond to disjoint unions of circuits in H. The symmetric difference between two T-joins is a union of circuits. Let us add edges to G until $\overline{G}$ is complete, the system $U(G)$ is a compact system for unions of circuits (cuts of the dual graph) of H. Let F be a (fixed) T-join of H. Let us define

$$y(e) = \begin{cases} x(e) & \text{if } e \notin F, \\ 1 - x(e) & \text{if } e \in F. \end{cases}$$

This transformation will be denoted by $y = Dx + h$. The matrix D is non-singular and h is a 0-1 vector.

Let w be a weight function for the edges of H, we obtain the value of a minimum weighted T-join by solving

$$\text{minimize } \sum_{e \in H} w(e)\, y(e)$$

subject to

$$y = Dx + h,$$

$$x \text{ satisfies } U(G).$$

It is an open question whether, for general graphs, Chinese Postman (or Optimum Matching) admits a compact formulation. Yannakakis (1988) proved that it is not possible by means of a *symmetric* system.

In Barahona (1988c) we showed that the Chinese Postman Problem in general graphs reduces to a sequence of $O(m^2 \log n)$ minimum mean cycle problems. This last problem can be formulated as a linear program of polynomial size. This gives a polynomial algorithm for matching that uses any polynomial algorithm for linear programming as a subroutine, see Khachiyan (1979), Karmarkar (1984) and Renegar (1988) for instance.

Acknowledgments. I am grateful to W. Pulleyblank for several helpful discussions about this material.

References

Ball, M.O., Derigs, U. (1983): An analysis of alternate strategies for implementing matching algorithms. Networks **13**, 517–549

Barahona, F. (1980): Application de l'Optimisation combinatoire à certains modèles de Verres de Spins: Complexité et Simulations. Thèse de Docteur-Ingenieur, Université de Grenoble

Barahona, F. (1981): Balancing signed toroidal graphs in polynomial time. Depto. de Matemáticas, Universidad de Chile

Barahona, F. (1982): On the computational complexity of Ising spin glass models. J. Phys. A **15**, 3241–3253

Barahona, F. (1983): The max cut problem in graphs not contractible to K_5. Oper. Res. Lett. **2**, 107–111

Barahona, F. (1987): Planar multicommodity flows, max cut and the chinese postman problem. Report No. 87454-OR, Institut für Operations Research, Universität Bonn

Barahona, F. (1988a): On via minimization. Research Report CORR 88-10, University of Waterloo, to appear in IEEE Trans. Circuits Syst.

Barahona, F. (1988b): On cuts and matchings in planar graphs. Report No. 88503-OR, Institut für Operations Research, Universität Bonn

Barahona, F. (1988c): Reducing matching to polynomial size linear programming. Research Report CORR 88-51, University of Waterloo

Barahona, F., Grötschel, M., Jünger, M., Reinelt, G. (1988): An application of combinatorial optimization to statistical physics and circuit layout design. Oper. Res. **36**, 493–513

Barahona, F., Maccioni, E. (1982a): On the exact ground states of three-dimensional Ising spin glasses. J. Phys. A **15**, L611–L615

Barahona, F., Mahjoub, A.R. (1986): On the cut polytope. Math. Program. **36**, 157–173

Barahona, F., Maynard, R., Rammal, R., Uhry, J.P. (1982b): Morphology of ground states of a two-dimensional frustration model. J. Phys. A **15**, 673–699

Bieche, I. (1979): Thèse de 3ème cycle. Université de Grenoble

Bieche, I., Maynard, R., Rammal, R., Uhry, J.P. (1980): On the ground states of the frustration model of a spin glass by a matching method of graph theory. J. Phys. A **13**, 2553–2576

Chen, R.W., Kajitani, Y., Chan, S.P. (1983): A graph theoretic via minimization algorithm for two-layer printed circuit boards. IEEE Trans. Circuits Syst. **30**, 284–299

Cornuéjols, G., Naddef, D., Pulleyblank, W.R. (1985): The traveling salesman problem in graphs with three-edge cutsets. J. Assoc. Comput. Mach. **32**, 383–410

Edmonds, J. (1965): Maximum matching and a polyhedron with (0,1)-vertices. J. Res. Nat. Bur. Standards **69B**, 125–130

Edmonds, J., Johnson, E.L. (1973): Matching, Euler tours and the chinese postman. Math. Program. **5**, 88–124

Galil, Z., Micali, S., Gabow, H. (1986): An $O(E\ V \log V)$ algorithm for finding a maximal weighted matching in general graphs. SIAM J. Comput. **15**, 120–130

Gerards, A.M.H. (1988): Graphs and polyhedra, binary spaces and cutting planes. Ph. D. Thesis, Tilburg University

Grötschel, M., Lovász, L., Schrijver, A. (1981): The ellipsoid method and its consequences in combinatorial optimization. Combinatorica **1**, 169–191

Hadlock, F. (1975): Finding a maximum cut of a planar graph in polynomial time. SIAM J. Comput. **4**, 221–225

Kasteleyn, P.W. (1961): The statistics of dimers on a lattice, the number of dimer arrangements on a quadratic lattice. Physica **27**, 1209–1225

Kasteleyn, P.W. (1967): Graph theory and crystal physics. In: Harary, F. (ed.): Graph theory and theoretical physics. Academic Press, London, pp. 43–110

Karmarkar, N. (1984): A new polynomial-time algorithm for linear programming. Combinatorica **4**, 373–395

Khachiyan, L.G. (1979): A polynomial algorithm in linear programming. Sov. Math. Dokl. **20**, 191–194

Lawler, E.L. (1976): Combinatorial optimization: Networks and matroids. Holt, Rinehart and Winston, New York, NY

Lipton, R.J., Tarjan, R.E. (1979): A separator theorem for planar graphs. SIAM J. Appl. Math. **36**, 177–189

Matsumoto, K., Nishizeki, T., Saito, N. (1986): Planar multicommodity flows, maximum matchings and negative cycles. SIAM J. Comput. **15**, 495–510

Pinter, R.Y. (1984): Optimal layer assignment for interconnect. J. VLSI Comput. Syst. **1**, 123–137

Renegar, J. (1988): A polynomial-time algorithm, based on Newton's method, for linear programming. Math. Program. **40**, 59–93

Seymour, P.D. (1981a): On odd cuts and planar multicommodity flows. Proc. Lond. Math. Soc., III. Ser. **42**, 178–192

Seymour, P.D. (1981b): Matroids and multicommodity flows. Eur. J. Comb. **2**, 257–290

Toulouse, G. (1977): Theory of the frustration effect in spin glasses I. Commun. Phys. **2**, 115–119

Truemper, K. (1986): A decomposition theory for matroids IV: Decomposition of graphs. University of Texas at Dallas

Wagner, K. (1964): Beweis einer Abschwächung der Hadwiger-Vermutung. Math. Ann. **153**, 139–141

Wagner, K. (1970): Graphentheorie. Bibliographisches Institut, Mannheim

Yannakakis, M. (1978): Node- and edge-deletion NP-complete problems. Proc. 10th ACM STOC, pp. 253–264

Yannakakis, M. (1988): Expressing combinatorial optimization problems by linear programs. Proc. 29th IEEE FOCS, pp. 223–228

Separator Theorems and Their Applications

Fan R. K. Chung

1. Introduction

Two subsets U and V of vertices in a graph G are said to be *separated* if no vertex in U is adjacent to any vertex in V. A subset S of vertices in a graph G is said to be a *separator* of G if by removing vertices in S from G, the remaining graph can be partitioned into two separated parts, say A and B, satisfying $|A| \leq |B| < 2|A|$. The concept of separators has been an extremely useful tool for dealing with many families of graphs (such as trees, planar graphs). For graphs with small separators, efficient and systematic methods can be developed for solving extremal and computational problems in so-called "divide-and-conquer" fashion. Namely, the original problem is divided into two or more smaller problems. The subproblems are solved by applying the method recursively, and the solutions to the subproblems are combined to give the solution to the original problem.

There are many different formulations and variations of separator theorems scattered about in the literature. For example, some very useful separator properties involve the trade-off of the separator size and the ratio of the two separated parts as well as additional requirements when the vertices are colored. We here intend to briefly survey various separator theorems. Then we will discuss some applications of these separator theorems to an extremal graph problem of finding optimal universal graphs. We then include references to many other applications in algorithmic design, data structure, and circuit complexity as well as in a number of other areas.

2. Separator Theorems for Trees

For a subset S of vertices in a graph G, we say S separates a subgraph H in G if there is no edge between vertices in H and vertices $V(G) - S - V(H)$.

We will start from the easiest but most fundamental separator theorems (see [CG1]).

Theorem 2.1. *Suppose α is any real number more than $\frac{1}{2}$ and T is a tree with at least $\alpha+1$ vertices. Then some vertex v separates a forest F in T satisfying*

$$\alpha \leq |V(F)| < 2\alpha.$$

Proof. If $\alpha \leq 1$, we choose F to be an end vertex in T which can be separated by removing one vertex. We assume $\alpha > 1$. If $|V(T)| \leq \alpha + 1$, the result is immediate. We may assume $|V(T)| > \alpha + 1$. Choose a leaf v_0 in T and let $\{v_0, v_1\}$ denote the edge incident to v_0. If all the connected components of $T - v_1$ have no more than α vertices, then by taking unions of some connected components, the desired forest F can be formed. Thus we may assume that some connected component T_1 has more than α vertices. If T_1 has fewer than 2α vertices, then we take $F = T_1$. We may assume $|V(T_1)| \geq 2\alpha$. Let v_2 be the vertex in T_1 adjacent to v_1 and consider the set S_1 of connected components in $T - v_2$ not containing v_1. The total number of vertices in S_1 is at least $2\alpha - 1 > \alpha$. As before, if all trees in S_1 have no more than α vertices, then F can be formed similarly. If some tree in S_1 has at least α but fewer than 2α vertices, again we are done. If some tree T_2 in S_1 has more than 2α vertices, then we let v_3 denote the vertex in T_2 adjacent to v_2 and we consider the connected components of $T - v_3$ not containing v_2, etc. By continuing in this matter, the theorem follows by induction.   □

As an immediate consequence of Theorem 2.1, we have the following (also see [CG1, LSH]).

Corollary 2.1. *Any n-vertex tree can be divided into two separated parts, each with no more than $\frac{2}{3}n$ vertices, by removing one vertex.*

Remark 2.1. If we replace "tree" by "forest", Theorem 2.1 and Corollary 2.1 are obviously true.

Remark 2.2. If we limit ourselves to binary trees, then we can achieve the same separation by removing one edge as follows:

Corollary 2.2. *For any real number $\alpha > \frac{1}{2}$, let T be a binary tree with at least $\alpha + 1$ vertices. Then some edge separates a forest F in T satisfying*

$$\alpha \leq |V(F)| < 2\alpha.$$

For trees with maximum degree $d+1$, the ratio of the two separated parts can be as large as d [Va2]. In a very similar way, the following can be proved.

Corollary 2.3. *For any real number $\alpha > \frac{1}{d}$, let T be a tree, with maximum degree $d+1$ having at least $\alpha + 1$ vertices. Then some edge separates a forest F satisfying*

$$\alpha \leq |V(F)| < d\alpha.$$

If we allow removing more than one vertex from a tree, the ratio of the two separated parts can be closer to $\frac{1}{2}$. In [CGP], it was shown that by removing w vertices, the ratio of two separated parts is no more than $1 + (\frac{2}{3})^{w-1}$. We here give an improved version.

Theorem 2.2. *Let* w *denote a positive integer and* T *denote a tree with at least* $\beta + w$ *vertices where* β *is a real value more than* $\frac{3^w}{2}$. *Then some set of* w *vertices separates a forest* F *in* T *satisfying*

$$\|V(F)\| - \beta| \leq \left(\frac{1}{3}\right)^w \beta.$$

Proof. For $w=1$, this is an immediate consequence of Theorem 2.1 by choosing α to be $\frac{2}{3}\beta$. Suppose it holds for any w', where $1 \leq w' < w$. By induction we can choose a set W' of w-1 vertices so that there is a forest F_1, formed by taking the union of some connected components in $T - W'$ satisfying

$$\|V(F_1)\| - \beta| \leq \left(\frac{1}{3}\right)^{w-1} \beta.$$

If $\|V(F_1)\| - \beta| \leq \left(\frac{1}{3}\right)^w \beta$, then Theorem 2.2 holds for w. We may assume

$$\left(1 - \left(\frac{1}{3}\right)^{w-1}\right)\beta \leq |V(F_1)| < \left(1 - \left(\frac{1}{3}\right)^w\right)\beta$$

or

$$\left(1 + \left(\frac{1}{3}\right)^w\right)\beta < |V(F_1)| \leq \left(1 + \left(\frac{1}{3}\right)^{w-1}\right)\beta.$$

We consider the following two cases.

Case 1.

$$\left(1 - \left(\frac{1}{3}\right)^{w-1}\right)\beta \leq |V(F_1)| < \left(1 - \left(\frac{1}{3}\right)^w\right)\beta.$$

Let F_2 be the forest formed by taking the union of all the connected components in $T - W' - F_1$. Let $\beta_1 = \frac{2}{3}(\beta - |V(F_1)|)$ and apply induction assumption for F_2 if $\beta_1 > \frac{1}{2}$. There is a vertex v so that a forest F_3 can be formed by taking the union of some connected components in $F_2 - v$ satisfying

$$\beta_1 \leq |V(F_3)| < 2\beta_1.$$

Let F_4 denote the forest in $T - W' - v$ which is the union of F_1 and F_3. Then we have

$$\begin{aligned}
|V(F_4)| &= |V(F_1)| + |V(F_3)| \\
&\geq |V(F_1)| + \beta_1 \\
&\geq |V(F_1)| + \frac{2}{3}(\beta - |V(F_1)|) \\
&\geq \frac{2}{3}\beta + \frac{1}{3}|V(F_1)| \\
&\geq \frac{2}{3}\beta + \frac{1}{3}\left(1 - \left(\frac{1}{3}\right)^{w-1}\right)\beta \\
&\geq \left(1 - \left(\frac{1}{3}\right)^w\right)\beta.
\end{aligned}$$

On the other hand,

$$|V(F_4)| \le |V(F_1)| + |V(F_3)|$$

$$\le |V(F_1)| + 2\beta_1$$

$$\le |V(F_1)| + \frac{4}{3}(\beta - |V(F_1)|)$$

$$\le \frac{4}{3}\beta - \frac{1}{3}|V(F_1)|$$

$$\le \frac{4}{3}\beta - \frac{1}{3}\left(1 - \left(\frac{1}{3}\right)^{w-1}\right)\beta$$

$$\le \left(1 + \left(\frac{1}{3}\right)^{w}\right)\beta.$$

Suppose $\beta_1 \le \frac{1}{2}$. Then $\beta - V(G_1) \le \frac{3}{4}$. We choose a vertex v' so that there is an isolated vertex v'' in $T - W' - F_1 - v'$. Let F_4' be the forest in $T - W' - v'$ which is the union of F_1 and v''. We have

$$|V(F_4')| = |V(F_1)| + 1 \ge \beta + \frac{1}{4} \ge \beta \quad \text{and}$$

$$|V(F_4')| = |V(F_1)| + 1 \le \left(1 - \left(\frac{1}{3}\right)^{w}\right)\beta + 1$$

$$\le \left(1 + \left(\frac{1}{3}\right)^{w}\right)\beta.$$

We can then take F to be F_4 or F_4'.

Case 2.

$$\left(1 + \left(\frac{1}{3}\right)^{w}\right)\beta < |V(F_1)| \le \left(1 + \left(\frac{1}{3}\right)^{w-1}\right)\beta.$$

Let $\beta_2 = \frac{2}{3}(|V(F_1)| - \frac{3}{4} - \beta)$ and apply the induction assumption for F_1 if $\beta_2 > \frac{1}{2}$. There is a vertex u so that a forest F_5 can be formed by taking a union of some connected components in $F_1 - u$ satisfying

$$\beta_2 \le |V(F_5)| < 2\beta_2.$$

Let F_6 denote the forest $F_1 - F_5 - u$ which is the union of some connected components in $T - W' - u$. We then have

$$|V(F_6)| = |V(F_1)| - |V(F_5)| - 1$$

$$\geq |V(F_1)| - 2\beta_2 - 1$$

$$\geq |V(F_1)| - \frac{4}{3}(|V(F_1)| - \frac{3}{4} - \beta) - 1$$

$$\geq \frac{4}{3}\beta - \frac{1}{3}|V(F_1)|$$

$$\geq \frac{4}{3}\beta - \frac{1}{3}\left(1 + \left(\frac{1}{3}\right)^{w-1}\right)\beta$$

$$\geq \beta - \left(\frac{1}{3}\right)^w \beta.$$

On the other hand, we have

$$|V(F_6)| = |V(F_1)| - |V(F_5)| - 1$$

$$\leq |V(F_1)| - \frac{2}{3}(|V(F_1)| - \frac{3}{4} - \beta) - 1$$

$$\leq \frac{1}{3}|V(F_1)| + \frac{2}{3}\beta - \frac{1}{2}$$

$$\leq \frac{1}{3}\left(1 + \left(\frac{1}{3}\right)^{w-1}\right)\beta + \frac{2}{3}\beta$$

$$\leq \beta + \left(\frac{1}{3}\right)^w \beta.$$

Suppose $\beta_2 \leq \frac{1}{2}$. Then $|V(F_1)| - \frac{3}{4} - \beta \leq \frac{3}{4}$. We can choose a vertex u' so that there is an isolated vertex u'' in $F_1 - u'$. Let F_6' denote the forest $F_1 - u' - u''$ which is the union of some connected components in $T - w' - u'$. We have

$$|V(F_6')| = |V(F_1)| - 1 \leq \beta + \frac{1}{2} \leq \left(1 + \left(\frac{1}{3}\right)^w\right)\beta$$

$$|V(F_6')| = |V(F_1)| - 1 \geq \left(1 + \left(\frac{1}{3}\right)^w\right)\beta - 1$$

$$\geq \left(1 - \left(\frac{1}{3}\right)^w\right)\beta.$$

F_6 or F_6' is the forest F we want. This completes the proof for Theorem 2.2.

Corollary 2.4. *Any n-vertex tree can be divided into two separated parts, each with no more than $\left(1 + \left(\frac{1}{3}\right)^w\right)\frac{n}{2}$ vertices, by removing w vertices, if $n > 3^w$.*

Theorem 2.3. *Any n-vertex tree can be divided into two separated equal parts by removing at most $\lfloor \frac{\log n}{\log 3} \rfloor + 1$ vertices.*

Proof. Using Theorem 2.2 and setting $w = \lfloor \log_3(n - \log_3 n) \rfloor$, $\beta = \frac{(n - \log_3 n)}{2}$, it is easily checked that $\beta \cdot (\frac{1}{3})^w < \frac{3}{2}$. Therefore by removing $w + 1$ vertices, the tree is divided into two separated equal parts.   □

A separator is called a bisector if it separates the graph into two equal parts. In [BCG], examples have shown that the bisectors in Theorem 2.3 are very close to the optimum.

Theorem 2.4. *In the complete ternary tree on 2t levels with $n = \frac{3^{2t}-1}{2}$ vertices, one cannot remove fewer than $\frac{\log n}{\log 3} - \frac{2\log n}{\log\log n}$ vertices to separate the remaining graph into two equal parts.*

Remark 2.3. Theorems 2.1 and 2.2 can all be generalized by considering a cost assignment to vertices instead of counting vertices. Here we state the generalized versions whose proofs are extremely similar and will not be given.

Theorem 2.1′. *Let F be any n-vertex forest with non-negative vertex costs. Let α be any real number that is greater than half of the maximum cost of the vertices and is smaller than the sum of the total vertex costs subtracting the minimum vertex cost. Then some vertex v separates a forest F′ so that the total cost of vertices in F′ is between α and 2α.*

Theorem 2.2′. *Let w denote a positive integer and F denote a forest with non-negative vertex costs. Let β be a real value that is at least $\frac{3^w}{2}$ times the maximum vertex cost and is smaller than the sum of the all vertex costs except for the w smallest vertex costs. Then some set of w vertices separates a subforest F′ and the total cost c(F′) of vertices in F′ satisfies*

$$|c(F') - \beta| \leq \left(\frac{1}{3}\right)^w \beta.$$

3. Separators for Colored Trees

Suppose the vertices of a tree are colored in k colors for some given integer k. It is often desirable to remove a small number of vertices so that the remaining graph is separated into two parts, each having about a half of the number of vertices in each color. This has the similar flavor of a very nice result of Goldberg and West [GW].

Theorem 3.1. *Suppose the beads on a string are colored in k colors. One can cut the necklace at k places so that the resulting strings of beads can be placed into two piles, each of which contains the equal (to within one) number of beads in each color.*

This necklace-splitting theorem can be combined with a decomposition lemma to derive the tree-splitting analogue (see [BL]). Since the subgraphs of a tree inherently have small separators, we can associate the separators in a complete binary tree structure. A complete binary tree of t levels has a vertex set consisting of all (0,1)-tuples of length $< t$ together with the root denoted by *. Each vertex u is adjacent to its two children $u0$ and $u1$.

Lemma 3.1. *Any n-vertex tree can be mapped into a complete binary tree C of $\lceil \frac{\log n}{\log 3/2} \rceil$ levels satisfying the following properties.*

(i) The mapping f is 1-1 from $V(G)$ into $V(C)$.
(ii) For each w in C, let S_w denote the set of all vertices in T that are mapped to descendants of w in C. Then $|S_{w0}| \leq |S_{w1}| < 2|S_{w0}|$.
(iii) For each w in C, let A_w denote the set of all vertices in T that are mapped to the ancestors of w in C and w itself. Then by removing A_w, the set S_w is separated from the rest of the graph in $T - A_w$.

Proof. This follows immediately by mapping the separator vertex of Theorem 2.1 to the root and the two separated parts to S_0 and S_1, recursively.

This decomposition tree induces a natural linear order for vertices in T which can be viewed as the order from left to right in the plane layout of C. To be precise, for two distinct binary tuples α and β of length $\leq t$, we say $\alpha < \beta$ if at the first place they differ, the corresponding coordinate of α is 0 and the corresponding place for β is 1.

Lemma 3.2. *Each initial segment, which consists of all u with $f(u) < w$ for some w in C, can be separated from the rest of the tree by removing vertices that mapped to the path $A(w)$ consisting of all ancestors of w.*

Combining Theorem 3.1 and Lemmas 3.1 and 3.2, we can then arrive at the following result for splitting colored trees (see [BL]).

Theorem 3.2. *Suppose the n vertices of a tree are colored in k colors. By removing $ck \log n$ vertices, the connected components can be partitioned into two parts, each containing the same number of vertices in each color.*

Proof. First we use Lemma 3.1 to decompose the tree T. That is to map the vertices of T into a complete binary tree C of $\lceil \frac{\log n}{\log 3/2} \rceil$ levels. Using the linear order introduced by the decomposition tree, the colored vertices can be viewed as a string of beads, so we are ready to use the necklace-splitting theorem. Theorem 3.1 ensures that by making no more than k cuts in the string, the k-colored beads can be almost equally partitioned into two parts. Now, for each cut, choose a vertex w next to the cut and let S_0 consist of all vertices in the path $A(w)$ consisting of all ancestors of w. It follows from Lemma 3.2 that by removing all vertices in S_0, the remaining graph in T can be partitioned into two parts so that the sum of differences of the number of vertices in each color is no more than $|S_0| + k$. By adding additional $|S_0| + k$ vertices to S_0 to form the separator S, the number of vertices of each color in the two parts can then be balanced. The separator S contains no more than $2k(\lceil \frac{\log n}{\log 3/2} \rceil + 1)$ vertices. This completes the proof of Theorem 3.2.

Although Theorem 3.2 is best possible within a constant factor, there is a trade-off between the size of the separator and the precision of bipartition. This is often crucial in proving optimality of various universal graphs in later sections [BCLR1, BCLR2].

Theorem 3.3. *For each constant $p < \frac{1}{2}$, there exists a constant q such that any n-vertex forest with w vertices of color A can be partitioned into two sets by the removal of q vertices so that each set has at least $\lfloor pn \rfloor$ vertices and at least $\lfloor pw \rfloor$ vertices of color A.*

The proof of Theorem 3.3 mainly follows from forming an appropriate decomposition tree and will not be included here. There is a generalized version that will be stated here without proof.

Theorem 3.4. *For each constant $p < \frac{1}{2}$ and each positive integer k, there exists a constant q such that any n-vertex forest with n_i vertices of color c_i, $i = 1, \ldots, k$, can be partitioned into two sets by the removal of q vertices so that for each i, each set has at least $\lfloor p(n_1 + \ldots + n_i) \rfloor$ vertices in color $1, 2, \ldots i$.*

4. Separator Theorems for Planar Graphs

Separator theorems for planar graphs and their applications were first described in the seminal papers of Lipton and Tarjan [LT1, LT2]. Here we will state without proof several major versions.

Theorem 4.1 [LT1]. *Let G be any n-vertex planar graph with non-negative vertex costs. Then the vertices of G can be partitioned into three sets A, B, C, such that no edge joins a vertex in A with a vertex in B, neither A nor B has vertex cost exceeding $\frac{2}{3}$ of the total cost, and C contains no more than $2\sqrt{2}\sqrt{n}$ vertices.*

Remark 4.1. The constant $2\sqrt{2}$ was improved by Djidjev [D1] to $\sqrt{6}$ and later by Gazit [G] to $\frac{7}{3}$.

Theorem 4.2 [LT1]. *Let G be any n-vertex planar graph with non-negative vertex costs. Then vertices of G can be partitioned into three sets A, B, C, such that no edge joins a vertex in A with a vertex in B, neither A nor B has vertex cost exceeding $\frac{1}{2}$ of the total cost, and C contains no more than $2\sqrt{2}/(1 - \sqrt{2/3})\sqrt{n}$ vertices.*

Remark 4.2. The constant $2\sqrt{2}/(1-\sqrt{2/3}) \approx 15.413$ was later improved in several papers; $\sqrt{6}/(1 - \sqrt{2/3}) \approx 13.348$ in [D1] and $8 + \frac{16}{81}\sqrt{6}/(1 - \sqrt{2/3}) \approx 10.637$ in [Ve]. The current best constant is $\frac{108}{19}\sqrt{3} \approx 9.845$ (see [C1]). On the other hand, the best known constant for the lower bound is $\frac{2}{\sqrt{3}} \approx 1.155$ by considering a plane graph with vertex set consisting of points $\{(x, y\sqrt{3/2}) : x, y \in z$ and $(x, y\sqrt{3/2})$ is in a regular hexagon of side $\sqrt{n/3}\}$ and the edges are between points of distance at most 1.

Remark 4.3. The separator can be constructed in $O(n)$ time [LT1].

The constant can be further improved if we consider the separation of a planar graph into two parts, each with roughly equal number of vertices. In other words, every vertex has a cost of 1.

Theorem 4.3 [C1]. *Let G be any n-vertex planar graph. The vertices of G can be partitioned into three sets A, B, C, such that no edge connects A with B, A and B each have $\leq \frac{n}{2}$ vertices, and C contains $\leq 3\sqrt{6}\sqrt{n}$ vertices.*

The constant $3\sqrt{6} \approx 7.348$ is an improvement over the results in [LT1], [D1] and [Ve].

We will give the proof for Theorem 4.3 which needs the following facts.

Lemma 4.1 [LT1]. *Let G be a planar graph of radius s, with non-negative vertex-costs summing to 1. Then the vertices of G can be partitioned into three parts A, B, C, such that there is no edge between A and B, neither A nor B has total vertex cost exceeding $\frac{2}{3}$, and C contains at most $2s + 1$ vertices.*

Lemma 4.2 [D1]. *Let G be an n-vertex planar graph of radius s. For any real number r, $\frac{1}{2} \leq r \leq 1$, there exists a set $S \subseteq V(G)$ with at most $3s + 1$ vertices such that by removing vertices in S from G the remaining graph is separated into three parts A, B, C, such that A, B each contain at most $(1 - r)n$ vertices, and C contains at most rn vertices.*

Lemma 4.3 [Ve]. *For any integer s, an n-vertex planar graph G contains a subgraph H of at least $n - \frac{n}{s}$ vertices so that any subgraph of H can be embedded into another planar graph of radius $s - 1$.*

Proof of Theorem 4.3. Let G denote a planar graph on n vertices. We will determine A, B and C iteratively as follows:

Step 0. Set $s = \lfloor \sqrt{\frac{n}{6}} \rfloor$ and use Lemma 4.3 to find a set S_0 with at most $\frac{n}{s}$ vertices and embed $G - S_0$ into a graph G' of radius $s - 1$. Set $A = B = \emptyset$, $C = S_0$ and $j = 1$.

In general, for $i = 1, 2, \ldots$, the step i can be described as follows.

Step i. Set $r = (\frac{n}{2} - |A|)/n'$ where $n' = |V(G')|$. For $j \leq 2$, use Lemma 2 to find a separator S' containing at most $3(s - 1)$ vertices which separate G' into A', B' and C' with $|A'| \leq |B'| \leq (1 - r)n'$ and $|B'| \leq |C'| \leq rn'$. Set A to be the smaller one of $A \cup C'$ and $B \cup B'$; B to be the larger one of $A \cup C'$ and $B \cup B'$; and C to be $C \cup S'$. If $j < 2$, apply Lemma 3 to the induced subgraph of A' and form a new graph G' of radius $s - 1$, set $j = j + 1$ and repeat Step i. If $j = 2$, set i to be $i + 1$, j to be 1. Then set s to be $\sqrt{n'/6}$ when applying Lemma 3 to A' and form G' of radius $s - 1$, add $\frac{n'}{s}$ vertices to C. Then go to the next step until G' is empty.

The correctness of this algorithm can be established by verifying the following facts in Step i. Suppose at the beginning of Step i A has p vertices, B has

q vertices and G' has w vertices. The new A and B have no more than the maximum of $p + |C'|$ and $q + |B'|$. Since $p + |C'| \leq p + rw \leq p + (\frac{n}{2} - p) \leq \frac{n}{2}$ and $q + |B'| \leq q + (1 - r)w \leq q + w - \frac{n}{2} + p \leq \frac{n}{2}$, the new A and B has no more than $\frac{n}{2}$ vertices. Furthermore $|A'| \leq \frac{w}{3}$ since $|A'| \leq |B'| \leq |C'|$ and $|A'| + |B'| + |C'| \leq w$. This implies the new G' has at most $\frac{w}{3}$ vertices. Because of our choice of j, for each i, Step i is repeated twice for $j = 1$ and 2 (except for possibly the last step). So altogether $|G'|$ is reduced by a factor of 9.

We can bound the separator $C = C(n)$ as follows:

$$C(n) \leq 2\sqrt{6}\sqrt{n} + \sum_{i \geq 1} 2\sqrt{6}\sqrt{\frac{n}{9^i}}$$

$$\leq 2\sqrt{6}\sqrt{n} + \left| C\left(\frac{n}{9}\right) \right|.$$

Therefore

$$|C(n)| \leq 3\sqrt{6}\sqrt{n}.$$

This completes the proof for Theorem 4.3.

Miller [M] further requires the planar separator to be a simple cycle.

Theorem 4.4 [M]. *If G is an embedded 2-connected planar graph with non-negative weights assigned to vertices and faces that sum to 1, and no face has weight $> \frac{2}{3}$, then there exists a simple cycle on at most $2\sqrt{2n}$ vertices so that neither the interior nor the exterior has total weight $> \frac{2}{3}$.*

For graphs with genus g, the separator is of size $c\sqrt{gn}$ [D2, GHT]. The best constant is given in [D3].

Theorem 4.5. *Any n-vertex graph of orientable genus g has a separator with $\sqrt{2g + 1}\sqrt{6n}$ vertices.*

5. Separator Theorems and Universal Graphs

Separator theorems have many applications in a broad range of areas. In this section, we will illustrate some applications in extremal graph theory, namely in the construction of universal graphs.

The problem of universal graphs is a fundamental problem that arises in various contexts in many topics such as universal circuit [Va2], data representation [CRS, RSS], VLSI design [CR, BCG] and simulations of parallel computer architecture [BCLR1, BCHLR]. A typical problem is the following:

How many edges must a graph have that contains all trees with n vertices?

Obviously, the complete graph on n vertices and $\binom{n}{2}$ edges has the required universal property. However, the objective is to determine the minimum number $f(n)$ of edges in such a universal graph $G(n)$, which contains all n-vertex trees.

It is easy to see that the universal graph $G(n)$ must contain at least $\frac{1}{2}n\log n$ edges since it must contain one vertex of degree $\geq n-1$, two vertices of degree $\geq \frac{n}{2}$, and, in general, i vertices of degree $\geq \frac{n}{i}$ so that its degree sequence dominates $n-1, \frac{n}{2}, \frac{n}{3}, \ldots$. On the upper bound for $f(n)$, we can improve upon $\binom{n}{2}$ of the complete graph by the following series of applications of the separator theorem for trees.

Construction 1. $G_1(n)$ is the union of $G_1\left(\frac{2}{3}n\right)$ and $G_1\left(\frac{n}{2}\right)$ together with a vertex u that is adjacent to all other vertices. By using Theorem 2.1, any n-vertex tree can

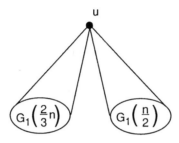

Fig. 1

be embedded into $G_1(n)$ by mapping the separator to u and the separated parts to $G_1\left(\frac{2}{3}n\right)$ and $G_1\left(\frac{n}{2}\right)$ respectively. On the other hand, the number of edges in $G_1(n)$ is bounded above by $f_1(n)$, which satisfies the following inequality[1]:

$$f_1(n) \geq f_1\left(\frac{2}{3}n\right) + f_1\left(\frac{n}{2}\right) + n.$$

It is easy to verify that

$$|E(G_1(n))| \leq f_1(n) \leq cn^{1.3}.$$

Although this is significantly better than the complete graph, it can be much improved by using Theorem 2.3.

Construction 2. $G_2(n)$ is the union of two copies of $G_2\left(\frac{n}{2}\right)$ together with $\frac{\log n}{\log 3}$ vertices, each of which is adjacent to all other vertices. A straightforward application of Theorem 2.3 shows that $G_2(n)$ contains all trees on n vertices and the number of edges in $G_2(n)$ is bounded by $f_2(n)$ which satisfies

$$f_2(n) \geq 2f_2\left(\frac{n}{2}\right) + n\log_3 n.$$

---

[1] Strictly speaking, we should use $G_1(\lfloor\frac{2}{3}n\rfloor)$ and $G_1(\lfloor\frac{n}{2}\rfloor)$. However, we will usually not bother with this type of detail since it has no significant effect on the arguments or results.

It can be easily checked that

$$|E(G_2(n))| \leq f_2(n) \leq cn \log^2 n.$$

The above bound can be further improved by using Theorem 2.2 appropriately (see [CGP]).

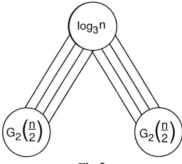

Fig. 2

Construction 3. $G_3(n)$ consists of $G_3(m)$ and $G_3(m)$ together with w vertices that are adjacent to all other vertices. By choosing $w = \frac{\log \log n}{\log 3/2}$ and

$$m = w \cdot n$$

we can then derive

$$|E(G_3(n))| \leq n \log n (\log \log n)^2.$$

In fact, it has been shown [CG3] that the minimum number $f(n)$ of edges in a graph that contains all n-vertex trees satisfies

$$\frac{1}{2}n \log n < f(n) < \frac{7}{\log 4}n \log n.$$

The construction assumption and proof are based on an elaborated induction and repeated usage of Theorem 2.1 that we will not discuss here. Our relatively simple descriptions of construction 1-3 merely illustrate various ways of using separator theorems and their effects.

A related problem is to determine the minimum number $f(n, d)$ of vertices in a graph that contains all n-vertex trees with maximum degree d. This problem can be solved by using the separator theorems for colored trees (Theorems 3.2, 3.3) and the decomposition lemmas. We here consider a simpler version of universal graphs for binary trees, that are trees with maximum degree 3 (see [BCLR2]).

Lemma 5.1. *The n vertices of a binary tree T can be mapped into a complete binary tree C on no more than $2^q - 1$ vertices ($2^q - 1 \leq n < 2^{q+1} - 1$) so that $6 \log \frac{n}{2^t} + 18$ vertices of T are mapped into a vertex of C at distance t from the root, and so that any two vertices adjacent in T are mapped to vertices at most 3 apart in C.*

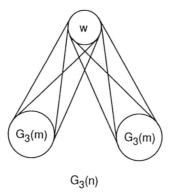

$G_3(n)$

Fig. 3

Proof. The idea is to recursively bisect T, placing the successive sets of bisector vertices within successively lower levels of C, until T is decomposed into single vertices. For example, the vertices placed at the root of C bisect T into two subgraphs T_1 and T_2. Similarly, vertices mapped to the left child of the root bisect T_1 and vertices mapped into the right child bisect T_2. In addition, at level i of C we map vertices of T (that have not already been mapped within levels $i-1$, $i-2$) that are adjacent to vertices mapped at level $i-3$ of C. This ensures that vertices adjacent in T will be mapped to vertices of C at most distance 3 apart.

To keep the number of vertices of T mapped to a level i vertex in C within the required bounds, we use separators for 3 colors, so-called 3-bisectors, as in Theorem 3.2. The following procedure describes how this is done.

Step 0. Initialize every vertex of T to color A, bisect T, and place the bisector vertices at the root (level 0) of C.

Step 1. For each subgraph created in the previous step, recolor every vertex adjacent to the bisector in the previous step with color 0, and place a 2-color bisector for the subgraph at the corresponding level 1 vertex of C.

Step 2. For each subgraph created in the previous step, recolor every vertex of color A adjacent to the bisector in the previous step with color 1, and place a 3-color bisector for the subgraph at the corresponding level 2 vertex of C.

Step t. $(\log |T| \geq t \geq 3)$. For each subgraph created in the previous step, place every vertex of color $t-1$ (mod 2) at the corresponding level t of C, recolor every vertex of color A that is adjacent to one of color $t-1$ (mod 2) with color t (mod 2), and place a 3-color bisector for the remaining subgraph at the corresponding level t vertex of C.

To ensure the accuracy of Step t, it suffices to show $n_t \leq 6\log \frac{n}{2^t} + 18$ for $3 \leq t < \log |T|$. Since we have

$$n_t \leq 3 \log \frac{n}{2^t} + \frac{1}{2} n_{t-3}$$

$$\leq 6 \log \frac{n}{2^t} + 18,$$

Lemma 5.1 is proved.

The analogous version for bounded-degree trees and planar graphs can be proved in a very similar way and the proofs are left to the reader. The main difference in proving these results is that vertices adjacent to previously mapped vertices are themselves only mapped at every $\log d$ instead of at every level.

Lemma 5.2. *The vertices of a tree T with maximum degree d can be mapped into a complete binary tree C on no more than $2^q - 1$ vertices $(2^q - 1 \leq n < 2^{q+1} - 1)$ so that $o(\log \frac{n}{2^t})$ vertices of T are mapped to a vertex of C at distance t from the root, and so that any two vertices adjacent in T are mapped to vertices at most distance $O(\log d)$ apart in C.*

Lemma 5.3. *The vertices of a planar graph G of maximum degree d can be mapped into a complete binary tree C on $2^q - 1$ vertices $(2^q - 1 \leq n < 2^{q+1} - 1)$ so that $O(\sqrt{n/2^t})$ vertices of G are mapped to a vertex of C at distance t from the root, and so that any two vertices adjacent in G are mapped to vertices at most distance $O(\log d)$ apart in C.*

A graph G is said to have a *k-bisector function f* if any subgraph of G on m vertices has a k-bisector of size no more than $f(m)$. The preceding lemmas are all special cases of the following.

Lemma 5.4. *Suppose G on n vertices with maximum degree d has a k-bisector function f. The vertices of G can be mapped into a complete binary tree C on no more than $2^q - 1$ vertices where $2^q - 1 \leq n < 2^{q+1} - 1$ so that $\alpha d f(\frac{n}{2^t})$ vertices of G are mapped to a vertex of C at distance t from the root, and so that any two vertices adjacent in G are mapped to vertices at most distance k apart in C if*

$$2f(xd^{3k}) \leq d^{3k-4} f(x) \text{ for all } x.$$

Although Lemma 5.4 looks somewhat complicated, it is merely a straightforward generalization of Lemmas 5.1 $\approx$ 5.3, and we omit the proof. We can now construct universal graphs using the decomposition lemmas.

Theorem 5.1. *The minimum universal graph for the family of all bounded-degree trees on n vertices has n vertices and $O(n)$ edges.*

Proof. Using Lemma 5.2, we consider the graph with vertices grouped into clusters corresponding to the vertices in the complete binary tree C. A cluster corresponding to a vertex of level t contains $O(\log \frac{n}{2^t})$ vertices. We connect all pairs of vertices in clusters with corresponding vertices within distance $O(\log d) = O(1)$ apart in C. By Lemma 5.2 the resulting graph is universal for the family of all trees with maximum degree d. The number $h(n)$ of edges in this graph is $O(n)$, since $h(n)$ satisfies the following recurrence inequality:

$$h(n) \leq 2h \left(\frac{n}{2} \right) + c(\log n)^2$$

where c is an appropriate constant depending on d.

The construction just described has $O(n)$ vertices. To obtain a universal graph with precisely n vertices, we modify the embedding of Lemma 5.1 so that the same number of vertices of T are wrapped to vertices in the same level of C. This is easy to do since we can always arbitrarily expand the bisector of any subtree to be within one of its maximum allowed value (which is the lesser of the number of vertices remaining and $O(\log \frac{n}{2^t})$ for vertices on level t of C). The exact value of the maximum bisector is the same for all vertices on a level and depends on the parity of the number of vertices in the subgraphs at that level. Hence, the size of the bisectors at each level depends only on n, and the universal graph can be assumed to have precisely n vertices.

Theorem 5.2. *The minimum universal graph for the family of all bounded-degree planar graphs on n vertices has n vertices and $O(n \log n)$ edges.*

Proof. The construction is by using Lemma 5.3 in similar fashion as in the proof of Theorem 5.1. The number of edges $h(n)$ satisfies

$$h(n) \leq 2h \left(\frac{n}{2} \right) + cn$$

and therefore the minimum universal graph has $O(n \log n)$ edges.   □

Theorem 5.3. *The minimum universal graph for a family of bounded-degree graphs on n vertices with bisector function $f(x) = n^\alpha$ has n vertices with: $O(n)$ edges if $\alpha < \frac{1}{2}$; $O(n \log n)$ edges if $\alpha = \frac{1}{2}$; $O(n^{2\alpha})$ edges if $\alpha > \frac{1}{2}$.*

Proof. The construction follows from Lemma 5.4 together with the fact that the number $h(n)$ of edges satisfies

$$h(n) \leq 2h \left(\frac{n}{2} \right) + (f(n))^2.$$

Theorem 5.4. *The minimum universal graph for the family of all bounded-degree outerplanar graphs on n vertices has n vertices and $O(n)$ edges.*

Proof. Since an outerplanar graph on n vertices has a bisector of size $O(\log n)$, the result follows from Theorem 5.3.

Recently, the author has shown that there is a universal graph with cn edges that contains all bounded-degree n-vertex trees as induced subgraphs [C2]. Many results on universal graphs and induced universal graphs for various classes of graphs can be found in [Bo, BCEGS, CG1, CG2, CGP, CCG, CGS, R]. In particular, the question of determining the minimum universal graphs that contain all n-vertex planar graphs remains open (it is between $n \log n$ and $n^{3/2}$ (see [BCEGS])).

6. Concluding Remarks

There are many aspects of separator theorems that we have not covered here. The references include various papers for many applications such as approximation algorithms [Ba, LT2], dynamic programming [LT2], pebbling [LT2], VLSI layout [BCLR1, BL, L], Boolean circuits [Va2], routing [LT2], nested dissection methods in numerical analysis [LRT, PR2] and parallel algorithms [GM, PR1] to find planar separators.

Very recently, N. Alon, P. Seymour and R. Thomas have generalized the separator theorem of Tarjan and Lipton to graphs excluding certain minors. A graph H is said to be a minor of a graph G if H can be obtained from a subgraph of G by contracting edges. The Kuratowski theorem asserts that a graph is planar if and only if it does not have K_5 or $K_{3,3}$ as minors. The separator theorem for graphs excluding minors can be stated as follows [AST]:

Let G be any n-vertex with non-negative vertex costs and the complete graph K_h on h vertices is not a minor of G. Then the vertices of G can be partitioned into three parts A, B and C, such that no edge joins a vertex in A with a vertex in B, neither A nor B has vertex cost exceeding $\frac{2}{3}$ of the total cost, and C contains no more than $h^{3/2}n^{1/2}$ vertices. Such a separator can be determined in time $O(h^{1/2}n^{1/2}m)$, where $m = n + |E(G)|$.

References

[AST]   Noga, A., Seymour, P., Thomas, R.: A separator theorem for non-planar graph. Preprint

[Ba]   Baker, B.S. (1983): Approximation algorithms for NP-complete problems on planar graphs. Proc. 24th IEEE FOCS, pp. 265–273

[BCEGS]   Babai, L., Chung, F.R.K., Erdős, P., Graham, R.L., Spencer, J. (1982): On graphs which contain all sparse graphs. In: Rosa, A., Sabidussi, G., Turgeon, J. (eds.): Theory and practice of combinatorics. North Holland, Amsterdam, pp. 21–26 (Ann. Discrete Math., Vol. 12)

[BCG]   Bollobás, B., Chung, F.R.K., Graham, R.L. (1983): On complete bipartite subgraphs contained in spanning tree complements. In: Erdős, P. (ed.): Studies in pure mathematics. Birkhäuser, Basel, pp. 83–90

[BCHLR]   Bhatt, S.N., Chung, F.R.K., Hong, J.-W., Leighton, F.T., Rosenberg, A.L. (1988): Optimal simulations by butterfly networks. Proc. 20th ACM STOC, pp. 192–204

[BCLR1]   Bhatt, S.N., Chung, F.R.K., Leighton, T., Rosenberg, A.L. (1986): Optimal simulations of tree machine. Proc. 27th IEEE FOCS, pp. 274–282

[BCLR2]   Bhatt, S.N., Chung, F.R,K., Leighton, T., Rosenberg, A.L.: Universal graphs for bounded degree trees and planar graphs.

[BL]   Bhatt, S.N., Leighton, F.T. (1984): A framework for solving VLSI graph layout problems. J. Comput. Syst. Sci. **28**, 300–343

[Bo]   Bondy, J.A. (1971): Pancyclic graphs. J. Comb. Theory, Ser. B **11**, 80–84

[C1]   Chung, F.R.K.: Improved separators for planar graphs.

[C2]   Chung, F.R.K.: Universal graphs and induced universal graphs.

[CCG]   Chung, F.R.K., Coppersmith, D., Graham, R.L. (1981): On trees which contain all small trees. In: Chartrand, G., Alavi, Y., Goldsmith, D.L., Lesniak-Foster, L., Lick, D.R. (eds.): The theory and applications of graphs. J. Wiley & Sons, New York, NY, pp. 265–272

[CG1]   Chung, F.R.K., Graham, R.L. (1978): On graphs which contain all small trees. J. Comb. Theory, Ser. B **24**, 14–23

[CG2]   Chung, F.R.K., Graham, R.L. (1979): On universal graphs. Ann. N.Y. Acad. Sci. **319**, 136–140

[CG3]   Chung, F.R.K., Graham, R.L. (1983): On universal graphs for spanning trees. J. Lond. Math. Soc., II. Ser. **27**, 203–211

[CGP]   Chung, F.R.K., Graham, R.L., Pippenger, N. (1978): On graphs which contain all small trees II. In: Hajnal, A., Sós, V. (eds.): Combinatorics I. North Holland, Amsterdam, pp. 213–223 (Colloq. Math. Soc. Janos Bolyai, Vol. 18)

[CGS]   Chung, F.R.K., Graham, R.L., Shearer, J. (1981): Universal caterpillars. J. Comb. Theory, Ser. B **31**, 348–355

[CR]   Chung, F.R.K., Rosenberg, A.L. (1986): Minced trees, with applications to fault-tolerant VLSI processor arrays. Math. Syst. Theory **19**, 1–12

[CRS]   Chung, F.R.K., Rosenberg, A.L., Snyder, L. (1983): Perfect storage representations for families of data structures. SIAM J. Algebraic Discrete Methods **4**, 548–565

[D1]   Djidjev, H.N. (1982): On the problem on partitioning planar graphs. SIAM J. Algebraic Discrete Methods **3**, 229–241

[D2]   Djidjev, H.N. (1981): A separator theorem. C.R. Ac. Bulg. Sci. **34**, 643–645

[D3]   Djidjev, H.N. (1985): A separator theorem for graphs of fixed genus. Serdica **11**, 319–321

[FP]   Friedman, J., Pippenger, N. (1987): Expanding graphs contain all small trees. Combinatorica **7**, 71–76

[G]   Gazit, H.: An improved algorithm for separating a planar graph. Preprint

[GHT]   Gilbert, J.R., Hutchinson, J.P., Tarjan, R.E. (1984): A separator theorem for graphs of bounded genus. J. Algorithms **5**, 391–407

[GM]   Gazit, H., Miller, G.L. (1987): A parallel algorithm for finding a separator in planar graphs. Proc. 28th IEEE FOCS, pp. 238–248

[GW]   Goldberg, C.H., West, D.B. (1985): Bisection of circle colorings. SIAM J. Algebraic Discrete Methods **6**, 93–106

[L]   Leiserson, C.E. (1980): Area-efficient graph layouts (for VLSI). Proc. 21st IEEE FOCS, pp. 270–281

[LRT]   Lipton, R.J., Rose, D.J., Tarjan, R.E. (1979): Generalized nested dissection. SIAM J. Numer. Anal. **16**, 346–358

[LSH]   Lewis, P.M., Stearns, R.E., Hartmanis, J. (1965): Memory bounds for recognition of context-free and context-sensitive languages. IEEE Computer Society Press, Washington, DC, pp. 191–202 (IEEE Conf. on Switching Theory and Logical Design)

[LT1]   Lipton, R.J., Tarjan, R.E. (1979): A separator theorem for planar graphs. SIAM J. Appl. Math. **36**, 177–189

[LT2]   Lipton, R.J., Tarjan, R.E. (1980): Applications of a planar separator theorem. SIAM J. Comput. **9**, 615–627

[M]   Miller, G.L. (1984): Finding small simple cycle separators for 2-connected planar graphs. Proc. 16th ACM STOC, pp. 376–382

[PR1]   Pan, V., Reif, J.H. (1985): Extension of parallel nested dissection algorithm to the path algebra problems. Technical Report TR-85-9, Department of Computer Science, State University of New York at Albany

[PR2]   Pan, V., Reif, J.H. (1985): Efficient parallel solution of linear systems. Proc. 17th ACM STOC, pp. 143–152

[R]   Rado, R. (1964): Universal graphs and universal functions. Acta Arith. **9**, 331–340

[RSS]   Rosenberg, A.L., Stockmeyer, L.J., Snyder, L. (1980): Uniform data encodings. Theor. Comput. Sci. **11**, 145–165

[Va1]   Valiant, L.G. (1977): Graph-theoretic arguments in low-level complexity. Department of Computer Science, University of Edinburgh

[Va2]   Valiant, L.G. (1981): Universality consideration in VLSI circuits. IEEE Trans. Comput. **30**, 135–140

[Ve]   Venkatesan, S.M. (1987): Improved constants for some separator theorems. J. Algorithms **8**, 572–578

Some of My Old and
New Combinatorial Problems

Paul Erdős

I published many papers on combinatorial problems [1] and to avoid repetitions I will mention only a few of my favourite old problems and a few problems which have been forgotten and will try to concentrate on some more recent ones. First I discuss some of the old problems.

1. In my old paper with Ko and Rado [2] many problems were stated, all but one of them have been solved. Here is the one which is still open: Let $|S| = 4n$, denote by $f(n)$ the largest integer for which there is a family $A_i \subset S$, $|A_i| = 2n$, $1 \le i \le f(n)$ for which $|A_i \cap A_j| \ge 2$. We conjectured that

(1)
$$f(n) = \left(\binom{4n}{2n} - \binom{2n}{n}^2 \right) / 2.$$

It is easy to see that $f(n) \ge \left(\binom{4n}{2n} - \binom{2n}{n}^2 \right) / 2$. Recently I offered 250 pounds for a proof or disproof of (1). More general conjectures have been stated by Peter Frankl and Cooper. Many papers have appeared on the Erdős-Ko-Rado theorem, here I refer only to a recent paper of Deza and Frankl and to a paper of Füredi. Both papers contain very extensive lists of references [3].

2. This is an old conjecture of Rado and myself [4]. A family of sets A_i, $1 \le i \le k$ is called a Δ-system if the intersection of any two of them equals the intersection of all of them. Peter Frankl uses for a Δ-system the picturesque and descriptive name "sunflower". Our conjecture with Rado states as follows: Let $f_k(n)$ be the smallest integer for which if $|A_i| = n$, $1 \le i \le f_k(n)$ is a family of $f_k(n)$ sets of size n, then there always is a subfamily of k sets which form a Δ-system. Our conjecture states that there is a constant C_k for which

(2)
$$f_k(n) < C_k^n.$$

I offered and offer 1000 dollars for a proof or disproof of (2). I would be satisfied with a proof for $k = 3$ which probably contains the whole difficulty. Rado and I proved

$$2^n < f_3(n) < 2^n n! \, .$$

Joel Spencer proved $f_3(n) \le (1 + \varepsilon)^n n!$ and Abbott and Hanson proved $f_3(n) > 10^{n/2}$. I would like to call the attention of the reader to the fact that $f_3(n) < n!$ for $n > n_0$ has not yet been proved.

Our conjecture would have many applications in number theory and combinatorics. Let G be an Abelian group of n elements. Let $S \subset G$ be a subset of $c \log n$ elements of G. Is it true that if c is sufficiently large, one can find three disjoint subsets S_1, S_2, S_3 of S so that the sum of the elements in S_i, $i = 1, 2, 3$ is the same? Coppersmith observed that (2) would immediately imply this and our theorem only gives this if $|S| > c(\log n)^2$, he can prove it by other methods for $|S| > c(\log n)^2 / \log \log n$. Coppersmith can prove that if $|S| > c \log n$ is false then there is an algorithm for matrix multiplication in $N^{2+\varepsilon}$ steps.

I was most interested in the following special case of the problem of Coppersmith: Let there be given $c \log n$ integers $\leq n$. Can one find three disjoint subsets of them whose sum is the same? A forthcoming paper of Sárközy and myself will deal with various extensions of this problem. Perhaps one could try to determine the smallest integer $r(n)$ for which if $1 \leq a_1 < a_2 < \ldots < a_{r(n)} \leq n$ there are always three disjoint groups of a's whose sum is the same. Perhaps this problem is hopeless but can be solved for small values of n. I can remark that perhaps my first serious conjecture going back to 1931 or 1932 (A.D. not B.C.) states: Let $a_1 < a_2 < \ldots < a_k \leq n$ be a sequence of integers for which all the sums $\sum_{i=1}^{k} \varepsilon_i a_i$, $\varepsilon_i = 0$ or 1, are distinct. Is it true that

$$(3) \qquad\qquad k < \log n + C \ ?$$

I offer 500 dollars for a proof or disproof of (3). Conway and Guy proved that there is such a sequence for which $a_k < 2^{k-2}$, but no such sequence is known for which $a_k < 2^{k-3}$. It seems certain that these conjectures have nothing to do with my conjecture with Rado.

Szemerédi and I considered the following problem: Let $|S| = n$, $A_i \subset S$, $1 \leq i \leq g(n)$. $g(n)$ is the smallest integer for which $g(n)$ subsets of S always contain a Δ-system of 3 members. We prove

$$g(n) < 2^n / 2^{c\sqrt{n}}$$

but conjecture $g(n) < (2 - \varepsilon)^n$ for some $\varepsilon > 0$. As far as I know this problem is still open [5].

3. Our next problem is the well known conjecture of Faber, Lovász and myself [6]. In September 1972 we conjectured that if G_i, $1 \leq i \leq n$ are n edge disjoint complete graphs of size n then the chromatic number of $\bigcup_{i=1}^{n} G_i$ is n. It is surprising that this simple conjecture is probably difficult. I offer 500 dollars for a proof or disproof. Many generalizations and extensions can be formulated. Hindman proved the conjecture for $n < 10$ and Seymour has some interesting preliminary results, he proved among others that if $\bigcup_{i=1}^{n} G_i$ has m vertices then it contains an independent set of size $\geq \frac{m}{n}$, which of course would follow if the chromatic number of $\bigcup_{i=1}^{n} G_i$ would be n. Winkler once told me that it is an embarrassment that our conjecture is still open.

In connection with Seymour's result perhaps I can mention an old conjecture of mine. Several years before Appel and Haken's proof of the four colour conjecture I conjectured that every planar graph of n vertices contains an independent set of $\left\lceil \frac{n+3}{4} \right\rceil$ vertices. Perhaps it would still be of interest to prove this without having to use computers. Albertson proved this with $\frac{2n}{9}$ instead of $\frac{n}{4}$. Akiyama told me the following conjecture (which does not follow from the four color theorem): Let G be a planar graph of n vertices. Then it contains a subset of $\left[\frac{n+1}{2}\right]$ vertices which contains no circuit. As far as I know this conjecture is still open.

4. This is an old problem of Lovász and myself [7]. Determine the smallest integer $m(n)$ for which there is a family of sets $|A_i| = n$, $1 \leq i \leq m(n)$, $A_i \cap A_j \neq \emptyset$ and the family can not be represented by a set $|S| = n - 1$ (i.e. if $|S| \leq n - 1$ there always is an A_j with $A_j \cap S = \emptyset$). We conjectured that

$$(4) \qquad\qquad m(n)/n \to \infty$$

but could not even prove $m(n) \geq 3n$ for $n > n_0$. We proved $m(n) < n^{\frac{3}{2}+\varepsilon}$. I often stated that our proof probably would give $m(n) < cn \log n$. I still believe this but could not never supply the details. I offer 250 dollars for a proof or disproof of (4).

Lovász and I used the probability method. It seems likely that in a finite geometry of $p^2 + p + 1$ elements one can find $cp \log p$ lines which can not be represented by a set $|S| = p$, but I could not complete the proof. Perhaps $cp \log p$ is best possible and if $\varepsilon > 0$ is sufficiently small any set of $\varepsilon p \log p$ lines can be represented by $\leq p$ points but as far as I know this is not even known for cp if c is large. Goldberg and I have a stronger formulation of (4): To every C there is an ε and a family of sets

$$|A_i| = n, \; 1 \leq i \leq Cn, \; A_i \cap A_j \neq \emptyset, \; 1 \leq i < j \leq Cn,$$

which can not be represented by a set $|S| < (1 - \varepsilon)n$. It would of course be of interest to determine the dependence of C on ε as accurately as possible.

5. Now finally I state an old problem of mine which has been partially solved by Frankl and Rödl [8].

Denote by $G(n; e)$ a graph of n vertices and e edges, or more generally let $G^{(r)}(n; e)$ denote an r-uniform hypergraph of n vertices and e hyperedges (i.e. r-tuples). We say that a sequence of hypergraphs $G^{(r)}(n_i; e_i)$, $n_i \to \infty$ has edge density α if α is the largest real number for which there is a subsequence of the n_i say $n_{i_j} \to \infty$ so that $G^{(r)}(n_{i_j}; e_{i_j})$ has an induced subhypergraph $G^{(r)}(m_i; e'_{i_j})$ with $e'_{i_j} = (\alpha + \sigma(1))\binom{m_i}{r}$. An old theorem of Stone, Simonovits and myself states that for $r = 2$ the only possible values of the density are $1 - \frac{1}{t}$, $1 \leq t \leq \infty$ [9]. This result follows immediately from a very old theorem of Stone and myself. Let $\alpha > \frac{1}{2}(1 - \frac{1}{t}) + \varepsilon$ and $G(n; e)$, $e \geq \alpha n^2$ a graph. Then $G(n; e)$ contains for $n > n_0(t, k)$ a complete $t + 1$ partite graph of k vertices of each color. For $r > 2$ I proved [10] that if $G^{(r)}(n; e)$, $e > \varepsilon n^r$ then our r-graph contains for $n > n_0(r, \varepsilon, t)$ a set of rt vertices $x_i^{(j)}$, $1 \leq i \leq t$, $1 \leq j \leq r$ so that our hypergraph contains all the t^r hyperedges $(x_{i_1}^{(1)}, x_{i_2}^{(2)} \ldots x_{i_r}^{(r)})$, $1 \leq i_j \leq t$. This theorem implies that every r-graph of positive density has density $\geq \frac{r!}{r^r}$ and the complete r partite r-graph has of course density $\frac{r!}{r^r}$. Many years ago I conjectured the so called jumping constant conjecture: There is an absolute constant c_r so that every sequence of hypergraphs of density $> \frac{r!}{r^r} + \varepsilon$ has in fact density $\geq \frac{r!}{r^r} + c_r$. This conjecture is still open. But then I overconjectured. I conjectured that for $r > 2$ the possible values of the densities form a well ordered set. This conjecture was disproved by Frankl and Rödl in their brilliant paper Hypergraphs do not jump. I offer 1000 dollars for the determination of all possible values of the densities of an r-graph. Perhaps for $r > 2$ all densities α, $\frac{r!}{r^r} \leq \alpha \leq 1$ can occur, but if my original jumping conjecture is true after all and there might in the future be a paper (perhaps authored again by Frankl and Rödl) entitled: Hypergraphs jump after all.

Now I state some recent problems.

6. First an attractive problem of Hajnal and myself [11]: Let H be a graph. Assume $n > n_0(H)$ and $G(n)$ is a graph of n vertices which does not contain an induced subgraph isomorphic to H. Is it then true that there is an $\varepsilon = \varepsilon(H)$ for which $G(n)$ contains an independent set or a complete subgraph of size n^ε? We proved this with $\exp(c\sqrt{\log n})$ instead of n^ε and proved it with n^ε for many special graphs. The simplest unsolved problem is if H is a path of four edges or a C_5. Lovász observed that the perfect graph conjecture implies that if no odd circuit of G and $\overline{G}$ contains a diagonal then G contains a complete graph or an independent set of size $\sqrt{n}$, but we can not even prove that it contains a complete graph or an independent set of size n^ε [11].

7. An old problem of Hajnal and myself states as follows: Let G be a graph of infinite chromatic number and let $2n_i + 1$, $1 \leq i < \infty$ be the length of the odd circuits contained in G. Is it true that $\sum_{i=1}^{\infty} \frac{1}{n_i} = \infty$? Perhaps in fact the sequence n_i has positive upper density, and maybe the upper density is 1.

Gyárfás, Komlós and Szemerédi and Gyárfás, Prömel, Szemerédi and Voigt [12] settled another related but probably much simpler conjecture of ours: Let $G(n; Cn)$ be a graph and $u_1, u_2, \ldots, u_t$ the length of the circuits occurring in our graph. Then $\sum \frac{1}{u_i} > \alpha \log C$.

8. Another old problem of ours states: Is it true that to every r and k there is an $f_k(r)$ so that every G of chromatic number $\geq f_k(r)$ contains a subgraph of chromatic number r all whose circuits have length $\geq k$. In other words a graph of sufficiently large chromatic number contains a subgraph of large girth and large chromatic number. Rödl [13] proved this for $k = 3$ and every r. It is surprising that this simple and attractive problem is still open. The bounds for the chromatic number given by Rödl's result are probably too large and better bounds would be very desirable.

Hajnal and I have a very nice problem for infinite graphs. Is it true that for every cardinal number m there is an $f(m)$ so that every graph of chromatic number $f(m)$ contains a triangle free induced subgraph of chromatic number m. Perhaps it contains for sufficiently large $f_r(m)$ a subgraph of chromatic number m which contains no odd circuit of size $\leq 2r + 1$. On the other hand it must contain a C_4 [14].

Now I state a few problems on random structures.

9. A well known result of Rényi, V.T. Sós and myself states [15] that every $G(n; (\frac{1}{2} + o(1))n^{3/2})$ contains a C_4 and the constant $\frac{1}{2}$ is best possible. Now Joel Spencer and I asked the following question. Choose the edges of a $K(n)$ with probability $\frac{1}{2}$. What is the largest C_4-free graph our random graph is expected to contain. Trivially almost surely the size of this graph will be between $(\frac{1}{4} + o(1))n^{3/2}$ and $(\frac{1}{2} + o(1))n^{3/2}$. We never could get any better estimates than this trivial one. Clearly this problem can be asked for other graphs instead of C_4.

10. S. Stein conjectured several years ago that if the elements of an n by n matrix are symbols and each symbol occurs at most $n - 1$ times then the determinant has an expansion term in which each symbol occurs at most once. This conjecture was known only with $c \log n$ instead of $n - 1$ [16]. Recently by a modification of the so called Local Lemma of Lovász [7] Spencer and I proved this with $cn, \frac{1}{10} > c > \frac{1}{16}$. The conjecture of Stein if true will no doubt need a new idea.

Spencer, Łuczak and I considered the following problem: Choose the edges of $K(n)$ with probability $\frac{1}{2}$. We proved that there are two constants $0 < c_1 < c_2 < 1$ so that with probability tending to 1 there is an induced subgraph of m vertices, $c_1 n < m < c_2 n$ each vertex of which has degree $\geq \frac{m}{2}$.

Very likely our result can be strengthened and there is a constant c so that with probability tending to 1 the largest induced subgraph of m vertices every vertex of which has degree $\geq \frac{m}{2}$ satisfies $m = (c + o(1))n$.

We also considered the following Turán type problem. Let m be the largest integer for which every $G(n; [\frac{n^2}{4}] + 1)$ contains an induced subgraph of m vertices every vertex of which has degree $\geq \frac{m}{2}$. We proved

$$(5) \qquad\qquad c_1 n^{\frac{1}{2}} < m < n^{2/3}.$$

Probably both the upper and lower bound in (5) are far from being best possible.

Pach and I considered a few years ago a related but slightly different problem [17]: Put $K(n) = G_1 \cup G_2$ then there always is a set of $m > \frac{cn}{\log n}$ vertices so that the graph induced by these vertices has either in G_1 or in G_2 in every vertex degree $\geq \frac{m}{2}$ We also showed that this no longer holds for $m > \frac{cn \log\log n}{\log n}$. In fact, we implicitly proved the lower bound in (5).

11. Here is a very nice extremal problem of T. Gallai: Let $G(n)$ be a graph which contains no wheel. Is it then true that the number of triangles of $G(n)$ is less than $(1 + o(1))\frac{n^2}{8}$. The Turán graph $K([\frac{n}{2}], [\frac{n}{2}])$ with a matching on one side shows that $\frac{n^2}{8}$ if true is best possible. It is not difficult to prove that the number of triangles is less than $\frac{n^2}{6} - cn^2$ for some small positive c. It is surprising that this harmless looking problem seems to present difficulties.

12. Gallai and I considered the following problem. Let $f(n)$ be the smallest integer for which the set of cliques of $G(n)$ can be represented by $f(n)$ points. We proved easily that $f(n) < n - \sqrt{n}$, this was improved by Tuza to $f(n) < n - (1 + c)\sqrt{n}$. We conjectured that

$$(6) \qquad\qquad f(n) = n - t_n$$

where t_n is the largest integer for which $r(t, 3) \leq n$ where $r(t, 3)$ is the Ramsey function, i.e. it is the smallest integer for which every graph of $r(t, 3)$ vertices either contains a triangle or an independent set of t vertices. It is well known that [18]

$$c_1 t^2 / (\log\ t)^2 < r(t, 3) < c_2 t^2 / (\log\ t).$$

We could not even prove that

$$\frac{n - f(n)}{\sqrt{n}} \to \infty.$$

13. I stated many problems and results on extremal graph theory, many in collaboration with Bollobás, Simonovits and others [19]. Here I just state two of them. Denote by $f(n; C_4)$ the smallest integer for which every $G(n; f(n; C_4))$ contains a C_4. Simonovits and I conjectured many years ago that it must contain $[c\sqrt{n}]$ C_4's. But in fact we could not even prove that it must contain two C_4's. It would be very nice if in general we could decide for which graphs does this phenomenon occur, i.e. let H be any graph and $f(n; H)$ is the smallest integer for which every $G(n; f(n; H))$ must contain H as a subgraph. Is it then true that our $G(n; f(n; H))$ must contain two (many) copies of H? It is well known that this holds for all K_4's.

14. Denote by $g(n)$ the largest integer for which every $G(n; \left[\frac{n^2}{4}\right] + 1)$ contains a triangle, the sum of the degrees of its vertices is $\geq g(n)$. Renu Laskar and I proved that

$$(1 + c_1)n < g(n) < (\frac{3}{2} - c_2)n.$$

Fan [20] proved

$$g(n) > \frac{21}{16}n$$

which is the current record. It would be desirable to determine $g(n)$ exactly. For several related problems see our papers with Caccetta and Vijayan [21].

15. Let H be a graph, $T(n; H)$ (the Turán number of H) is the largest integer for which there is a $G(n; T(n; H))$ which does not contain a subgraph isomorphic to H. Let n be large and denote by $f(n; H)$ the number of graphs of n vertices which do not contain a subgraph isomorphic to H. Assume first that H is bipartite and is not a tree. I conjectured that

(7) $$f(n; H) = 2^{(1 + o(1)) T(n; H)}$$

(7) if true will probably not be easy to prove. Kleitman and Winston proved that [22]

(8) $$f(n; C_4) < 2^{cn^{3/2}}.$$

If (7) is true then (8) should hold with $c = \frac{1}{2}$. To get an asymptotic formula for $f(n; H)$ (and not only for its logarithm) is probably hopeless. The determination of $T(n; H)$ is also usually very difficult and has been investigated a great deal and has been started by Turán. Here I mention only an old conjecture of Simonovits and myself which states that

(9) $$T(n; H) < cn^{3/2}$$

if and only if H does not contain a subgraph each vertex of which has degree > 2. We are not at all sure if (9) is true and I offer 250 dollars for a proof and 100 dollars for a counterexample. Turán determined $T(n; K_r)$ for every r.

Rothschild, Kleitman and I [23] obtained an asymptotic formula for $f(n; K_3)$ and an asymptotic formula for the logarithm of $f(n; K_r)$. Kolaitis, Prömel and Rothschild in a very difficult paper recently obtained an asymptotic formula for $f(n; K_r)$ [24].

Hajnal and I wanted recently an asymptotic formula or a good inequality for $f(n; K_r)$ if r can depend on n, e.g. if $r < c \log \log n$. As far as we know this problem has not yet been investigated. Prömel and I have some preliminary results.

16. Here are four old problems of mine. Is it true that every triangle free graph of $5n$ vertices can be made bipartite by the omission of at most n^2 edges? Faudree, Pach, Spencer and I in a recent paper proved a weaker result [25].

Is it true that if a graph of $5n$ vertices contains more than n^5 C_5's then it must contain a triangle? If true this is clearly best possible. E. Győri proved this with $1.03n^5$.

An old forgotten problem of mine (which perhaps deserves to be remembered) states as follows: Let $G(10n)$ be a graph of $10n$ vertices. Assume that every set of $5n$ vertices induces more than $2n^2$ edges. Then the graph contains a triangle. If true it is easy to see that it is best possible.

Is it true that if G is a bipartite graph of n white and $n^{2/3}$ black vertices and has more than Cn edges then for sufficiently large C our graph contains a C_6. It is not difficult to prove that it must contain a C_8.

17. To end the paper I state a few recent problems:

17.1 This is a problem of Tuza and myself posed when we both attended the meeting at Positano. Let $G(n; [\frac{n^2}{4}] + 1)$ be a graph which contains no K_4. Denote by $f(n)$ the largest integer for which $\overline{G}$ (the complement of G) has $f(n)$ edges $e_1, e_2, \ldots, e_{f(n)}$, so that $G + e_r$ contains a K_4, $1 \le r \le f(n)$. We proved $f(n) > cn^2$ and conjectured

(10) $$f(n) = (1 + o(1)) \frac{n^2}{16}.$$

A simple example shows $f(n) \le (1 + o(1)) \frac{n^2}{16}$.

17.2 Let $G(n; e)$ be a graph of n vertices and e edges. Assume that every edge is contained in a triangle. Gallai defines $e(\max)$ as the largest integer for which there are $e(\max)$ edges no two of which are in a triangle and $e(\min)$ is the smallest integer for which the omission of $e(\min)$ edges will make $G(n; e)$ triangle free. Gallai asked for the investigation of these quantities. Tuza and I conjectured that

$$(11) \qquad\qquad e(\min) + e(\max) \le \frac{n^2}{4}.$$

Gallai and Tuza found many graphs of different structure for which there is equality in (11), which perhaps suggests that the proof of (11) (if true) will not be quite simple. Very recently Lehel suggested the following strengthening of (11). Denote by $e_b(\min)$ the smallest integer for which the omission of $e_b(\min)$ edges will make $G(n; e)$ bipartite. Is it true that

$$(12) \qquad\qquad e_b(\min) + e(\max) \le \frac{n^2}{4}.$$

I tried unsuccessfully to give a counterexample to (12). I asked: Is it true that for every $\varepsilon > 0$ there is a graph of e edges for which

$$(13) \qquad\qquad e(\min) > (\frac{1}{2} - \varepsilon)e \text{ and } e(\max) > (\frac{1}{2} - \varepsilon)e.$$

We could make no progress with (13) and perhaps for ε small enough such graphs do not exist.

Simonovits and I proved

$$(14) \qquad\qquad c_1 e^{2/3} < \min_{G}(e(\min) + e(\max)) < c_2 e^{2/3}$$

where in (14) the minimum is to be taken over all graphs with e edges where every edge occurs in at least one triangle. We hope to publish these problems and results with proofs in a finite time (not posthumously for Gallai and myself I hope).

Recently Tuza told me one of his older conjectures which states as follows: Let G be a graph of e edges $e_1 = e(\max), e_2$ is the largest subset of edges which contains no traingle. Then

$$(15) \qquad\qquad \frac{e}{3} + \frac{e_1}{2} + \frac{e_2}{6} \le e_2.$$

If G has not triangle (i.e $e_2 = e$) (15) is trivial and Tuza conjectures that if G has a triangle then the inequality is strict.

17.3 To end the paper let me state a nice recent problem of Füredi and Seymour. Larman conjectured several years ago that if $|\mathcal{S}| = n$ and F is a family $A_1, A_2, \ldots$ of subsets of $\mathcal{S}$ every two of which have at least r elements in common (i.e $|A_i \cap A_j| \ge r$), then the family F can be decomposed into at most n subfamilies $F_1, F_2, \ldots$ where for every k, $1 \le k \le n$ two A's in F_k have at least $r + 1$ elements in common. As far as I know this conjecture is open even for $r = 1$. Now Füredi and Seymour ask: Let $\mathcal{S}$ be a set of n elements $x_1, \ldots, x_n$ and F an intersecting family of subsets $A_1, A_2, \ldots$. Can one find n pairs (x_i, x_j) so that every A contains at least one of these pairs? If the conjecture is false they ask for the smallest $f(n)$ for which there are always $f(n)$ pairs (x_i, x_j) so that every A contains at least one of these $f(n)$ pairs. They easily proved $f(n) < cn^{4/3}$ and of course believe that $f(n)$ is much smaller.

References

[1] Here is a partial list of my papers on combinatorial problems:

Problems and results in graph theory and combinatorial analysis. In: Nash-Williams, C., Sheenan, J. (eds.): Proc. 5th British Combinatorial Conf. Utilitas Mathematica, Winnipeg, pp. 169–192 (1976, Congr. Numerantium 15)

Problems and results on finite and infinite graphs. In: Fiedler, M. (ed.): Recent advances in graph theory. Academia Praha, Prague, pp. 183–190 (1975)

Extremal problems in graph theory. In: Fiedler, M. (ed.): Theory of graphs and its applications. Academic Press, New York, NY, pp. 29–36 (1964)

Problems and results on finite and infinite combinatorial analysis. In: Hajnal, A., Rado, R., Sós, V. (eds.): Infinite and finite sets, Vol. 1. North Holland, Amsterdam, pp. 403–424 (1975, Colloq. Math. Soc. Janos Bolyai, Vol. 10)

Some unsolved problems in graph theory and combinatorial analysis. In: Welsh, D.J.A. (ed.): Combinatorial mathematics and its applications. Academic Press, London, pp. 97–109 (1971)

Problems and results in chromatic graph theory. In: Harary, F. (ed.): Proof techniques in graph theory. Academic Press, New York, NY, pp. 27–35 (1969)

Extremal problems on graphs and hypergraphs. In: Berge, C., Ray-Chaudhuri, D. (eds.): Hypergraph Semin. at the Ohio State University 1972. Springer Verlag, Berlin, Heidelberg, pp. 75–83 (1974, Lect. Notes Math., Vol. 411)

Topics in combinatorial analysis. In: Mullin, R.C., Reid, K.B., Roselle, D.P., Thomas, R.S.D. (eds.): Proc. 2nd Southeastern Conf. on Combinatorics, Graph Theory and Computing. Utilitas Mathematica, Winnipeg, pp. 2–20 (1971, Congr. Numerantium 3)

Some new applications of probability methods to combinatorial analysis and graph theory. In: Hoffman, F., Kingsley, R.A., Levow, R.B., Mullin, R.C. (eds.): Proc. 5th Southeastern Conf. on Combinatorics, Graph Theory and Computing. Utilitas Mathematica, Winnipeg, pp. 39–54 (1974, Congr. Numerantium 10)

Some recent progress on extremal problems in graph theory. In: Hoffman, F., Mullin, R.C., Levow, R.B., Roselle, D.P., Stanton, R.G., Thomas, R.S.D. (eds.): Proc. 6th Southeastern Conf. on Combinatorics, Graph Theory and Computing. Utilitas Mathematica, Winnipeg, pp. 3–14 (1975, Congr. Numerantium 14)

Some recent problems and results in graph theory, combinatorics and number theory. In: Hoffmann, F., Lesniak-Foster, L., Mullin, R.C., Reid, K.B., Stanton, R.G. (eds.): Proc. 7th Southeastern Conf. on Combinatorics, Graph Theory and Computing. Utilitas Mathematica, Winnipeg, pp. 3–14 (1976, Congr. Numerantium 17)

Problems and results in combinatorial analysis. In: Hoffman, F., Lesniak-Foster, L. (eds.): Proc. 8th Southeastern Conf. on Combinatorics, Graph Theory and Computing. Utilitas Mathematica, Winnipeg, pp. 3–12 (1977, Congr. Numerantium 19)

Problems and results in combinatorial analysis and combinatorial number theory. In: Hoffman, F., MacCarthy, D., Mullin, R.C., Stanton, R.G. (eds.): Proc. 9th Southeastern Conf. on Combinatorics, Graph Theory and Computing. Utilitas Mathematica, Winnipeg, pp. 29–40 (1978, Congr. Numerantium 21)

Some extremal problems on families of graphs. In: Holton, D.A., Seberry, J. (eds.): Combinatorial mathematics. Springer Verlag, Berlin, Heidelberg, pp. 13–21 (1978, Lect. Notes Math., Vol. 686)

With D.J. Kleitman: Extremal problems among subsets of a set. Discrete Math. 8, 289–294 (1974, and Proc. 2nd Chapel Hill Conf., pp. 144–170)

Problems and results in graph theory and combinatorial analysis. In: Bermond, J.-C., Fournier, J.-C., Las Vergnas, M., Sotteau, D. (eds.): Problèmes combinatoires et

Théorie des Graphes. CNRS, Paris, pp. 127–129 (1978, Colloq. Int. CNRS, Vol. 260)

Problems and results in combinatorial analysis. In: Motzkin, T.S. (ed.): Combinatorics. AMS, Providence, RI, pp. 77–89 (1971, Proc. Symp. Pure Math., Vol. 19)

Old and new problems in combinatorial analysis and graph theory. Proc. 2nd International Conf. on Combinatorial Mathematics, New York, NY, pp. 177–187 (1979)

[2] Erdős, P., Ko, C., Rado, R. (1961): Intersection theorems for systems of sets. Q. J. Math. Oxf. II. Ser. **12**, 313–320

[3] Deza, M., Frankl, P. (1983): Erdős-Ko-Rado theorem 22 years later. SIAM J. Algebraic Discrete Methods **4**, 419–431

Füredi, Z. (1988): Matchings and covers in hypergraphs. Graphs Comb. **4**, 115–206

Frankl, P. (1987): The shifting technique in extremal set theory. In: Whitehead, C. (ed.): Surveys in combinatorics. Cambridge University Press, Cambridge, pp. 83–110 (Lond. Math. Soc. Lect. Note Ser., Vol. 123)

[4] Erdős, P., Rado, R. (1960): Intersection theorems for systems of sets I. J. Lond. Math. Soc., II. Ser. **35**, 85–90

Erdős, P., Rado, R. (1969): Intersection theorems for systems of sets II. J. Lond. Math. Soc., II. Ser. **44**, 467–479

Erdős, P., Rado, R., Milner, E. (1974): Intersection theorems for systems of sets III. J. Aust. Math. Soc., Ser. B **18**, 22–40

[5] Erdős, P., Szemerédi, E. (1978): Combinatorial properties of systems of sets. J. Comb. Theory, Ser. A **24**, 308–313

[6] Our conjecture was first mentioned in my paper "Problems and results on finite and infinite combinatorial analysis" (see [1])

[7] Erdős, P., Lovász, L. (1975): Problems and results on 3-chromatic hypergraphs and some related questions. In: Hajnal, A., Rado, R., Sós, V. (eds.): Infinite and finite sets, Vol. 2. North Holland, Amsterdam, pp. 609–624 (Colloq. Math. Soc. Janos Bolyai, Vol. 10)

[8] Frankl, P., Rödl, V. (1984): Hypergraphs do not jump. Combinatorica **4**, 149–159

[9] Erdős, P., Simonovits, M. (1966): A limit theorem in graph theory. Stud. Sci. Math. Hung. **1**, 51–57

Erdős, P., Stone, A. (1946): On the structure of linear graphs. Bull. Am. Math. Soc., New Ser. **52**, 1087–1091

[10] Erdős, P. (1965): On extremal problems of graphs and generalized graphs. Isr. J. Math. **2**, 189–190

[11] Erdős, P., Hajnal, A. (1977): On spanned subgraphs of graphs. In: Beiträge zur Graphentheorie und deren Anwendungen. Technische Hochschule Ilmenau, pp. 80–96 (International Colloq. 1977, Oberhof)

[12] Gyárfás, A., Komlós, J., Szemerédi, E. (1984): On the distribution of cycle length in graphs. J. Graph Theory **8**, 441–462

Gyárfás, A., Prömel, H.J., Szemerédi, E., Voigt, B. (1985): On the sum of the reciprocals of cycle lengths in sparse graphs. Combinatorica **5**, 41–52

[13] Rödl, V. (1977): On the chromatic number of subgraphs of a given graph. Proc. Am. Math. Soc. **64**, 370–371

[14] Erdős, P., Hajnal, A. (1966): On the chromatic number of graphs and set-systems. Acta Math. Hung. **17**, 61–99

[15] Erdős, P., Rényi, A., Sós, V. (1966): On a problem of graph theory. Stud. Sci. Math. Hung. **1**, 215–235

Erdős, P., Rényi, A. (1962): On a problem in graph theory. Publ. Math. Inst. Hung. Akad. Sci. **7**, 215–227 (In Hungarian)

[16] Erdős, P., Hickerson, D.R., Norton, D.A., Stein, S.K. (1988): Has every latin square of order n a partial latin transversal of order $n-1$? Am. Math. Mon. 95, 428–430

[17] Erdős, P., Pach, J. (1983): On a quasi-Ramsey problem. J. Graph Theory 7, 137–147

[18] Erdős, P. (1961): Graph theory and probability II. Can. J. Math. 13, 346–352
Ajtai, M., Komlós, J., Szemerédi, E. (1980): A note on Ramsey numbers. J. Comb. Theory, Ser. A 29, 354–360

[19] Bollobás, B. (1978): Extremal graph theory. Academic Press, London (Lond. Math. Soc. Monogr., Vol. 11)
Simonovits, M. (1983): Extremal graph theory. In: Beineke, L.W., Wilson, R.J. (eds.): Selected topics in graph theory, Vol. 2. Academic Press, London
Simonovits, M. (1984): Extremal graph problems, degenerate extremal problems and supersaturated graphs. In: Bondy, J.A., Murty, U.S.R. (eds.): Progress in graph theory. Academic Press, London
Erdős, P., Simonovits, M. (1983): Supersaturated graphs and hypergraphs. Combinatorica 3, 181–192

[20] Fan, G. (1988): Degree sum for a triangle in a graph. J. Graph Theory 12, 249–263

[21] Cacetta, L., Erdős, P., Vijayan, K. (1985): A property of random graphs. Ars Comb. 19A, 287–294

[22] Kleitman, D.J., Winston, K.J. (1982): On the number of graphs without 4-cycles. Discrete Math. 41, 167–172

[23] Erdős, P., Kleitman, D.J., Rothschild, B. (1976): Asymptotic enumeration of K_n-free graphs. In: Teorie combinatorie. Accademia Nazionale dei Lincei, Rome (International Colloq. 1973, Rome, Atti Conv. Lincei, Vol. 17)

[24] Kolaitis, P.G., Prömel, H.J., Rothschild, B. (1987): K_{l+1}-free graphs: Asymptotic structure and a 0-1 law. Trans. Am. Math. Soc. 303, 637–671

[25] Erdős, P., Faudree, R., Pach, J., Spencer, J. (1988): How to make a graph bipartite. J. Comb. Theory, Ser. B 45, 86–98

Packing Paths, Circuits, and Cuts – A Survey

András Frank

1. Introduction

The main motivation of the topic treated in this paper comes partly from routing problems in VLSI design and partly from the intrinsic development of graph theory. Therefore our purpose is twofold. First we want to cover a relatively broad area from graph theory and, doing so, many results will be listed which do not seem to have too much to do with VLSI applications. On the other hand we try to put some emphasis on results that are closer to routing problems.

The paper, however, is not merely a list of theorems. Sometimes only an outline of a proof is given to provide an impression of the idea. In other cases, when the original proof is not easily available (because it is either too complicated or appeared in "hidden" journals or proceedings) we include a relatively detailed proof.

At this point I would like to express my thanks to Professor A.V. Karzanov for his comments and information on the fundamental contribution to this topic made by Soviet mathematicians.

Throughout the paper we work with a connected graph $G = (V, E)$ that may be directed or undirected. Here V denotes the node set of G. For $X \subseteq V$ put $\overline{X} := V - X$. $A \subset B$ means that $A \subseteq B$ but $A \neq B$. For $s, t \in V$ a set $T \subseteq V$ is called a $t\bar{s}$-set if $t \in T$ and $s \notin T$. We do not distinguish between an element v and the one-element set $\{v\}$. The set of edges between A and $V - A$ is called a cut and is denoted by $\nabla(A)$. If both A and $V - A$ induce a connected subgraph then $\nabla(A)$ is called a bond. It is well-known that any cut can be partitioned into bonds.

If G is directed, an element e of E with tail u and head v is denoted by $e = uv$. For a directed edge we sometimes use the shorter term arc. If G is undirected, an element e of E with endpoint u and v is denoted by $e = uv(= vu)$. (Such a notation is not precise because G may have parallel edges, but no ambiguity can arise from this sloppiness.)

For a graph G (directed or undirected) $d_G(X, Y)$ denotes the number of edges with one end in $X - Y$ and one end in $Y - X$. We use $d_G(X)$ for $d_G(X, V - X)$. (When it is not ambiguous we leave out the subscript G.) For two graphs $G = (V, E)$ and $H = (V, F)$ (where E, F are disjoint but may contain elements that are parallel) $G + H$ denotes the graph $(V, E \cup F)$. In a directed graph $\varrho_G(X)$ ($\delta_G(X)$) denotes the number of edges entering (leaving) X.

An undirected (directed) graph is called *Eulerian* if $d(v)$ is even ($\varrho(v) = \delta(v)$) for every node v.

Where S is a finite set, $X \subseteq S$ and $h : S \to R$ is a function we use the notation $h(X) := \sum_{x \in X} h(x)$.

In an undirected graph $G = (V, E)$ *splitting off* two edges uv and vz means the operation of replacing uv and vz by a new edge uz. Similarly, in a directed graph splitting off two arcs uv and vz is an operation that replaces uv and vz by a new arc uz. If $u = z$, we leave out the resulting loop uz.

The following two equalities will prove extremely useful. The first concerns directed graphs while the second is for undirected graphs. For $A, B \subseteq V$

$$(1.1) \qquad \varrho_G(A) + \varrho_G(B) = \varrho_G(A \cap B) + \varrho_G(A, B) + d_G(A, B)$$

$$(1.2) \qquad d_G(A) + d_G(B) = d_G(A \cap B) + d_G(A \cup B) + 2d_G(A, B)$$

The proof consists of showing that the contribution of any of the edges to the two sides of the equality is the same.

An obvious consequence of (1.1) and (1.2) is the submodular property of ϱ and d:

$$(1.1') \qquad \varrho_G(A) + \varrho_G(B) \geq \varrho_G(A \cap B) + \varrho_G(A \cup B)$$

$$(1.2') \qquad d_G(A) + d_G(B) \geq d_G(A \cap B) + d_G(A \cup B)$$

Sometimes more complicated relations are needed. Suppose that the node set V is partitioned into 5 sets; A, M, N, X, Y. Then

$$(1.3) \quad d(X \cup M) + d(Y \cup M) + 2d(A, N) = d(X \cup N) + d(Y \cup N) + 2d(A, M).$$

The proof is an easy exercise.

The starting point of the whole theory is Menger's (1927) theorem. In what follows s and t are two specified nodes of the graph or digraph $G = (V, E)$ in question.

Theorem 1.1. *a, In a digraph (graph) there are k arc-disjoint (edge-disjoint) st-paths if and only if every $t\bar{s}$-set has at least k entering edges.*
b, In a digraph (graph) if there is no arc (edge) from s to t, there are k openly disjoint st-paths if and only if the paths from s to t cannot be covered by less than k nodes distinct from s and t.

(A set of paths is called openly disjoint if they are disjoint except for their end nodes).

Actually here we have four theorems according to whether directed or undirected and edge-(arc-)disjoint or openly disjoint st-paths are considered. Menger originally proved the undirected, openly disjoint version.

Although this theorem is included in almost every book concerning graphs, here we exhibit a proof since its basic idea, splitting off a pair of adjacent edges and the use of submodularity, is extensively used throughout the whole paper.

Proof. Let us first consider the arc-disjoint case. Let the minimum in question be $k > 0$. Call a $t\bar{s}$-set T *tight* if $\varrho(T) = k$.

Lemma. *If A and B are tight, then both $A \cap B$ and $A \cup B$ are tight, furthermore $d(A, B) = 0$.*

Proof. By (1.1) we have $k + k = \varrho(A) + \varrho(B) = \varrho(A \cap B) + \varrho(A \cup B) + d(A, B) \geq k + k + d(A, B)$ from which the lemma follows. ☐

We use induction on the number of edges. Let $e = uv$ be an edge with $v \neq t$. (If there is no such an edge, the theorem is trivial.) We can assume that e enters a tight set, for otherwise, by deleting e, we are done by induction. By the lemma there is a unique minimal tight set T that is entered by e. Now there is an edge vz with $z \in T$, for otherwise $\varrho(T - v) < k$. There is no tight set Z containing z but not containing u and v since then $Z \cap T$ is tight by the lemma contradicting the minimal choice of T. There is no tight set Z containing u and z but not v since then $d(Z, T) > 0$ contradicting the lemma again.

Therefore if we split off uv and vz, no $t\bar{s}$-set can arise with indegree less than k. By induction the resulting graph includes k edge-disjoint paths from s to t. Replacing back the new edge uz by uv and vz we obtain k edge-disjoint paths in G.

From the directed edge-version the other three cases of the Menger theorem follow by elementary construction. Namely, in case a, replace each edge by a pair of oppositely directed arcs and observe that if there is a set of k arc-disjoint paths in the resulting digraph, then there is one that does not use both arcs assigned to an original edge. The same construction yields the undirected openly-disjoint version from the directed one.

To see the directed openly-disjoint version construct a new digraph D' from D as follows. Replace each node v of D ($\neq s, t$) by a pair of new nodes v' and v''. Let $v'v''$ be an arc of D' and for an arc uv of D let $u''v'$ be an arc of D'. Arc-disjoint st-paths in D' correspond to openly disjoint paths in D. Moreover, if there are k arcs in D' covering all st-paths, then these arcs can be assumed to be of type $v'v''$ and this set of arcs corresponds to a set of k nodes of D covering all st-paths. ☐

There exist other versions of Menger's theorem. For example, given a graph and two disjoint subsets S, T of its node set, there are k disjoint paths between S and T if and only if there are no $k-1$ nodes covering all such paths. By elementary construction this result easily follows from the original Menger theorem.

Since this paper is about paths and circuits let us close this introductory first section by mentioning a recent application of the Menger theorem.

Theorem 1.2 (Egawa, Kaneko and Matsumoto 1988). *In an undirected graph there are k edge-disjoint circuits passing through two specified nodes s and t if and only if every cut separating s and t contains at least $2k$ edges and after deleting any node distinct from s and t every cut separating s and t contains at least k edges.*

2. Disjoint Paths Problem

In this section we address the following problem, called the *disjoint paths problem*. Let us given a connected graph $G = (V, E)$ or a digraph and k pairs of nodes $(s_1, t_1), (s_2, t_2), \ldots, (s_k, t_k)$. Find k pairwise disjoint paths connecting the corresponding pairs (s_i, t_i). If we are interested in finding edge-disjoint paths we speak about the *edge-disjoint paths problem*.

First, let us concentrate on undirected edge-disjoint paths. Sometimes it is convenient to mark the terminal pairs to be connected by an edge. The graph $H = (U, F)$ formed by the marking edges is called a *demand graph* while the original graph $G = (V, E)$ is the *supply graph*. (Of course, H may not be connected). In this terminology the edge-disjoint paths problem is equivalent to seeking for $|F|$ edge-disjoint circuits in $G + H$ each of which contains exactly one edge of F.

A natural necessary condition is the cut criterion:

$$\text{CUT-CRITERION } d_G(X) \geq d_H(X) \text{ for every } X \subseteq V.$$

Since any cut of G can be partitioned into bonds cut criterion holds if we require the inequality above only for subsets X for which both X and $V - X$ induce a connected subgraph.

We call $d_H(X)$ the *congestion* and the difference $s(X) := d_G(X) - d_H(X)$ the *surplus* of cut $\nabla(X)$ The cut criterion is equivalent to saying that the surplus of every cut is non-negative. A cut $\nabla(X)$ is called *tight* or *saturated* if $s(X) = 0$.

The cut criterion is sufficient if the demand graph consists of a set of parallel edges (in which case we are back at the undirected edge-version of Menger's theorem), or if H is a star (that is, the demand edges share a common endpoint). (This immediately follows from Menger).

The cut criterion is not sufficient, in general, as the following simple example shows (Figure 2.1).

REDUCTION PRINCIPLE. Let us introduce a simple device by which the edge disjoint paths problem can be reduced to a case when every degree in $G + H$ is at most 4. First replace each demand edge by a path of three edges such that the middle edge is a demand edge, the two other edges are supply edges. As a result, no demand edge is incident to a node of degree bigger than 2. Next let v be a node with degree at least 5. Replace this node and the incident edges as is shown in the picture 2.2.

It is easy to see that the edge-disjoint paths problem is solvable in the original graph if and only if it is solvable in the new graph. Applying this reduction at one node v as long as the degree of v is bigger than 4, we see that v is replaced by a subgraph displayed in Figure 2.2a.

The problem we obtain by eliminating all nodes of degree at least five is not only equivalent to the original problem but its size is a polynomial of the original size. Indeed, every node has been replaced by $O(d(v)^2)$ new nodes of degree four. For applications of the reduction principle, see Sections 3 and 4.

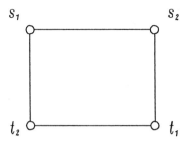

Fig. 2.1

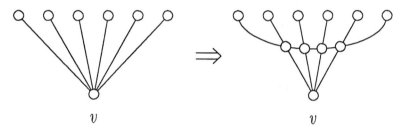

Fig. 2.2

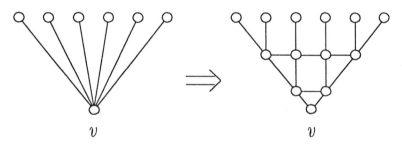

Fig. 2.2a

There is a natural relaxation of the edge-disjoint paths problem called *multicommodity flow*, or for short, the *multiflow* problem. Let G be undirected. The problem is to assign non-negative variables to paths connecting the prescribed terminal pairs $s_1t_1, s_2t_2, \ldots, s_kt_k$ so that for each terminal pair s_it_i the sum of variables assigned to paths connecting s_i and t_i is at least one and the sum of variables assigned to paths passing through any edge of G is at most one. (In the general multiflow problem one may have capacities on the edges).

Obviously a solution to the edge-disjoint paths problem is a 0-1 solution to the multiflow problem and vice versa. This is why we say that the edge-disjoint path problem *has a fractional solution* when its multiflow relaxation has a solution. Notice that the problem in Figure 2.1 has a fractional solution (assign $1/2$ to the 4 paths of length 2).

One way to formulate the multiflow problem as a linear program is the following. Let A be a 0,1 matrix the rows of which correspond to the edges of G the colums correspond to the good circuits. An entry (i, j) is 1 if the edge corresponging to i is in the circuit corresponding to j and 0 otherwise. Similarly let B be a 0,-1 matrix the rows of which correspond to the edges of H the columns correspond to the good circuits. An entry (i, j) is -1 if the edge corresponding to i is in the circuit corresponding to j and 0 otherwise. (The structure of B is simple: every column has exactly one non-zero entry). The multiflow problem is equivalent to the following linear inequality system. $Ax \leq \underline{1}$, $Bx = -\underline{1}$, $x \geq 0$, where $\underline{1}$ and $-\underline{1}$ is appropriately sized vectors of 1's and -1's, respectively.

By Farkas' lemma this system has no solution if and only if there is a vector w in R_+^E and a vector z in R^F such that $\Sigma(w(e) : e \in E) - \Sigma(z(f) : f \in F) < 0$ and such that $\Sigma(w(e) : e \in C - f) - z(f) \geq 0$ holds for every demand edge f and every circuit C for which $C \cap F = \{f\}$. Obviously, if there is such a w and z, then z can be chosen so as to satisfy $z(f) = dist_w(u, v)$ where $f = uv$ and $dist_w(u, v)$ is the minimum w—weight of a path in G connecting the end nodes of demand edge f.

Theorem 2.0 *The multiflow problem has a solution if and only if*

$$\text{DISTANCE CRITERION } \Sigma(dist_w(u, v) : uv \in F) \leq \Sigma(w(e) : e \in E)$$

holds for every vector $w \in \mathbb{R}_+^E$.

By chosing d to be 1 on the edges of a cut and 0 otherwise we see that the distance criterion implies the cut criterion. But not the other way round! In the next figure one can check by inspection that the cut criterion holds true but the distance criterion does not: choose w to be 1 everywhere.

This example also shows that the cut criterion is not sufficient in general even if $G + H$ is Eulerian. The next example, due to Éva Tardos, shows that even the stronger distance criterion is not sufficient (Figure 2.4).

Actually this is not surprising in the view of the following.

Theorem 2.1 (R. Karp 1972). *The undirected (edge-) disjoint paths problem (when k can vary) is NP-complete.*

The disjoint paths problem remains NP-complete for planar G and even for gridgraphs (a gridgraph is an induced subgraph of a rectilinear grid) (Richards), (Kramer and Leeuwen).

Even, Itai and Shamir (1976) proved that the problem is NP-complete in the special case when the demand graph consists of two sets of parallel edges. Recently, Middendorf and Pfeiffer (1989) proved that both the edge-disjoint

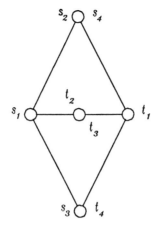

Fig. 2.3

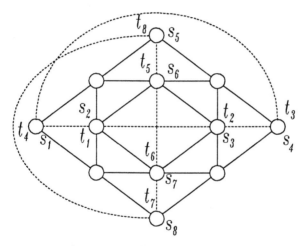

Fig. 2.4

and the node-disjoint paths problem is NP-complete if $G + H$ is planar even if every degree in $G + H$ is restricted to be at most 3. They also showed that the half-integer multicommodity flow problem is NP-complete. This implies that the edge-disjoint paths problem is NP-complete even if $G + H$ is Eulerian.

To consider the arc-disjoint paths problem in directed graphs let $D = (V, A)$ be a digraph and let (s_i, t_i) $(i = 1, 2, \ldots, k)$ be ordered pairs of terminals. The problem is to find arc-disjoint paths from s_i to t_i.

Let $H = (U, F)$ denote the demand digraph, where $F = \{(t_i s_i : i = 1, 2, \ldots, k\}$. Then the problem can be reformulated as follows: Find k arc-disjoint circuits in $G + H$ each of which contains exactly one demand edge.

Again a natural necessary condition is available:

DIRECTED CUT CRITERION $\varrho_G(X) \geq \delta_H(X)$ for every $X \subseteq V$.

If $s_1 = \ldots = s_k$ and $t_1 = \ldots = t_k$, then the directed cut criterion is sufficient as well (directed arc version of Menger's theorem). It remains true, via an elementary construction, if we require only $s_1 = \ldots = s_k$.

For general digraphs one has the following negative result.

Theorem 2.2 (Fortune, Hopcroft and Wyllie 1980). *The (arc-) disjoint paths problem is NP-complete for $k = 2$.*

To close this section we formulate a necessary condition for the disjoint path problem (in an undirected graph).

NODE-CUT CRITERION. The counterpart of the cut condition requires that a subset S of nodes must not separate more than $|S|$ terminal pairs.

This condition is sufficient if the terminal pairs share a common node (a node-version of the Menger theorem) but not in general. Another special case when the node-cut criterion is sufficient is the following.

Theorem 2.3 (N. Robertson and P. Seymour 1986). *Suppose that G is planar and the terminals are on the outer face. This disjoint paths problem has a solution if and only if the node-cut condition holds and there are no two "crossing" terminal pairs (that is, any two pairs (s_1, t_1) and (s_2, t_2) are in this order on the outer face: s_1, t_1, s_2, t_2).*

(The proof of this result is easy).

3. $G + H$ is Eulerian

In Section 2 we saw how submodularity can be used for proving Menger's theorem. Let us start this section by claiming a simple lemma that makes possible some more sophisticated uses of submodularity.

Let $G = (V, E)$ and $H = (V, F)$ be two graphs for which the cut criterion holds, that is $d_G(X) \geq d_H(X)$ for every $X \subseteq V$. Call a subset X of nodes tight if $d_G(X) = d_H(X)$.

Lemma 3.1. *a, If A and B are tight and $d_H(A, B) = 0$, then both $A \cap B$ and $A \cup B$ are tight and $d_G(A, B) = 0$. b, If A and B are tight and $d_H(A, \overline{B}) = 0$, then both $A - B$ and $B - A$ are tight and $d_G(A, \overline{B}) = 0$.*

Proof. By applying (1.2) to G and H we have
$d_H(A) + d_H(B) = d_G(A) + d_G(B) = d_G(A \cap B) + d_G(A \cup B) + 2d_G(A, B) \geq d_H(A \cap B) + d_H(A \cup B) + 2d_G(A, B) = d_H(A) + d_H(B) + 2(d_G(A, B) - d_H(A, B))$
from which part a, follows. We obtain part (b) if (a) is applied to A and $V - B$. □

In this section we outline the edge-disjoint paths problem when $G + H$ is Eulerian. It was already mentioned that the edge-disjoint paths problem can be formulated in terms of packing of circuits. When $G + H$ is Eulerian, the problem is equivalent to finding a partition of the edge set of $G + H$ into circuits each of which contains at most one edge from H. Such a partition will be called *good*. Figure 2.3 shows that the cut criterion is not sufficient in general even if G is planar.

However, there are important special cases when the cut criterion is sufficient. In one class of examples the supply graph G is planar and there are additional restrictions on H. In another class G is arbitrary but the demand graph H is rather restricted.

First let us survey the results concerning planar G.

Theorem 3.2 (Okamura and Seymour 1981). *Suppose that G is planar, $G + H$ is Eulerian, and each terminal is on one face of G. Then the cut criterion is necessary and sufficient for the solvability of the edge-disjoint paths problem.*

Proof (Okamura and Seymour 1981). By induction on the number of edges of G. Let G be embedded in the plane. We can assume that G is 2-connected. Then every face is bounded by a circuit. Let C denote the circuit bounding the infinite face and let the subscripts of the nodes $v_1,\ldots,v_h$ of C reflect the cyclic order. Assume that the terminals are on C.

Choose an edge e of C which is in a tight cut (if there is no tight cut any $e \in E(C)$ will do) and renumber the nodes of C such that $e = v_h v_1$. Let A be a minimal tight set containing v_i but not v_h. Choose a demand edge $f = v_i v_j$ ($i < j$) such that $v_i \in A$, $v_j \notin A$ and j is as big as possible. (If there is no tight set at all, any demand edge will do).

Delete e from G and replace f by $v_1 v_i$ and $v_j v_h$. We are going to show that the cut criterion holds with respect to the resulting $\overline{G}$ and $\overline{H}$. This will imply the theorem since $\overline{G}$ has one less edge than G and the other hypotheses of the theorem hold for $\overline{G}$ and $\overline{H}$. So by induction we have the edge-disjoint paths in $\overline{G}$. This provides the required edge-disjoint paths in G if we observe that glueing together the path between v_1 and v_i and the path between v_j and v_h and the edge e we obtain a path between v_i and v_j.

If the cut criterion, indirectly, fails to hold for $\overline{G}$ and $\overline{H}$, then there is a set B which is tight with respect to G and H and, among the four nodes v_1, v_i, v_j, v_h, B contains exactly (i) v_1, (ii) v_h, (iii) v_1 and v_h.

By Lemma 3.1 if A and B are tight and $d_H(A, B) = 0$, then both $A \cap B$ and $A \cup B$ are tight and $d_G(A, B) = 0$.

By the choice of f in each case we have $d_H(A, B) = 0$ so Lemma 3.1 applies. Thus $A \cap B$ is tight which in Cases (i) and (iii), contradicts the minimal choice of A. Lemma 3.1 also implies that $saved_G(A, B) = 0$ showing that Case (ii) cannot occur either (in Case (ii) $d_G(A, B) > 0$ because of edge e).   □

Remark. Around the same time when Okamura and Seymour proved their theorem S. Lins (1981) showed that the maximum number of edge-disjoint non-separating circuits in an Eulerian graph embedded into the projective plane is equal to the minimum cardinality of a non-separating cut. This theorem in the present context is nothing but the theorem of Okamura and Seymour's theorem in the special case when all the terminals are distinct and they are positioned around the specific face in the cyclic order $s_1, s_2, \ldots, s_k, t_1, t_2, \ldots, t_k$. However, it can be shown by a simple trick that this special case implies the Okamura-Seymour theorem. Indeed, if there are two terminals s_1, s_2 sitting at the same node u, then add two new nodes v_1, v_2 and two new edges uv_1, uv_2 to the graph and move terminal s_1 to v_1 and s_2 to v_2. Applying this operation we can ensure that the terminals are distinct. The requirement on the cyclic order of the terminals is equivalent to saying that any two terminal pairs $s_i t_i$ and $s_j t_j$ crosses each other, that is, their cyclic order is s_i, s_j, t_i, t_j If this is not the case, then there are two non-crossing terminal pairs $s_i t_i$ and $s_j t_j$ such that one of s_i and t_i, say s_i, and one s_j and t_j, say s_j, are consecutive in the cyclic order (this is an easy exercise). Now modify the graph and the position of s_i and s_j as is depicted in Figure 3.1.

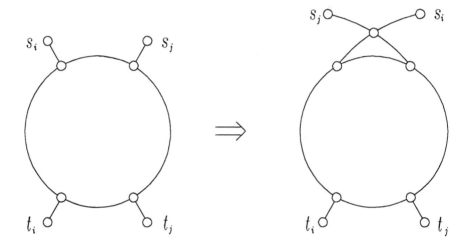

Fig. 3.1

It is easily seen that the cut criterion is satisfied for the new problem if it is satisfied for the old and if the required paths exist in the new problem, then so do they in the old. Furthermore the number of crossing terminal pairs is one bigger in the new problem. Applying this technique as long as there are non-crossing terminal pairs finally we obtain a problem which is equivalent to the original one and the terminals satisfy Lins' requirement.

H. Okamura generalized the theorem of herself and Seymour in two directions. The first one is:

Theorem 3.3 (Okamura 1983). *Suppose that G is planar, $G + H$ is Eulerian, and there are two faces C_1, C_2 such that each demand edge connects two nodes of either C_1 or C_2. Then the cut criterion is necessary and sufficient for the solvability of the edge-disjoint paths problem.*

The proof below (due to Gábor Tardos (Tardos 1984)) is a slightly simplified version of Okamura's original proof.

Proof. Again, we can assume that G is 2-connected. We say that a set K of nodes crosses a face if K contains a node of the face but not all. If there is a tight set crossing only one of the two specified faces C_1 and C_2, then the reduction step used in the proof of Okamura and Seymour's theorem can be applied. (Notice that the crucial equality $d_H(A, B) = 0$ in that proof cannot spoil down since every terminal pair is either on C_1 or on C_2.)

So assume that every tight set crosses both C_1 and C_2. Assume that a terminal pair st is in C_1 and that C_1 is the outer face of G. (It will cause no confusion that we use the same term C_i to denote the graph-circuit of G bounding the face C_i.) The nodes s and t divide C_1 into two paths P and Q connecting s and t.

First, delete the edges of P from G and remove the demand edge st from H. For the resulting G_1 and H_1 the hypotheses of the theorem hold and then we are done if the cut criterion is satisfied. So assume this is not the case. Then there is a set K which is tight with respect to G and H such that $s, t \notin K$ and K intersects P.

Second, delete the edges of Q from G and remove the demand edge st from H. Analogously to the first case, we are in trouble only if there is a set L tight with respect to G and H such that $s, t \notin L$ and L intersects Q.

Let $Z := V - (K \cup L)$. Since both K and L cross C_2, in the subgraph induced by Z there is no path connecting s and t. Therefore there is a partition of Z into two sets A and N with $s \in N$, $t \in A$ such that $d_G(A, N) = 0$ Let us introduce the following notation: $M := K \cap L$, $X := K - L$, $Y := L - K$. If M is non-empty, then at least one of A and N, say A, is disjoint from C_2. Theorefore $d_H(A, M) = 0$ and this is also true if $M = \emptyset$.

We will apply formula (1.3) from Section 1:

$$d(X \cup M) + d(Y \cup M) + 2d(A, N) = d(X \cup N) + d(Y \cup N) + 2d(A, M)$$

Now $X \cup M$ and $Y \cup M$ are tight and $d_H(A, M) = 0 = d_G(A, N)$, thus we have

$$0 + 0 = s(X \cup M) + s(Y \cup M) = s(X \cup N) + s(Y \cup N) + 2[d_G(A, M) + d_H(A, N)] \geq 0.$$

Therefore each term is 0, in particular, $d_H(A, N) = 0$. But this is impossible since the demand edge st leads between A and N. $\qquad\square$

Okamura's other generalization of Okamura and Seymour's theorem is as follows.

Theorem 3.4 (Okamura 1983). *Let G be planar, $G + H$ Eulerian, C a specified face of G and s is a node of C. Suppose that each terminal pair has either both members on C or one member at s. Then the cut criterion is necessary and sufficient for the solvability of the edge-disjoint paths problem.*

There is a recent result by A. Schrijver of similar vein concerning path-packing problems in a planar graph.

Theorem 3.5 (Schrijver 1988b). *Let G be planar, $G + H$ Eulerian and let C_1 and C_2 be two specified inner faces of G. Assume that the demand edges $s_1t_1, \ldots, s_kt_2$ are such that each s_i is on C_1 and each t_i is on C_2 and their cyclic order is the same. Then the cut criterion is necessary and sufficient for the solvability of the edge-disjoint paths problem. (Notice that if C_1 is chosen to be the outer face of G then the cyclic orders should be opposite.)*

In Theorems 3.3 and 3.5 G is planar $G + H$ is Eulerian and the terminals are on two specified faces. Figure 2.3 shows that if we do not impose some extra conditions on the terminals, then the cut condition is not sufficient, in general. In the example in Figure 2.3 even no fractional solution exists. Thus one may suspect that under the circumstances above the existence of a fractional solution already implies solvability. However this is not the case as is shown in Figure 2.4.

Here is yet another fundamental result concerning planar graphs.

Theorem 3.6 (Seymour 1981). *Suppose that $G + H$ is planar and Eulerian. Then the cut criterion is necessary and sufficient for the solvability of the edge-disjoint paths problem.*

Proof (Z. Zubor 1989). We can assume that every edge $e \in E$ is in a tight cut since otherwise e can be moved from E into F without destroying the cut criterion. By the reduction principle we can assume that in $G + H$ every degree is 2 or 4. Suppose that $G + H$ is a counter–example with a minimum number of nodes of degree 4. Define

$$w : E \cup F \to \{+1, -1\} \quad \text{by}$$

$$w(e) := \begin{cases} +1 & \text{if } e \in E \\ -1 & \text{if } e \in F. \end{cases}$$

The cut criterion is equivalent to: $d_w(X) \geq 0$ for every $X \subseteq V$. We need the following observation of A. Sebő (1987b).

Claim. *Let $A \subseteq V$ be tight, i.e. $d_w(A) = 0$, and define*

$$w'(e) = \begin{cases} w(e) & \text{if } e \notin \nabla(A) \\ -w(e) & \text{if } e \in \nabla(A). \end{cases}$$

Then $d_{w'}(X) \geq 0$ for every $X \subseteq V$.

Proof. We have $d_{w'}(X) = d_w(A \oplus X) - d_w(A) = d_w(A \oplus X) \geq 0$. ($A \oplus X$ denotes $(A - X) \cup (X - A)$.) □

By *interchanging along* a cut C we mean an operation that replaces F by $F \oplus C$ and E by $E \oplus C$. By the Claim the theorem holds for $G + H$ if and only if it holds after interchanging along a tight cut.

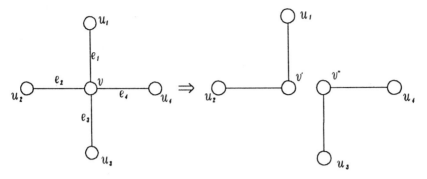

Fig. 3.2

Let vu_1 be a demand edge. Assume that the four edges $e_i = vu_i$ $(i = 1, 2, 3, 4)$ incident to v are indexed in cyclic order so that $e_1 \in F, e_2 \in E$. Modify slightly the "splitting off" operation as follows. Replace v by v' and v'' so that v' is connected to u_1 and u_2 and v'' is connected to u_3 and u_4 (Figure 3.2).

Let $G' = (V', E')$ and $H' = (V', F')$ denote the resulting graphs. If there were a solution to the edge–disjoint paths problem in $G' + H'$, there would be one in $G + H$. Thereby there is a bond $\nabla'(A)$ for which $d_{G'}(A) < d_{H'}(A)$. We can assume that $v' \in A$. Since the cut criterion holds for $G + H$ we have

(∗) $v'' \notin A$ and an edge e_i $(i = 1, 2, 3, 4)$ belongs to $\nabla'(A)$ precisely if $e_i \in F$.

These are two cases.

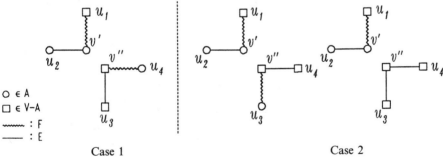

O ∈ A
□ ∈ V–A
⌇⌇⌇ : F
──── : E

Case 1   Case 2

Fig. 3.3

Case 1. $e_4 \in F$. By (∗) $u_2, u_4 \in A$ and $u_1, u_3 \notin A$. Both A and $V' - A$ induce a connected subgraph of $G' + H'$ contradicting the planarity of $G' + H'$.

Case 2. $e_4 \in E$ By (∗) $u_2 \in A$ and $u_1, u_4 \notin A$. Now $A - v'$ is tight in $G + H$. By interchanging along $\nabla(A - v')$ (and re–indexing the e_i's) we are at Case 1. □

It is a challenging open problem to find a unified theorem that implies all the "planar" results above.

Actually Seymour proved a result more general than Theorem 3.6. If we take planar dual, then the role of circuits and cuts is interchanged, in particular Eulerian turns into bipartite. It turns out that planarity can be left out from the hypotheses:

Theorem 3.6' (Seymour 1981a). *Suppose that $G + H$ is bipartite. There are $|F|$ edge-disjoint cuts in $G + H$, each containing exactly one element of F if and only if every circuit of $G + H$ contains as many edges from G as from H.*

In Section 8 (Theorem 8.1) we will prove this result along with some generalizations. Also a strongly polynomial-time algorithm will be provided for the more general weighted case. P. Seymour found another generalization of Theorem 3.6.

Theorem 3.7 (Seymour 1981b). *Suppose that $G + H$ is Eulerian and no subgraph of it can be contracted to K_5 (complete graph on 5 nodes). Then the cut criterion is necessary and sufficient for the solvability of the edge-disjoint paths problems.*

Let us now turn to another class of graphs when, supposing $G + H$ Eulerian, the cut criterion is sufficient.

For a given demand graph $H = (V, F)$, H' will denote the graph arisen from H by replacing each (maximal) set of parallel edges by one edge. Let us call a graph a *double star* if there are at most two nodes that cover all the edges. In what follows K_n denotes the complete graph on n nodes and C_5 denotes a circuit on 5 nodes. Let $K_2 + K_3$ denote a graph on 5 nodes with components K_2 and K_3. Similarly $3K_2$ denotes a graph consisting of three disjoint edges.

Theorem 3.8. *Suppose that $G + H$ is Eulerian and H' is either a double-star or K_4 or C_5. Then the cut criterion is necessary and sufficient for the solvability of the edge-disjoint paths problem.*

The case when H' is $2K_2$ was proved by Rothschild and Whinston (1966b), sharpening earlier results of T. C. Hu (1963). From this the theorem easily follows for double-stars (see below). The K_4 case was proved by P. Seymour (1980a) and M. Lomonosov (1979), independently. The C_5 case is due to Lomonosov (1979).

The theorem is sharp in the sense that if H' is different from each of the three graphs in the theorem, then there is a G and H such that $G + H$ is Eulerian, the cut criterion holds but there is no solution to the edge-disjoint paths problem. To see this, observe that the example in Figure 2.3 shows that H' must not contain $K_2 + K_3$ as a subgraph. The example in Figure 3.4 shows that H' must not contain $3K_2$. This is due to Papernov (1976).

It is an easy exercise to show that if a graph contains neither of these two forbidden graphs, then it is either a double star or K_4 or C_5.

Proof of Theorem 3.8. Suppose the $G + H$ is counterexample with a minimum number of edges. Obviously $G + H$ is connected. We need some preparation that is useful for each case.

Claim 1. There are no edges $e \in E$ and $f \in F$ that are parallel.

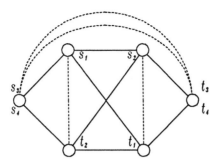

Fig. 3.4

Proof. Deleting e and f does not destroy the cut criterion and then a good circuit-partition of the smaller graph along with circuit $\{e, f\}$ would form a good circuit-partition of $G + H$. $\qquad\square$

Lemma 3.1 implies:

Claim 2. Let vz be an edge of G where v is not a terminal node. Let A and B be two (distinct) maximal tight $z\bar{v}$–sets. Then $d_H(A, \bar{B}) > 0$ and $d_H(A, B) > 0$. $\qquad\square$

Let $A_1, A_2, \ldots, A_k$ be the minimal tight $z\bar{v}$–sets. Suppose that there is an edge uv which does not occur in any tight cut. Then splitting off uv and vz does not destroy the cut criterion. So the resulting graph has a good circuit-partition. But this provides a good circuit-partition of $G + H$ which is impossible. Therefore every edge vu enters some A_i.

Proof. If $k = 1$, then every neighbour of v is in $\nabla(A_1)$ but then $\nabla(A_1 + v)$ would violate the cut condition. If $k = 2$ and $d_G(v, A_1) \geq d_G(v, A_2)$, say, then $d_G(A_1 + v) > d_G(A_1)$ (because of edge vz) and therefore $\nabla(A_1 + v)$ violates the cut criterion. $\qquad\square$

Turning to the different cases of the theorem let us first assume that $H' = 2K_2$ and the demand edges are between s_i and t_i ($i = 1, 2$).

We claim the there is a node v which is not a terminal. Indeed, if no such a node exists, then, by Claim 1, G must be a four-circuit with possible parallel edges. But this cannot be a counterexample as is seen by inspection (or by the theorem of Okamura and Seymour).

Therefore there is an edge vz of G where $z = S_1$ and v is not a terminal node. By Claim 3 there are at least three maximal tight $z\bar{v}$-sets A_1, A_2, A_3. By Claim 2 $d_H(A_i, A_j) > 0$ ($1 \leq i < j \leq 3$) but this is impossible since $s_1 \in A_1 \cap A_2 \cap A_3$.

Next we show how the double star case reduces to the case $H' = 2K_2$. Let s_1 and s_2 be the two nodes covering the edges of a double-star H. First subdivide each demand edge $s_1 t_i$ by a node t'_1 such that $s_1 t'$ belongs to the demand graph and $t'_i t_i$ to the supply graph. Then contract the nodes t'_i into one node. Finally do the same with the demand edges incident to s_2. This way we obtain a new problem which is equivalent to the original one and the demand graph consists of two sets of parallel edges.

Suppose that $H' = K_4$ and the four terminal nodes are $s_1, \ldots, s_4$. By Claim 1 there is no edge in G connecting two terminal nodes.

Let us denote s_1 by z. If there is no tight set containing s_1, s_2 and not containing s_3, s_4, then let vz be any edge in G with vs_i. If there is one, then the intersection Z of such sets is tight by Lemma 3.1. We claim that there is a $v \in Z - \{s_1, s_2\}$ such that vz is an edge of G. For otherwise, $d_G(Z) = d_G(z) + d_G(Z - z) \geq d_H(z) + d_H(Z - z) = d_H(Z) + 2d_H(z, Z - z) = d_G(Z) + 2d_H(z, Z - z) > d_G(Z)$, a contradiction.

By Claim 3 there are at least three maximal tight $z\bar{v}$–sets A_1, A_2, A_3. By Claim 2 $d_H(A_i, A_j) > 0$ $(1 \leq i < j \leq 3)$. But this is possible only if each of A_1, A_2 and A_3 contains a terminal node which is not in the union of the two others. Assume that A_2 contains s_2. Then A_2 is a tight set containing s_1, s_2 and not s_3, s_4 and v. This contradicts the choice of v and the definition of Z and thus the case of K_4 is settled.

Finally let us assume that $H' = C_5$. If $|V| = 5$, then, by Claim 1 G is a subgraph of a 5-circuit with possible parallel edges. But then the Okamura-Seymour theorem shows that $G + H$ cannot be a counterexample. So let vz be an edge of G where v is not a terminal.

By Claim 3 there are at least three maximal tight $z\bar{v}$-sets A_1, A_2, A_3. By Claim 2 (*) $d_H(A_i, A_j) > 0$ and $d_H(A_i, \overline{A_j}) > 0$ $(1 \leq i < j \leq 3)$. Then each A_i contains 2 or 3 terminals. The complement of A_i is also tight so we can assume that there are three tight sets B_1, B_2, B_3 for which (*) holds and each of them contains exactly two terminals. Now if $B_1 \cap B_2 \cap B_3$ contains a terminal node, then each of B_1, B_2, B_3 contains a terminal node which is not in the union of the two others. But then these three terminals must form a triangle in H' which is impossible.

Suppose now that $B_1 \cap B_2 \cap B_3$ contains no terminal node. Since $B_i \cap B_j$ contains a terminal node $(1 \leq i < j \leq 3)$ the other two terminal nodes must be outside $\cap B_i$ and then we must have again a triangle in H', a contradiction. $\quad \square$

Each of Theorem 3.2 through 3.8 has a fractional version as an easy consequence. For example:

Theorem 3.8' (Papernov 1976). *Let G be arbitrary and H as in Theorem 3.8. Then the cut criterion is necessary and sufficient for the solvability of the multiflow problem.*

There is a very useful device by which the reverse implication can also be proved. The idea, noticed by van Hoesel and Schrijver (1990), is as follows. (For more details, see (Schrijver 1988a)).

Proof of Theorem 3.8 from Theorem 3.8'. Let x be a solution to the multiflow problem and P a path for which $x(P) > 0$. Let v be any inner node of P and uv and vz the two edges of P incident to v. We claim that uv and vz can be split off without violating the cut criterion. Indeed, if the cut criterion does not hold after the splitting, there is a tight cut of G that contains both uv and vz. But this is impossible since a simple argument shows that any tight cut and any path Q with $x(Q) > 0$ have at most one edge in common.   □

One can similarly proceed to derive Theorems 3.2–3.6 from their corresponding fractional version. However, in order to maintain planarity, certain care is required while chosing the pair of edges to be split off:

Proof of Theorems 3.2–3.6 from the corresponding fractional versions. First, by the reduction principle described in Section 2 we assume that in $G + H$ every node has degree four. Let x be a solution to the corresponding multiflow problem (in either of Theorems 3.2–3.6). If there is a path P and an inner node v of P such that $x(P) > 0$ and the two edges uv and vz of P are in the same face of G, then splitting off these edges preserves not only the cut criterion but also the planarity. If no such a path exists (that is, for every inner node v of any path P with $x(P) > 0$ goes "across" v), then for every terminal pair (s, t) there can be only one path P with $x(P) > 0$ connecting s and t. Consequently, x is 0-1 valued, that is, x itself is a solution to the corresponding edge-disjoint paths problem.   □

Of course, the reduction method above can be considered useful only if there is a direct way to prove the "fractional" theorems. In Section 8 we indicate such a method.

By applying the splitting off technique to directed graphs a directed counterpart of the theorem of Rothschild and Whinston can be proved. By a *(directed) star* we mean a directed graph in which either all the edges enter the same node or all the edges leave the same node.

Theorem 3.9 (Frank 1985). *Suppose that $G + H$ is an Eulerian digraph and H is the union of two stars. Then the directed cut criterion is necessary and sufficient for the solvability of the undirected edge-disjoint paths problem.*

The following figure shows some small H which are not the union of two stars, the directed cut condition holds but there is no solution.

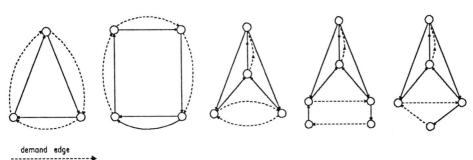

demand edge
- - - - - - - - - - - →

Fig. 3.5

Let us conclude this section by citing two recent results of Karzanov concerning undirected G and H.

Theorem 3.10 (Karzanov 1987). *Suppose that $G + H$ is Eulerian and the demand edges form a K_5. Then the distance criterion is necessary and sufficient for the solvability of the edge-disjoint paths problem. (In other words, if there is a fractional solution, there is an integral one.)*

Theorem 3.11 (Karzanov 1989a). *Suppose that $G + H$ is Eulerian, G is planar and the demand edges are on three faces of G. Then the distance criterion is necessary and sufficient for the solvability of the edge-disjoint paths problem.*

4. Further Necessary Conditions

The purpose of this section is to introduce some further necessary conditions concerning the (edge-) disjoint path problem. They belong to two classes. The first one is a kind of topological obstruction while the second is based on parity arguments.

Let G and H be undirected. We know that the cut-criterion is sufficient when H is a star. Suppose now that H consists of two disjoint edges. The following characterization appears in three different papers: E.A. Dinits and A.V. Karzanov (1979), P. Seymour (1980) and C. Thomassen (1980).

Theorem 4.1. *Let G be a graph such that no cut edge separates both of the two terminal pairs (s_1, t_1) and (s_2, t_2). There is no two edge-disjoint paths between the corresponding terminal pairs if and only if some edges of G can be contracted so that the resulting graph G' is planar, the four terminals have degree two while the other nodes are of degree 3 and the terminals are positioned on the outer face in this order: s_1, s_2, t_1, t_2.*

Figure 4.1 shows a typical example where the two edge-disjoint paths do not exist.

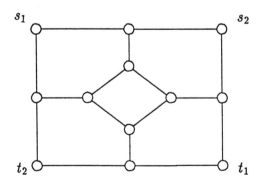

Fig. 4.1

Recall that if one wants k_i paths between s_i and t_i ($i = 1, 2$), then the problem becomes NP-complete. The necessity of the condition in the theorem depends on three observations. Namely, edge-contraction does not destroy solvability, a node of degree three can be used by at most one path, and two curves in the plane connecting antipodal pairs of points of a circle must intersect each other.

Actually this theorem immediately follows from the following node-version by considering the line graph.

Theorem 4.2 (Thomassen 1980a, Seymour 1980). *Let G be a graph such that no node separates s_1 from t_1 and s_2 from t_2. There are no disjoint paths between s_1 and t_1 and between s_2 and t_2 if and only if G arises from a planar graph G', where the four terminals are one the outer face in this order s_1, s_2, t_1, t_2, by placing an arbitrary graph into some faces of G' bounded by two or three edges.*

For directed graphs the two disjoint paths problem is NP-complete. However, Thomassen (1985) found a complete description of acyclic digraphs have no solution to the 2-disjoint paths problem. The core of his result os as follows.

Theorem 4.3. *Let us be given an acyclic digraph $D = (V, A)$ (with no cut-node and parallel edges) and terminal pairs (s_1, t_1), (t_2, t_2) such that $|V| \geq 5$, $\varrho(v), \delta(v) \geq 2$ for each non-terminal node v and $\varrho(s_1) = \varrho(s_2) = \delta(t_1) = \delta(t_1) = \delta(t_2) = 0$. If there are no disjoint paths from s_1 to t_1 and from s_2 to t_2, then D is planar and has a plane representation in such a way that s_1, t_2, t_1, s_2 are on the outer face occuring in that cyclic order.*

Notice that in these theorems the hypotheses are purely graphical and topological arguments come only in the characterization. But one can be interested in disjoint paths in a graph embedded in a plane with certain holes such that the paths must satisfy a certain homotopy requirement. (That is, the topological way how the paths have to go around the holes is spedified.) This general problem is precisely the central topic of A. Schrijver's article in this volume.

Let us turn to the other class of necessary conditions and consider the edge-disjoint paths problem in an undirected graph. In the preceding section we have considered special classes when $G + H$ is Eulerian, that is, when $d_{G+H}(X)$ is even for every subset of V. Let us now call a set X *odd* (or the cut ∇_G odd) if $d_{G+H}(X)$ is odd. It is useful to observe that the number of odd nodes is always even and that a set X is odd if and only if X contains an odd number of odd nodes.

The crucial observation concerning odd cuts is that, given an odd set X and any solution to the edge-disjoint paths problem, an odd number of edges of $\nabla_G(X)$, in particular at least one edge, can not be used by the paths in the solution. (Actually, we have already relied on an special case of this idea when we argued after Theorem 4.1 that no two edge-disjoint paths can go through a node of degree three.)

Thus this parity argument provides a kind of force that intuitively prevents a solution to use too many edges. On the other hand, in a tight cut all of the edges are necessarily used. Or more generally, for a set X with surplus $s(X)$ $(= d_G(X) - d_H(X))$ there may be at most $s(X)$ edges in $\nabla_G(X)$ which are not used by a solution. Thus this surplus argument provides a kind of force that intuitively prevents a solution to use too few edges.

These two forces of opposite directions are the basis of each of the following necessary conditions. For example, it is necessary that

(4.1) $\nabla_G(D)$ cannot be covered by two tight cuts for any odd set D.

Observe that (4.1) is not satisfied by the graph in Figure 2.1. We mention three cases when (4.1) is sufficient, as well.

Suppose first that H consists of two sets of parallel edges, that is, there are two terminal pairs. We call a set X of nodes (and the cut $\nabla(X)$)) *separating* if $\nabla(X)$ separates both terminal pairs. Two separating sets X and Y are called *parallel* if either $X - Y$ or $X \cap Y$ is separating. Otherwise they are *non-parallel*. We say that a set X *crosses* C if $X \cap C$ and $C - X$ are non-empty.

Theorem 4.4 (Seymour 1981a). *Suppose that $G + H$ is planar and H consists of two sets or parallel edges. The edge-disjoint paths problem has a solution if and only if the cut criterion holds and* (**) $d_{G+H}(S \cap T)$ *is even for any two tight non-parallel separating sets S, T.*

Proof. We can assume that G is 2-connected. Assume that there are k_i demand edges connecting s_i and t_i, the terminal pair (s_i, t_i) is on face C_i $(i = 1, 2)$ and that C_i is the outer face. We can assume that both $k_1 \geq k_2 > 0$, since otherwise Menger's theorem applies. Let us recall that if the cut criterion does not hold, then there is bond $\nabla(K)$ violating it. Because of planarity K divides any facial circuit of G into at most two paths.

The nodes s_1 and t_1 divide C_1 into two paths P and Q connecting s_1 and t_1. First, delete the edges of P from G and remove one demand edge connecting s_1 and t_1 from H.

Assume first that the resulting G_1 and H_1 satisfy the cut criterion. One can observe that if X and Y violate (**) for G_1 and H_1, then X and Y violate (**) with respect to G and H. Then, by induction, there is a solution with respect to G_1 and H_1 and this solution along with path P yields a solution with respect to G and H. So we can assume that there is a set K violating the cut criterion with respect to G_1 and H_1. Then $s_1, t_1 \notin K$ intersects P and the surplus $s(K) \leq 1$. Then $\nabla(K)$ necessarily separates s_2 and t_2 and K crosses C_2.

Similarly, delete the edges of Q from G and remove one demand edge connecting $s_1 t_1$ from H. Analogously to the first case, we are in trouble only if there is a set L with surplus $s(L) \leq 1$ such that $s_1, t_1 \notin L$ and L intersects Q.

Let $Z := V - (K \cup L)$. Since both K and L cross C_2, in the subgraph of G induced by Z there is no path connecting s_1 and t_1. Therefore there is a partition of Z into two sets A and N with $s_1 \in N$, $t_1 \in A$ such that $d_G(A, N) = 0$. Let us introduce the following notation: $M := K \cap L$, $X := K - L$, $Y := L - K$. If M is non-empty, then at least one of A and N, say A, is disjoint from C_2. Therefore $d_H(A, M) = 0$ and this is also true if $M = \emptyset$.

We will apply formula (1.3) to both G and H. Exploiting that $d_H(A, M) = 0 = d_G(A, N)$, we have

$$1 + 1 \geq s(X \cup M) + s(Y \cup M) = s(X \cup N) + s(Y \cup N) +$$

(4.2) $\qquad 2[d_G(A, M) + d_H(A, N)] \geq 0 + 0 + 2[0 + 1] = 2.$

Therefore equality holds everywhere and, in particular, $s(X \cup M) = s(Y \cup M) = 1$, $s(X \cup N) = s(Y \cup N) = 0$, $d_G(A, M) = 0$, $d_G(A, N) = 1$. The last equality shows that $k_1 = 1$. Since $k_1 \geq k_2 > 0$, we have $k_2 = 1$. (This means that the two edges leaving K are common edges of C_1 and C_2).

Since $s(K) = s(L) = 1$ it follows that $d_G(K) = d_G(L) = 2$ and that $d_{G+H}(K) = 3$. Now $M = K \cap L$ must be empty for if a node v is in $K \cap L$, then there is a path in K from v to P. But such a path leaves L along an edge that is not in C_1. So we would have $d_G(L) \geq 3$. See Figure 4.2. We see that $K = X$ and $L = Y$.

Since M is empty $d_H(N, M) = 0$. (4.2) can be applied with interchanging A and N. We obtain that $Y \cup A$ is tight. Let $S =: V - (Y \cup A)$ $(= K \cup N)$ and $T := V - (Y \cup N) = (K \cup A)$. Now S and T violate (**) since S and T are tight, $K = S \cap T$ and $d_{G+H}(K)$ is 3, an odd number. $\qquad \square$

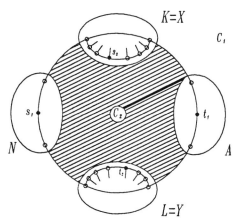

Fig. 4.2

Remark. The original proof of Seymour relies on the concept of T-cuts. The proof outlined above has the advantage that it can be extended to obtain the following generalization of Seymour's theorem.

Theorem 4.4a (Frank 1988). *Suppose that $G + H$ is planar and the edges of H are in two faces of G. The edge-disjoint paths problem has a solution if and only if the cut criterion holds and $d_{G+H}(S \cap T)$ is even for any two tight sets S, T.*

A direct conseqeuence of Theorem 4.4 is that the problem has a solution if the cut criterion holds with strict inequality on any separating cut. In Theorem 4.6 we shall see that in an extension of Okamura and Seymour's theorem, when no parity restriction is imposed on the nodes of the outer face, a similar type of results holds. The statement is not true if H has three disjoint edges as is shown by the following example of E. Korach where there is no tight cut at all. (Compare Theorem 4.4a to this example: here there are only three demand edges but they are on three faces.) M. Middendorf and F. Pfeiffer recently showed that the edge-disjoint paths problem is NP-complete if $G + H$ is planar.

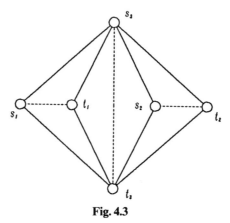

Fig. 4.3

The following theorem deals with a very special graph but it will find a nice application in the next section.

Theorem 4.5 (Frank and Tardos 1984). *Suppose that G is a circuit with parallel edges. The edge-disjoint paths problem has a solution if and only if the cut-criterion and (4.1) hold.*

Proof. If $G+H$ is Eulerian, then the cut criterion itself is sufficient by the Okamura-Seymour theorem (Theorem 3.2). So assume that the set $T = \{a_1, a_2, \ldots, a_{2h}\}$ of odd nodes is non-empty. The idea behind the proof is that we want to reduce the problem to Theorem 3.2 be eliminating the odd nodes. In order to do so first add the following h new demand edges to H: $a_1 a_2, a_3 a_4, \ldots, a_{2h-1} a_{2h}$. Let H_1 denote the extended demand graph. Obviously, $G + H_1$ is Eulerian, so we are done by Theorem 3.2 if the cut criterion holds in $G + H_1$. If this is not the case, then there is a bond $\nabla_G(X_1)$ which is tight in $G+H$ where $X_1 = \{a_i, a_{i+1}, \ldots, a_j\}$ ($1 < i < j$), i is even and j is odd.

Second, add $a_2 a_3, a_4 a_5, \ldots, a_{2h} a_1$ as new demand edges to H obtaining this way H_2. If the cut criterion does not hold in $G + H_2$, then there is a bond $\nabla(X_2)$ which is tight in $G+H$ where $X_2 = \{a_k, a_{k+1}, \ldots, a_l\}$ ($1 < k < l$), k is odd and l is even. But now the component of $G - (\nabla_G(X_1) \cup \nabla_G(X_2))$ containing a_1 contains and odd number of odd nodes and therefore it violates (4.1). □

This idea of pairing off the odd nodes can be used to prove the following consequence of the Okamura-Seymour theorem when no parity restriction is imposed at the nodes of the outer face.

Theorem 4.6. *Suppose that G is planar, the terminals are on the outer face and the degree of every node not on the outer face is even. If the cut criterion holds in a strong form, that is, $d_G(X) > d_H(X)$ for every $\emptyset \neq X \subset V$, then the edge-disjoint paths problem always has a solution.*

Proof. If there are no odd nodes, we are done by Theorem 3.2. Otherwise let $T = \{a_1, a_2, \ldots, a_{2h}\}$ be the set of odd nodes. Extend the demand graph by the following new edges: $a_1 a_2, \ldots, a_{2h-1} a_{2h}$. Observe that the cut criterion continuous to hold and apply Theorem 3.2. □

There is a pretty consequence of Theorem 4.6. Suppose that given a big triangle R in a triangular grid which is bounded by lattice lines. R defines a graph G_R in the natural way.

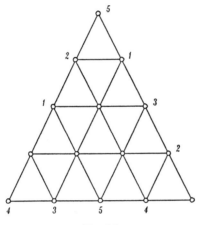

Fig. 4.4

Corollary. *If the terminals are on the boundary of R and are distinct, then the edge-disjoint paths problem has a solution.*

Actually, we can have a complete characterization for the case considered in Theorem 4.6:

Theorem 4.7 (Frank 1985). *Suppose that G is planar, the terminals are on the outer face and the degree of every node not on the outer face is even. The edge disjoint paths problem has a solution if and only if $\Sigma(s(C_i) \geq 1/2q$ for every family $(C_1, C_2, \ldots, C_l)$ of cuts $(l \leq |V|)$ where q denotes the number of components in $G - C_1 - C_2 \ldots - C_l$ which are odd (in $G + H$) and s(C) is the surplus of C.*

Note that this theorem provides a characterization for the edge-disjoint paths problem when the supply graph G is outerplanar.

Proof. (Outline) To show the necessity of the condition suppose that there is a solution and let $Q_1, Q_2, \ldots, Q_q$ be the odd components in question. For each Q_i

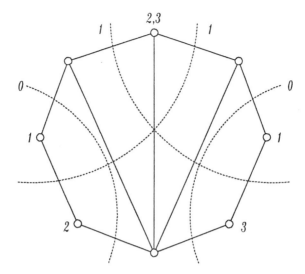

Fig. 4.5

at least one edge in $\nabla_G(Q_i)$ is not used by the solution. Because any edge of G may belong to (at most) two $\nabla_G(Q_i)'s$, we see that at least $1/2q$ edges must not be used. On the other hand all of these edges are in $\cup C_i$ therefore $\Sigma s(C_i) \geq 1/2q$.

(For example there is no solution to the edge-disjoint paths problem in the graph depicted in the following figure. The four cuts violate the necessary condition of the theorem since their sum of surpluses is 2 while the removal of them gives rise to $q = 8$ odd components and $2 \not\leq 8/2$.) The sufficiency of the condition can be proved with the same idea we have used for proving Theorem 4.6. The only difference is that this time finding the appropriate pairing of the odd nodes needs a little more care.

Namely we proceed as follows. If there is no tight cut, then we are back at Theorem 4.6. For simplicity suppose that G is 2-connected and let C denote the outer circuit of G. Call a tight set X *minimal* if $V(C) \cap X$ is minimal for inclusion. The basic step of the pairing algorithm is that we find a tight set X which is minimal and find the odd nodes $a_1, a_2, \ldots, a_j$ (in this order along C) in $V(C) \cap X$. (It can be shown that j is even). Now extend the demand graph H by the following new terminal pairs: $a_1 a_2, \ldots, a_{j-1} a_j$. The crucial observation is that the original problem has a solution if and only if the new one has. Therefore we can keep going on this pairing operation. If in the course of the procedure a cut arises which violates the cut condition with respect to the current (enlarged) demand graph, then this cut and the minimal tight cut used by the procedure in the previous steps violate the condition of Theorem 4.7. Let us consider a possible run of the pairing procedure on following example:

The odd nodes are indicated by solid points. X_1, X_2, X_3 are the current minimal tight sets. In the fourth step $\nabla_G(X_4)$ violates the cut condition with respect to the enlarged demand graph. The four cuts $C_i : \nabla_G(X_i)$ $(i = 1, 2, 3, 4)$ violate the condition in the theorem since the sum of their original surplus is 2 while the number of odd components in $G - (U C_i)$ is 8. See Figure 4.7.   □

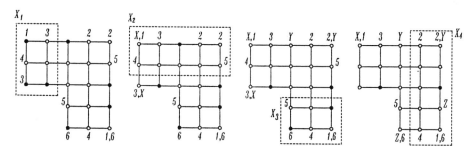

Fig. 4.6

In Theorem 4.7 the parity restriction on the inner nodes cannot be dropped as is shown by the example in Figure 4.1.

Our last result to demonstrate the use of parity conditions is due to P. Seymour. Let G be again arbitrary. The cut criterion is sufficient if H is a star. The next two simplest demand graphs are $2K_2$ and K_3. As another application of the "parity-versus-surplus" principle we exhibit a characterization when H is a K_3 with parallel edges.

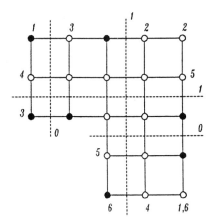

Fig. 4.7

Theorem 4.8 (Seymour 1980b). *If the demand graph H consists of three sets of parallel edges between nodes s_1, s_2 and s_3, the edge-disjoint paths problem has a solution if and only if the cut-criterion holds and*

(4.3) $q(V_1 \cup V_2 \cup V_3) \leq s(V_1) + s(V_2) + s(V_3)$

for every choice of disjoint sets V_i with $s_i \in V_i$ $(i = 1, 2, 3)$ where $s(X)$ denotes the surplus and $q(X)$ denotes the number of components C in $G - X$ for which $d_G(C) + d_H(C)$ is odd.

Proof. Let $X = V_1 \cup V_2 \cup V_3$. If there is a solution, then a least $q(X)$ edges of $\nabla_G(X)$ are not used. On the other hand at most $s(V_1) + s(V_2) + s(V_3)$ edges are avoided by the paths in the solution. Therefore (4.3) is necessary.

To see the sufficiency we invoke Mader's edge-disjoint A-paths theorem (see Theorem 7.3). Extend G by adding three new nodes a_1, a_2, a_3 and $k_{i,i+1} + k_{i,i-1}$ parallel edges between a_i and s_i (where $i = 1, 2, 3$) and the subscript are meant modulo 3). Let $A = \{a_1, a_2, a_3\}$. If there are $h := k_{1,2} + k_{2,3} + k_{2,3}$ edge-disjoint A-paths, then their restriction to G provides the desired edge-disjoint paths in G. So suppose there are no h edge-disjoint A-paths. By Mader's theorem there are three disjoint sets U_i with $a_i \in U_i$ $(i = 1, 2, 3)$ for which

(4.4) $h > \text{value}(U_1, U_2, U_3)$

where $\text{value}(U_1, U_2, U_3) := 1/2(\Sigma d_G(U_i) - q(\cup U_i))$.

We can assume that (i) each U_i induces a connected subgraph. Indeed, if the subgraph induced by U_1, say, has a component C not containing a_1, then $\text{value}(U_1 - C, U_2, U_3) \leq \text{value}(U_1, U_2, U_3)$.

We can assume that (ii) each component C of $V - \cup U_i$ is connected to every U_i $(i = 1, 2, 3)$. Indeed, if a component C is not connected to U_3, say, but it is connected to U_1, then $\text{value}(U_1 \cup C, U_2, U_3) \leq \text{value}(U_1, U_2, U_3)$. (Notice that if U_1 is connected, then so is $U_1 \cup C$.) If $s_i \in U_i$ for each $i = 1, 2, 3$, then the sets $V_i := U_i - a_i$ $(i = 1, 2, 3)$ violate (4.4). If $s_1 \notin U_1$, say, then $U_1 = \{a_1\}$ because of (i) and then $q(\cup U_i) \leq 1$ because of (ii).

But now we have $h > \text{value}(U_1, U_2, U_3) \geq 1/2(d(U_i) + 1) \geq h + 1/2$, a contradiction. (Notice that the cut-criterion was used in the last inequality.)   □

5. Problems on Grids

We devote this section to disjoint paths problems when the supply graph G is a subgraph of a rectilinear grid.

We are given a closed rectangle T (bounded by lattice lines). T defines a finite subgraph G of the plane grid in the natural way which has $m * n$ nodes when m horizontal and n vertical grid lines intersect T. Assume that the terminals are on the perimeter of T, that is on the outer face of G. Then the edge-disjoint paths problem is a special case of the problem solved in Theorem 4.7. However, exploiting the simpler structure of G we can obtain simpler theorems. By a *column (row)* of G we mean a cut consisting only of horizontal (vertical) edges. Obviously there are $n - 1$ columns and $m - 1$ rows.

Let us first restrict ourselves to the case when each terminal is on the upper or on the lower horizontal lines bounding T. We call this case *two-sided*. See Figure 5.1. Figure 2.1 shows that the cut criterion is not sufficient even for the two-sided problem. However a tiny restriction makes the cut criterion sufficient:

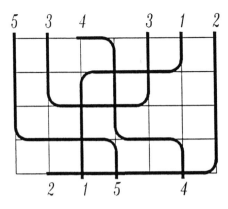

Fig. 5.1

Theorem 5.1. *In the two-sided case if the terminals are distinct and at least one of the assumptions (i) and (ii) holds, then the edge-disjoint paths problem has a solution if and only if the cut criterion holds for every column where*

(i)   at least one corner point of T is not a terminal,

(ii) at least one terminal pair is such that both members of it are on the same horizontal line.

Part (i) was proved in (Frank 1982) with the help of a polynomial time algorithm. A slight modification of the same idea led to (ii) (Abos, Frank, Tardos 1982).

Notice that if there is an inner node v on the upper boundary of G such that neither v nor its opposite node u on the lower boundary is a terminal, then the vertical line connecting u and v can be left out. Similarly we can leave out horizontal lines as long as the cut criterion holds for columns. This way we can preassume that $m \le k$ and $n \le 2k$. In applications it is desirable that the paths do not have too many bends since each bend corresponds to a via in a layout realization.

Theorem 5.2 (Preparata and Lipski 1984). *Assume that one member of each terminal pair is on the upper boundary line of T, the other member is on the lower boundary line and that there is no terminal on the $l = d/2$ right-most vertical lines where d denotes the maximum congestion of a column. If $m \ge d$ (that is, the cut criterion holds for the columns) then there is a solution to the edge-disjoint paths problem such that each path has one of the following shape (Figure 5.2).*

Let us turn to the case when the terminals are still distinct (with the possible exception that there may be two terminals sitting at one corner) but they are arbitrarily positioned on the boundary of T. Let c be an arbitrary column and let $\{r_1, r_2, \ldots, r_t\}$ be the set of tight rows ($t \ge 0$). Let $T_1, T_2, \ldots, T_{t+1}$ be the components of $G - (c \cup r_1 \cup \ldots \cup r_t)$ that are on the left-hand side of c. The *parity*

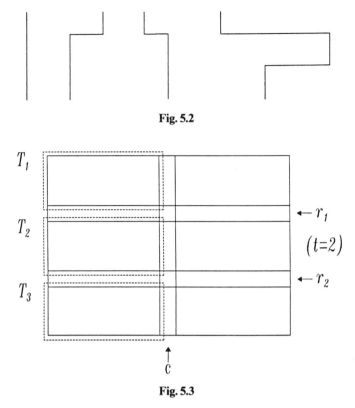

Fig. 5.2

Fig. 5.3

congestion of c is the number of odd sets T_i. The parity congestion of a row is defined analogously. By the parity-versus-surplus principle we see that the

REVISED CUT CRITERION: the parity congestion of a row or column cannot exceed the surplus is necessary for the solvability.

We have

Theorem 5.3 (Frank 1982). *We are given a rectangle in a rectilinear grid and k pairs of distinct terminals on its boundary. The edge-disjoint paths problem has a solution if and only if the revised cut criterion holds for every row and column.*

In Figure 5.4 two examples are shown differing only in the position of terminal "1". The first example has a solution but the second does not since the column c indicated in the picture violates the revised cut criterion.

A further advantage of Theorem 4.7 is that it makes possible to handle certain capacitated cases. For example, suppose that each vertical and horizontal line has a positive integer capacity (not necessarily the same). Let the capacity of an edge e of the grid-graph be the capacity of the line containing e. Instead of seeking for edge-disjoint paths we require that no edge is contained in more paths than its capacity. Obviously Theorem 4.7 can be applied since the sum of capacities of the edges incident to an inner node is even.

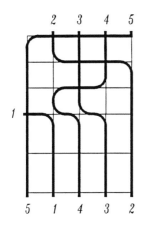
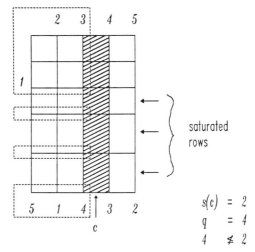

$s(c) = 2$

$q = 4$

$4 \nleq 2$

Fig. 5.4

In applications sometimes one needs regions of the grid more general than rectangles. As long as the terminals are on the outer face the problem is still a special case of the problem answered by Theorem 4.7. But the revised cut criterion is not sufficient in general as was shown by (Lai and Sprague 1987). See the problem in Figure 4.6.

With the help of the pairing method described in the proof of Theorem 4.7 we can reduce the problem to the Eulerian case. When the boundary region is x-convex, that is, any horizontal line intersects it in a segment, then there is an extremely simple algorithm due to M. Kaufmann.

We close the section by an application of Theorem 4.5. (The material is taken from (Frank and Tardos 1984). Let O and I be two closed rectangles bounded by lattice lines such that I is in the inside of O. The graph we consider is the subgraph of the rectilinear grid between O and I. The k terminal pairs to be connected are on the perimeter of I. The problem is to find edge-disjoint paths connecting the corresponding terminal pairs which, in addition, do not touch. That is, if a path bends at a certain node v, then v must not be used by other paths. This constraint is imposed in order to model two-layer routing problems where one layer is used for horizontal segments, the other for vertical ones and a bend corresponds to a via hole between the two layers.

Figure 5.5 shows an instance of the problem along with a solution.

A version of this problem was solved by LaPaugh (1980) when only the inner rectangle I is given and the problem is to find a surrounding rectangle O of minimum area such that the paths exist between O and I.

We need the following well known result. Let $\mathcal{F}$ be a family of closed intervals of a segment S. The *density* of $\mathcal{F}$ is the maximum number of intervals covering a point of S.

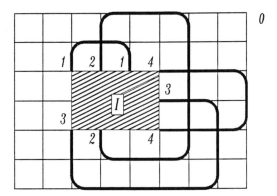

Fig. 5.5

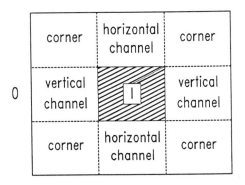

Fig. 5.6

Lemma (Gallai 1962). $\mathcal{F}$ *can be partitioned into d classes consisting of pairwise disjoint segments if and only if the density of* $\mathcal{F}$ *is at most d. (Furthermore the partition can be found in* $O(|\mathcal{F}| \log |\mathcal{F}|)$ *time.)*

The four straight lines of the boundary segments of I divide $O - I$ into four channels and four corners.

The *width* of the channels above and below I (resp. left and right to I) is the number of their horizontal (resp. vertical) lines.

Observe that each path has two different homotopies. Suppose for a moment that the homotopies have already been specified. Then they define four interval systems as is shown in Figure 5.7a and b.

If (∗) the density of each of these interval systems is at most the width of the corresponding channel, then by Gallai's lemma the intervals can be placed on the available lines, and the resulting segments belonging to the same homotopy can be connected in the corners so as to form the desired paths (Figure 5.7 c).

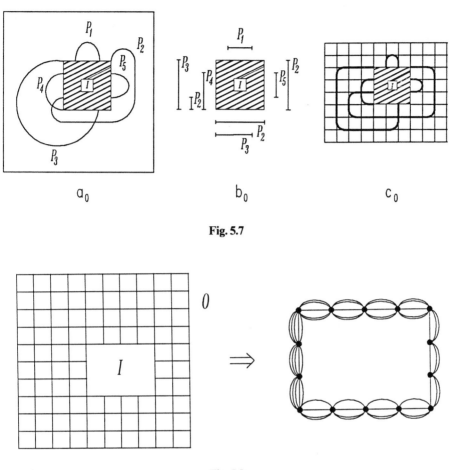

Fig. 5.7

Fig. 5.8

Therefore the only problem we are encountered is to find the homotopies that satisfy (∗). But this problem can immediately be solved if we apply Theorem 4.5 to the graph obtained by contracting all the edges in the four corners, the vertical edges in the two horizontal channels and the horizontal edges in the two vertical channels.

6. When the Disjoint Paths Problem is Tractable

In this section we survey restrictions of the (edge-) disjoint paths problem when either a polynomial time algorithm is available or a sufficient condition (or both).

Theorem 6.1 (Robertson-Seymour 1986b). *For fixed k the undirected (edge-) disjoint paths problem can be solved in polynomial time.*

Actually this theorem is the central topic of Robertson and Seymour's paper in this volume. As they remark the algorithm is completely out of the range of practical usability when $k > 2$. For acyclic digraphs an analogous result holds.

Theorem 6.2 (Fortune, Hopcroft and Wyllie 1980). *In acyclic digraphs the (arc-) disjoint paths problem can be solved in polynomial time if k is fixed.*

Unlike the undirected case, the algorithm of Fortune, Hopcroft and Wyllie is quite reasonable for small k. The idea behind it is reformulated in (Thomassen 1985) as follows. Let the set $S = \{s_1, \ldots, s_k\}$ of source nodes and the set $T = \{t_1, \ldots, t_k\}$ of target nodes be disjoint. We introduce an auxiliary digraph G^* the nodes of which correspond to the k−tuples of distinct nodes of G. There is an arc in G^* from $X = \{x_1, \ldots, x_k\}$ to $Y = \{y_1, \ldots, y_k\}$ if and only if there is a $j \in \{1, \ldots, k\}$ such that $x_i y_i$ for $i \neq j$ and G contains an arc $x_j y_j$ and contains no directed path from x_i to x_j $(i \neq j)$. It can be shown that there are k pairwise disjoint paths in G from s_i to t_i $(i = 1, \ldots, k)$ if and only if G^* has a path from S to T.

A by-product of Schrijver's disjoint homotopic paths theory (see his paper in this volume) is the following.

Theorem 6.3. *The disjoint paths problem is solvable in polynomial time when G is planar and the terminals are on a bounded number of faces of G.*

Note that in this case no restriction is put on the size of the demand graph. The status of the corresponding edge-disjoint paths problem is not known.

Theorem 6.4 (Sebő 1988). *The integer multicommodity flow problem is solvable in polynomial time if G + H is planar and there is a bounded number of demand edges (with arbitrary big demand values).*

The same question remains open if only the number of faces of G covering the terminal nodes is bounded.

Next we list results where connectivity assumptions prove to be sufficient for the (edge-) disjoint paths problem.

Let us call a graph *k-linked on the edges (or weakly k-linked)* if for any choice of k pairs of terminals there are k edge-disjoint paths connecting the corresponding terminal pairs. Let $g(k)$ denote the minimal number m such that every m-edge-connected graph is k-linked on the edges.

C. Thomassen has a nice conjecture asserting that $g(2k + 1) = g(2k) = 2k + 1$.

Theorem 6.5. $g(3) = 3$, $g(4) = 5$, $g(5) \leq 6$, $g(6) \leq 8$, $g(7) \leq 9, g(3k) \leq 4k$, and $g(3k + 1) \leq g(3k + 2) \leq 4k + 2$ $(k \geq 2)$.

Here $g(3) = 3$ is due to (Okamura 1984), $g(4) = 5$ to (T. Hirata-K. Kubota-O. Saito 1984) and to (Mader 1985), the other results are due to (Okamura 1988).

Surprisingly for directed graphs the analogous situation is much simpler. The following was observed by Shiloach (1979). Let us call a digraph $D = (V, A)$ *k-linked on the arcs* if for any choice of k pairs $\{(s_1, t_1), \ldots, (s_k, t_k)\}$ of (not necessarily distinct) terminals there are arc-disjoint paths p_i from s_i to t_i $(i = 1,...,k)$. Obviously such a digraph is strongly k-arc connected (that is every non-empty proper subset of nodes has k entering arcs.)

Theorem 6.6. *A strongly k-arc connected digraph is k-linked on the arcs.*

Proof. Add a new node r to D and new arcs rs_i $(i = 1, 2,...,k)$ and apply Edmonds' disjoint arborescence theorem (Edmonds 1973).   □

We call a graph *k-linked* if for any choice of k pairs of terminals there are k openly disjoint paths connecting the corresponding terminal pairs.

Theorem 6.7 (Jung 1970), (Larman and Mani 1970). *A 2^{3k} connected graph is k-linked.*

It is not known if 2^{3k} can be replaced by a linear bound. The natural $2k + 2$ is not enough as can be seen from a K_{3k-1} with edges $x_1 y_1,..., x_k y_k$ removed (an example due to (Strange and Toft 1983)).

In certain cases the cut condition is not strong enough to ensure the existence of all required paths but the demands can almost be met:

Theorem 6.8 (Korach and Penn 1985). *Suppose that $G + H$ is planar and that the demand edges are on k faces of G. If the cut criterion holds, there are edge-disjoint paths connecting all but $k - 1$ terminal pairs so that for one face F, specified in advance, all the terminal pairs on F are connected while for each other face F' the terminal pairs on F' with one possible exception are connected.*

Actually, Korach and Penn proved a more general result. There is an important corollary to Theorem 6.8.

Corollary 6.9. *Suppose that $G + H$ is planar and H consists of k demand edges (s_i, t_i) endowed with integer demands d_i. The supply edges e have integer capacities $c(e)$ so that the cut criterion holds. Then there are d_1 paths connecting s_1 and t_1 and $d_i - 1$ paths connecting s_i and t_i $(i = 2, 3,..., k)$ so that each supply edge e is used by no more than $c(e)$ among these $\sum d_i - k + 1$ paths.*

Another result of similar flavour is the following.

Theorem 6.10 (Itai and Zehavi 1984). *In a graph s_i, t_i are terminal pairs $(i = 1, 2)$ such that there are k edge-disjoint paths connecting s_i and t_i $(i = 1, 2)$. Then for each m, $0 \leq m < k$ there are k edge-disjoint paths $P, S_1, S_2,..., S_m, Q_1, Q_2,..., Q_{k-m-1}$ such that each S_i connects s_1 and t_1, each Q_j connects s_2 and t_2 and P connects either s_1 and t_1 or s_2 and t_2.*

7. Maximization

In combinatorial optimization sometimes we are interested in the existence of a certain configuration (e.g., is there a perfect matching in a graph) other times we need the biggest (or smallest) configuration (e.g. find the biggest matching). Not surprisingly, the corresponding feasibility and maximization problem often correlate and typically (though not always) the maximization problem is more difficult.

For example, in the matching case Tutte's theorem on the existence of a perfect matching is a direct consequence of the so-called Berge-Tutte formula on the maximum cardinality of a matching. Conversely, the Berge-Tutte formula can be derived from Tutte's theorem via an elementary construction.

There are however other cases when a good answer to the feasibility problem exists but the corresponding maximization problem is NP-complete. For example, suppose that G has a perfect matching. Then the problem of finding a stable set of $|V|/2$ elements is tractable, but to find a maximum cardinality stable set is NP-complete.

As far as (edge-) disjoint paths problems are concerned we have studied so for problems of feasibility type. In the *maximization problem* we want to find the maximum total number M of (edge-) disjoint paths connecting the corresponding terminal pairs $s_1 t_1, s_2 t_2, \ldots, s_k t_k$ (allowing many paths between one terminal pair).

In what follows we discuss, among others, some feasibility problems where the corresponding maximization problem is solvable but the derivation needs some work.

Let $V_1, V_2, \ldots, V_t$ be a family $\mathcal{P}$ of disjoint subsets of V such that each demand edge connects different V_i's. By a *multicut* defined by $\mathcal{P}$ we mean the set of edges uv of G such that $u \in V_i$, $v \notin V_i$ for some i. The capacity m of a multicut is defined to be $1/2 \Sigma d(V_i)$. Let m_1 denote the minimum cardinality of a cut separating each terminal pair. Obviously, $m_1 \geq m \geq M$. If $k = 1$, then $m_1 = M$ by Menger's theorem.

First, we will present two theorems for $k = 2$. In the first one, due to B. Rothschild and A. Whinston (1966a), we assume that the degree of every non-terminal node is even, in the second one, due to M. Lomonosov (1983), we assume that G together with the two edges $s_1 t_1, s_2 t_2$ is planar.

For both cases let c_i denote the cardinality of a minimum cut separating s_i and t_i but not separating $s_{3-i} t_{3-i}$. Let c_{12} denote the minimum cardinality of a cut separating both terminal pairs. (Then $c_{12} = m_1$). Obviously $c_1 + c_2 \geq c_{12}$.

Theorem 7.1 (Rothschild-Whinston 1966a). *If $k = 2$ and $d_G(v)$ is even for each non-terminal node, then $m_1 = M$.*

Proof. Assume first that c_{12} is even.

Case 1. $d(s_1)$ and $d(t_1)$ have the same parity (equivalently $d(s_2)$ and $d(t_2)$ have the same parity). Define a demand graph H to consist of c_1 parallel edges between s_1 and t_1 and $c_{12} - c_1$ parallel edges between s_2 and t_2. By Theorem 3.8 we are done since $G + H$ is Eulerian and the cut criterion holds.

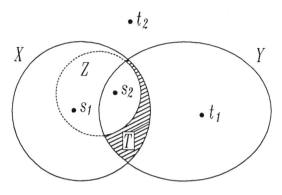

Fig. 7.1

Case 2. Precisely one member, say s_1 and s_2, of both terminal pairs has odd degree. If both c_1 and c_2 are even (resp. odd), define H to consist of c_1 edges between s_1 and t_1, $c_{12} - c_1$ edges between s_2 and t_2 and just one edge between s_1 and s_2 (resp. t_1 and t_2). If one of c_1 and c_2 is odd, say c_1, then let H consist of c_1 edges between s_1 and t_1, $c_{12} - c_1 + 1$ edges between s_2 and t_2 and one edge between t_1 and t_2. With some care one can check that in each case $G + H$ is Eulerian and the cut criterion holds. Therefore Theorem 3.8 implies the existence of the desired c_{12} paths.

If c_{12} is odd, then one member, say s_1 and s_2, of both terminal pairs has odd degree. Furthermore any set X for which $\nabla(X)$ separates both terminal pairs and $d(X) = c_{12}$ contains exactly one of s_1 and s_2. Therefore if we add a new edge $e = s_1 s_2$ to G, then the new c'_{12} is one bigger that c_{12}. So it is even. For even c'_{12} we have proved already the existence of c'_{12} edge-disjoint paths in $G + e$ between the two terminal pairs. If we take back the newly added edge e, we still have the c_{12} paths in G, as required.   □

Let us call a cut C *critical* if it has a minimum number of edges from G among all the cuts that separate the same terminal pairs as C does. Recall the definition of separating and parallel sets (before Theorem 4.4)

Theorem 7.2 (Lomonosov 1983). *Suppose that $k = 2$ and $G + \{s_1 t_1, s_2 t_2\}$ is planar. Then M is either $c_{12} - 1$ or c_{12}. $M = c_{12} - 1$ if and only if there is an odd cut $\nabla(T)$ which does not separate either of the two terminal pairs and which can be covered by three separating critical cuts $\nabla(X), \nabla(Y), \nabla(Z)$. Moreover X, Y, Z, T can be chosen in such a way that $Z \subset X$, X and Y are non-parallel and $T = (X \cap Y) - Z$. (See Figure 7.1)*

Proof. The original proof of Lomonosov consists of a direct construction and is rather complicated. Here we derive the result from Theorem 4.4 of Seymour. We can assume that $c_1 > 0$, $c_2 > 0$ and $c_{12} > \max(c_1, c_2)$ since otherwise the situtation is trivial.

Let H consist of $c_1 - 1$ parallel edges between s_1 and t_1 and $c_{12} - c_1$ parallel edges between s_2 and t_2, then the cut criterion holds and there is no tight separating cut. By Theorem 4.4 we see that $M \geq c_{12} - 1$.

The necessity of the condition in the second half of the theorem is straightforward.

Case 1. $c_1 + c_2 = c_{12}$.

Let H consist of c_i edges between s_i and t_i $(i = 1, 2)$. We are done if there is a solution to the feasibility problem concerning G and H. Suppose there is none. Then we have

Claim. *There are tight non-parallel separating sets X, Y and tight sets $Z_1 \subseteq X - Y$, $Z_2 \subseteq X \cap Y$ such that $d_{G+H}(X \cap Y)$ is odd and Z_1 separates one of the two terminal pairs while Z_2 separates the other.*

Proof. By Theorem 4.4 there are tight non-parallel separating sets X and Y so that $d_{G+H}(X \cap Y)$ is odd (and then automatically $d_{G+H}(X - Y)$, $d_{G+H}(Y - X)$ and $d_{G+H}(V - (X \cup Y))$ are all odd). Assume that $s_1 \in X - Y$, $t_1 \in Y - X$, $s_2 \in V - (X \cup Y)$ and $t_2 \in X \cap Y$.

Let Z_1 be a minimal tight set separating s_1 and t_1 such that $s_2, t_2 \notin Z_1$. Then $d_G(Z_1) = c_1$ and Z_1 contains one of s_1 and t_1, say s_1. Since $d_H(X, Z_1) = 0$, by Lemma 3.1 $Z_1 \cap Y$ is tight. Therefore $Z_1 \subseteq X$ by the minimal choice of Z_1. Since $d_H(Z_1, V - Y) = 0$, by Lemma 3.1 $Z_1 - Y$ is tight so we obtain that $Z_1 \subseteq X - Y$.

It can be seen analogously that there is a tight cut $\nabla(Z_2)$ separating s_2 and t_2 for which s_1, $t_1 \notin Z_2$ and Z_2 is either in $X \cap Y$ or in $V - (X \cup Y)$. We can assume that $Z_2 \subseteq X \cap Y$ for otherwise we can work with $V - Y$ on place of Y. □

Let $Z = Z_1 \cup Z_2$ and $T = X \cap Y - Z$. Since $c_{12} \leq d_G(Z) \leq d_G(Z_1) + d_G(Z_2) = c_1 + c_2 = c_{12}$ we have $d_G(Z) = c_{12}$. Furthermore $d_G(T) = d_{G+H}(T) = d_{G+H}(X \cap Y) - d_{G+H}(Z_2) + 2d_{G+H}(Z_2, T)$. Since $d_{G+H}(X \cap Y)$ is odd and $d_{G+H}(Z_2)$ is even, $d_G(T)$ is odd and therefore the theorem is proved for Case 1.

Case 2. $c_1 + c_2 > c_{12}$.

Let H' consist of c_1 edges between s_1 and t_1 and $c_{12} - c_1$ parallel edges between s_2 and t_2. If there is a solution to the feasibility problem concerning $G + H'$, then we are done. Otherwise, since the cut criterion holds, there are separating non-parallel tight sets X' and Y' such that $s_1 \in X' - Y'$, $s_2 \notin X' \cup Y'$ and $d_{G+H'}(X' \cap Y')$ is odd. Suppose that X' and Y' are minimal such sets. (To avoid confusion we will call these sets H'−tight.)

Next, let H'' consist of $c_1 - 1$ edges between s_1 and t_1 and $c_{12} - c_1 + 1$ parallel edges between s_2 and t_2. ($c_1 > 0$ since otherwise we are at Case 1). Again the cut criterion holds, so if the feasibility problem has no solution then (by Theorem 4.4) there are separating non-parallel H''-tight set X'', Y'' such that $s_1 \in X'' - Y''$, $s_2 \notin X'' \cap Y''$ and $d_{G+H}(X'' \cap Y'')$ is odd. Suppose that X'' and Y'' are minimal such sets.

It is not possible that $X' = X''$ and $Y' = Y''$ since $d_{G+H'}(X' \cap Y')$ is odd, therefore $d_{G+H''}(X' \cap Y')$ is even and $d_{G+H''}(X'' \cap Y'')$ is odd. Assume that

$X' \neq Y'$ By symmetry we can assume that $X' \nsubseteq X''$, that is $Z := X' \cap X'' \subset X'$. Let $T = X' \cap Y' - Z$. Since $d_G(X'') = c_{12}$, X'' is H'-tight. By Lemma 3.1 Z is H'-tight and separating. Since Z is a proper subset of X' and X' was chosen minimal, $d_{G+H'}(Z \cap Y')$ must be even. But then $d_G(T) = d_{G+H'}(T) = d_{G+H'}(X' \cap Y') - d_{G+H'}(Z \cap Y') + 2d_{G+H}(Z \cap Y', T)$ from which we conclude that $d_G(T)$ is odd and sets X', Y', Z, T satisfy the requirements in the theorem.   □

There are cases when the maximization form is easier to handle:

Theorem 7.3a (Lovász 1976b, Cherkasskij 1977). *If the demand edges form a complete graph induced by A ($A \subseteq V$) and $d_G(v)$ is even for $v \in V - A$, then $m = M$.*

Proof. For $a \in A$ let c_a denote the maximum number of egde-disjoint paths connecting a and $A - a$. By Menger's theorem this is the minimum cardinality of a cut separating a and $A - a$. Let us denote $\frac{1}{2}c(A) = \Sigma(c_a : a \in A)$. Obviously $c(A) \geq m \geq M$. We are going to show, by induction on the number of edges, that $c(A) = M$.

We can assume that G is connected. If $A = V$ there is nothing to prove. Call a set $X \subset C$ *critical* with respect to $a \in A$ if $X \cap A = \{a\}, |X| \geq 2$ and $d(X) = c_a$.

Case 1. There is a set X critical with respect to a certain $a \in A$. Contract the elements of X into one node a'. In the contracted graph there are $c(A)$ edge-disjoint A'-paths where $A' = A - a + a'$. In X there are $d(X)$ edge disjoint paths from a to the edges in $\nabla(X)$. Pasting together the two sets of paths we obtain $c(A)$ edge-disjoint A-paths in G.

Case 2. There are no critical sets. Choose any two edges e, f which are incident to a node v not in A. Because there are no critical sets, splitting off e and g does not reduce $c(A)$ and we are done by induction.   □

Generalizing this result to non-Eulerian graphs W. Mader (1978b) found the following (much more difficult) characterization for M.

Theorem 7.3. *Let $G = (V, E)$ be a graph and A a specified subset of nodes. The maximum number of edge-disjoint paths connecting distinct elements of A is $\min 1/2(\Sigma(d(V_i) - q(\cup V_i)))$ where the minimum is taken over all collections of disjoint subsets $V_1, V_2, \ldots, V_{|A|}$ for which $|V_i \cap A| = 1$. (Here $d(X)$ denotes the edges leaving X and $q(X)$ denotes the number of components C of $G - X$ for which $d(C)$ is odd.)*

To formulate a node-disjoint version of Theorem 7.3 suppose that A is independent.

Theorem 7.4 (Mader 1978c). *The maximum number of openly node-disjoint paths connecting distinct members of A is equal to $\min(|V_0| + \Sigma(\lfloor 1/2b(V_i, V_0)\rfloor : i = 1, 2, \ldots, k)$ where the minimum is taken over all collections of disjoint subsets $V_0, V_1, \ldots, V_k$ of $V - A$ ($k \geq 0$) (where only V_0 can be empty) such that*

$G - V_0 - \cup(E(V_i) : i = 1, \ldots, k)$ *contains no path connecting distinct nodes of A. In the formula* $b(V_i, V_0)$ *denotes the number of nodes of* V_i *which have a neighbour outside* V_0.

In the next figure the solid points belong to A. There are two openly disjoint A-paths and the only family where the minimum is attained is shown in the picture.

$$b(V_1, V_0) = b(V_2, V_0) = 3 \qquad\qquad V_0 = \emptyset$$

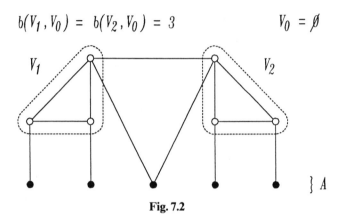

Fig. 7.2

This result can be regarded as a common generalization of Menger's theorem and the Berge-Tutte theorem. An immediate corollary of Theorem 7.4 is a result of T. Gallai (1961):

Corollary 7.5. *The maximum number of disjoint paths having end nodes in* T *is* $\min_{K \subseteq V}(|K| + \Sigma \lfloor 1/2 |C \cap T| \rfloor)$ *where the sum is taken over the components* C *of* $G - K$.

Let us turn back to edge-disjoint paths. A common generalization of Theorems 7.1 and 7.2 has been found recently:

Theorem 7.6 (Karzanov 1985b). *Let* $H = (T, F)$ *denote the demand graph. If the maximal independent sets of* H *can be partitioned into two classes such that both classes consist of disjoint sets (which is equivalent to saying that the complement of* H *is the line graph of a bipartite graph) and, in addition, if* $d_G(v)$ *is even for* $v \in V - T$, *then* $m = M$.

As far as the maximization problem is concerned for digraphs we mention the following counter-part of Theorem 7.3a.

Theorem 7.7 (Frank 1989). *In an Eulerian digraph* $D = (V, A)$ *the maximum number of arc-disjoint paths connecting distinct nodes of a specified subset* A *of* V *is equal to the minimum of* $\Sigma \varrho(V_i)$ *over all families of disjoint subsets* $V_1, V_2, \ldots, V_{|A|}$ *of* V *for which* $|V_i \cap A| = 1$ $(i = 1, 2, \ldots, |A|)$.

The proof goes along the same line as that of Theorem 7.3a.

Professor Karzanov kindly informed me that this theorem was proved much earlier by Lomonosov. Karzanov presented Lomonosov's proof in: Combinatorial Methods for Flow Problems (Inst. for System Studies, Moscow 1979, issue 3), 6-69, in Russian.

8. T-joins and T-cuts

Let $G = (V, E)$ be an undirected graph. Let T be a subset of nodes with even cardinality. A cut $\nabla(K)$ is called a T-cut if $|K \cap T|$ is odd. A T-join F is a set of edges that has an odd number of edges incident to a node v if and only if v is in T. G has a T-join if and only if every component of G has an even number of nodes from T (easy exercise). Obviously any T-join and T-cut has an odd number of edges in common, in particular at least one. Therefore the maximum number of disjoint T-cuts cannot exceed the minimum cardinality of a T-join. The complete graph on 4 nodes, when $T = V$, shows that we do not have equality in general. However,

Theorem 8.1 (Seymour 1981). *In bipartite graphs the maximum number of disjoint T-cuts is equal to the minimum cardinality of a T-join.*

This theorem implies that (Lovász 1976b) in an arbitrary graph G one half of the maximum number of half-disjoint T-cuts is equal to the minimum cardinality of a T-join. (Here half-disjoint means that each edge can be in at most 2 T-cuts.) Indeed, subdivide each edge of G by a new node and apply Theorem 8.1. A weaker version of this result, stating that the minimum cardinality of a T-join is equal to the maximum of a fractional packing of $T-$cuts, was proved algorithmically in (Edmonds and Johnson 1973).

Because minimum T-joins and T-cuts packings have a great number of applications we say some words about the algorithmic aspects.

A weighted generalization of Theorem 8.1 is the following.

Theorem 8.1w. *Let $d : E \to \mathbb{Z}_+$ be a non-negative integer-valued function with the even-circuit property, that is, the d-weight of every circuit of G is even. Then the minimum weight of a T-join is equal to $\max(\sum(y(A) : A \subseteq V, |A \cap T| \text{ odd}), y$ non-negative and integer-valued, $d(uv) \geq \sum(y(B) : |B \cap T| \text{ odd}, |B \cap \{u, v,\}| = 1$ for every $uv \in E))$.*

If we choose each weight to be 1, we are back at Theorem 8.1. On the other hand this weighted version can easily be derived from Theorem 8.1. The problem is to find algorithmically the minimum and maximum in question.

Let $m(uv)$ denote the minimum $d-$weight of a path in G between u and v, $(v, u \in V)$. Obviously $m(uv) \leq d(uv)$, m satisfies the triangle inequality, and m has the even-circuit property if d has.

In order to construct a minimum weight T-join Edmonds and Johnson (1973) associated the T-join problem with the following minimum weight perfect matching problem. For each pair $u, v \in T$ compute $m(uv)$ between u and v. Construct then a minimum m-weight perfect matching M in the complete graph on T. Finally, look at the union F of the minimum weight paths in G that connect the pairs of nodes determined by M. It is easy to see that these $|M|$ paths are pairwise edge-disjoint and F is a minimum weight T-join. Because the mimimum weight perfect matching problem is solvable in strongly polynomial time (as was shown first by Edmonds (1965)), so is the minimum weight T-join problem.

To construct the optimal packing of T-cuts several algorithms have been devised (Korach 1982, Karzanov 1986, Barahona 1987).

Here we exhibit the newest algorithm that seems to be conceptionally the simplest. Its basic idea, due to A. Sebő (1988), is that not only the minimum weight T-join problem reduces so handily to a minimum weight perfect matching problem, but also its dual reduces relatively easily to the dual of the associated matching problem.

To obtain an integral optimal solution to the dual matching problem we need the following observation of Barahona and Cunningham (1988). Let C_T be the complete graph on a set T of even cardinality and let m be a non-negative integer-valued weighting on the edges of C_T with the even-circuit property.

Barahona and Cunningham proved the following.

Lemma 8.2. *If m satisfies the even-circuit property, then the minimum weight of a perfect matching in C_T is equal to $\max(\sum(y(A) : A \subseteq T, |A|$ odd$), y(A) \geq 0$ for $|A| \geq 3$, y integer-valued and $m(uv) \geq \sum(y(B) : |B|$ odd, $|B \cap \{u,v\}| = 1$ for every edge $uv))$.*

This follows easily from Theorem 8.1w of Seymour. The main point in Barahona and Cunningham's paper is the observation that a natural modification of Edmonds' algorithm provides not only the minimum weight perfect matching but also the integer-valued dual y occuring in the lemma. In other words, the minimum and the maximum in the lemma can be computed in strongly polynomial time.

The modification consists of two parts. First, the minimum weight of a perfect matching is considered rather than the maximum. Second, unlike Edmonds' algorithm, where an alternating forest is grown, Cunningham and Barahona's algorithm grows only one alternating tree at a time. This ensures that the current dual variables are automatically integer-valued if the even-circuit property holds.

It is obvious from the algorithm that the family $\mathscr{F} := \{A : y(A) > 0\}$ is laminar. Our second observation is the following.

Lemma 8.3. *If, in addition to the assumptions in Lemma 8.2, m satisfies the triangle inequality, then y can be chosen non-negative.*

Proof. Let us start with an optimal y occuring in Lemma 8.2. If this is non-negative, we are done, so suppose that $y(z) < 0$ for some $z \in T$. For any set $A \in \mathscr{F}$ containing z increase $y(T - A)$ by $y(A)$ and then revise $y(A)$ to be 0. This way we get another optimal solution such that no member of $\mathscr{F}$ contains z. (Such a change keeps $\mathscr{F}$ laminar).

Denote $p(uv) := \sum(y(A) : |A \cap \{u,v\}| = 1)$. The dual constraint in Lemma 8.2 requires that $p(uv) \leq m(uv)$ for every $u,v \in T$. There must be an edge uz incident to z for which $p(uz) = m(uz)$ since otherwise by increasing $y(z)$ by 1 we would get a better y.

Let A be a maximal member of $\mathscr{F}$ containing u. For $v \notin A$ we have

$$(*) \qquad\qquad p(uv) = p(uz) + p(vz) - 2y(z)$$

Let $\Delta := \min(-y(z), y(A))$ and revise y by increasing $y(z)$ by Δ and decreasing $y(A)$ by Δ.

We claim that the revised dual solution is feasible. To see this all we have to show is that $m(vz) - p(vz) \geq \Delta$ for every $v \notin A$. Actually, this inequality turns out to be strict. Indeed, using $(*)$ and the triangle inequality we get $m(vz) \geq m(uv) - m(uz) = m(uv) - p(uz) \geq p(uv) - p(uz) = p(vz) - 2y(z) > p(vz) + \Delta$, as required.

Therefore we have another optimal dual solution. Repeat this procedure as long as there is a point z in T with $y(z)$ negative. We claim that after at most $2|T|$ iterations y becomes non-negative. Indeed, at every iteration the number of points v with negative $y(v)$ plus $|\mathscr{F}|$ reduces and this sum is at most $2|T|$. □

Let $\mathscr{F} \subseteq 2^V$ be a laminar family and $y : \mathscr{F} \to \mathbb{Z}_+$ a function. We call the pair $(y, \mathscr{F})$ a *weighted laminar family* on S as follows. Let $\mathscr{F}_S := \{X = F \cap S : \text{for some } F \in \mathscr{F}\}$ and let $y_S(X) := \sum(y(F) : F \in \mathscr{F}, X = S \cap F)$ for $X \in \mathscr{F}_S$. We will say that $(y, \mathscr{F})$ is an *extension of* $(y_S, \mathscr{F}_S)$ on V.

Let m be a metrics on V. We say that a $w-$laminar family $(y, \mathscr{F})$ on S is *feasible* if $m(uv) \geq \sum(y(F) : |F \cap \{u,v\}| = 1, F \in \mathscr{F})$ holds for every $u,v \in S$.

Lemma 8.4 (Sebő 1988). *Every feasible $w-$laminar family $(y, \mathscr{F})$ on S can be extended (in polynomial time) to a $w-$laminar family on V.*

Obviously, if we apply the lemma to the w-laminar family $(y, \mathscr{F})$ on T obtained in Lemmas 8.2 and 8.3, we obtain an optimal solution to the T-packing problem in Theorem 8.1w.

Originally, the lemma was proved, using a different method, by A. Sebő (1988). The present proof has a slight advantage that it provides a conceptionally simpler algorithm.

Proof. We are going to prove only that $(y, \mathscr{F})$ can be extended on a set $S + t$ ($t \in V - S$) because then, element by element, we can extend $(y, \mathscr{F})$ on V. So suppose that $V = S + t$.

It is well known that a laminar family $\mathscr{F}$ can be represented with the help of an arborescence $D = (V', A)$ (with $V' \cap V = \emptyset$) and a mapping from V to V' as follows. There is a one to one correspondence between the edges of D and the members of $\mathscr{F}$ with the property that for every edge e of D the corresponding member of $\mathscr{F}$ consists precisely of those elements of S whose map is reachable in D from the head of E. We denote the map of an element $u \in V$ by u'.

For an edge f of D let $y(f) := y(F)$ where F is the member of $\mathscr{F}$ corresponding to f. For $u', v' \in V'$ let $y(u'v')$ denote the y-length of the unique (undirected) path in D between u' and v'. In this representation the feasibility of $(y, \mathscr{F})$ means that $m(uv) \geq y(u'v')$ for every $u, v \in S$ and the lemma follows from the following

Claim. *Either there is a node t' of D for which*

$$(*) \qquad\qquad m(tu) \geq y(t'u') \text{ for every } u \in S$$

or there is an edge $e = s'z'$ of D and an integer $0 < h < y(s'z')$ so that subdividing e by t' and defining $y(s't') := h$, $y(t'z') := y(s'z') - h$ $(*)$ holds true.

Proof. Let r' denote the root of D. If $m(tu) \geq y(r'u')$ holds for every $u \in S$, then $t' := r'$ will do. So suppose that $y(r'u') - m(tu) > 0$ for some $u \in S$ and let $u_0 \in S$ be an element for which $M = y(r'u_0') - m(tu_0)$ is maximum.

Let $s'z'$ be an edge of the path from r' to u_0' in D for which $y(r's') \leq M < y(r'z')$. If $y(r's') = M$, then choose $t' := s'$. If $y(r's') < M$, then subdivide $s'z'$ by a new node t' and choose $h := M - y(r's')$. With this choice we have $y(t'u_0') = m(tu_0)$ and we claim that $(*)$ is satisfied. To see this let D' denote the subdivided arborescence and let $u \in S$ be arbitrary.

Case 1. The path in D' from r' to u' contains t'. Then $y(r'u_0') = y(r't') + y(t'u_0')$. By the maximal choice of u_0 we have $y(r'u_0') - m(tu_0) \geq y(r'u') - m(tu)$. Therefore $m(tu) \geq y(r'u') + m(tu_0) - y(r'u_0') = y(r'u') - y(r't') = y(t'u')$.

Case 2. The path in D' from r' to u' does not contain t'. Then $y(u'u_0') = y(t'u_0') + y(t'u')$. We have $m(tu) \geq m(uu_0) - m(tu_0) = m(uu_0) - y(t'u_0') \geq y(u'u_0') - y(t'u_0') = y(t'u')$. $\square$

This one element extension can be carried out in $O(n)$ steps, therefore the complete extension needs no more than $O(n^2)$ steps.

There is a version of Theorem 8.1 that ensures a maximum packing of T-cuts with a special structure. For a subset X of nodes let $q_T(X)$ denote the number of T-odd components in $G - X$ (a component is T-*odd* if it contains an odd number of nodes in T).

Theorem 8.5 (Frank, Sebő and Tardos 1984). *In a bipartite graph $G = (V_1, V_2; E)$ the minimum cardinality of a T-join is equal to $\max \Sigma q_T(X_i)$ taken over all partitions $\{X_i\}$ of V_1.*

This result immediately implies Theorem 8.1: for an optimal partition $\{X_i\}$ take the T-cuts defined by the T-odd-components of $G - X_i$ $(i = 1, 2 \ldots)$.

Before deriving these results let us mention an easy but useful lemma.

Lemma 8.6 (Mei-Gu Guan). *A T-join F is of minimum cardinality if and only if no circuit of G uses more edges from F than from $E - F$ (or in other words, there is no circuit of negative total weight in G where the edges of F have weight -1 the other edges have weight 1).*

We call a circuit of negative total weight a *negative circuit*.

To prove the non-trivial direction of Theorem 8.1 one starts with a minimum T-join F and wants to find $|F|$ disjoint T-cuts each of which containing one element of F. Obviously F is a forest since the edge-set of a circuit could be left out of F without changing the parity of the degrees. Here comes an observation: a cut that contains exactly one element of F is automatically a T-cut. Therefore Theorem 8.1 follows from the following.

Theorem 8.1′. *In a ± 1 edge-weighted bipartite graph there is no negative circuit if and only there are edge-disjoint cuts such that each contains one negative edge and each negative edge is contained in one cut.*

(This is exactly Theorem 3.6′ except that the wording is different.)

Actually Theorem 8.1′ follows from Theorem 8.1, as well. Indeed, the "if" part is easy. To see the "only if" part let F denote the set of edges of weight -1. Let T consist of those nodes that have an odd number of edges incident to F. Then F is a T-join and, by the lemma, F is a minimum T-join. By Theorem 8.1 there are $|F|$ disjoint T-cuts each of which necessarily contains exactly one element of F.

Using the same idea, Theorem 8.1w transforms into:

Theorem 8.1′w. *Let $G = (V, F^+ \cup F^-)$ be a graph and $d : F^+ \cup F^- \to \mathbb{Z}$ an integer-valued weight-function for which $d(e) \geq 0$ if $e \in F^+$ and $d(e) < 0$ if $e \in F^-$ and d satisfies the even-circuit property. There is no negative-circuit in G if and only if there is an integer-valued vector $y : 2^V \to \mathbb{Z}_+$ so that $y(A) > 0$ implies that $d_{F^-}(A) = 1$, that $\sum(y(A) : |A \cap \{u, v\}| = 1) \leq d(u, v)$ for every $uv \in F^+$ and $\sum(y(A) : |A \cap \{u, v\}| = 1) = d(uv)$ for every $uv \in F^-$.*

Note that the algorithm given after Theorem 8.1w can be used to construct either a negative circuit or a packing y. Namely, define $T = \{v \in V, d_{F^-}(v) \text{ odd}\}$ and find a minimum weight T-join F with respect to the weight function $d' |d(e)|$. If $d'(F) < d'(F^-)$, then the symmetric difference $F^- \oplus F$ contains a circuit of negative d-weight. If $d'(F) = d'(F^-)$, then the vector y in Theorem 8.1w will do for Theorem 8.1′w.

For later purpose we phrase here a fractional version of Theorem 8.1′.

Theorem 8.1″. *Let $\overline{G} = (V, \overline{E})$ be a graph where $\overline{E}$ is partitioned into two sets E and F. Let $y : E \to R$ be a rational vector for which $y(e) \geq 0$ if $e \in E$ and $y(e) \leq 0$ if $e \in F$. There is an assignment of non-negative variables $z(B)$ to cuts B containing exactly one edge from F for which $y(e) \geq \Sigma(z(B) : e \in B)$ for every $e \in E$ and $-y(e) \leq \Sigma(z(B) : e \in B)$ for every $e \in F$ if and only if there is no circuit of G with negative y-weight.*

If G is planar, we can take the dual graph and then Theorem 8.1 transforms into Theorem 3.6.

We can reformulate also Theorem 8.5, as follows.

Theorem 8.5′. *In a* ± 1 *edge-weighted bipartite graph* $G = (V_1, V_2; E)$ *there is no negative circuit in* G *if and only if there is a partition* $\{X_1, X_2, \ldots, X_k\}$ *of* V_1 *such that no component of* $G - X_i$ *is entered by more than one negative edge.* $(i = 1, \ldots, k)$

Proof ("only if" part).   We use induction on the number of nodes. If there are two nodes x and y in the same class V_1 with no negative path connecting them, then identify x and y into a single new node z. By induction there is a partition with the desired property. After splitting up z the same partition of V_1 satisfies the requirements.

So assume that there is a negative path between any two nodes in the same class. We are done by the following lemma of A. Sebő. See (Frank, Sebő and Tardos).

Lemma 8.7. *Let* $G = (V_1, V_2; E)$ *be a simple bipartite graph with at least three nodes and* w *a* ± 1-*weighting on* E. *Suppose that there is no negative circuit but there is a negative path between every two nodes in the same class* V_i. *Then* G *is a tree and* w *is identically* -1.

Proof. Let P be a path of minimum weight m that has as few edges as possible. Let t be an end node of P and tx the first edge of P. Then (*) any starting segment of P has negative weight, in particular, $w(tz) = -1$. By induction the next claim implies the lemma.

Claim. *tx is the only edge of* G *incident to* t.

Proof. Suppose ty is another edge. $w(ty)$ must be positive since, if $y \in P$, then $P[t, y] \cup ty$ is a negative circuit, if $y \notin P$, then $P \cup ty$ is a path of weight $m - 1$.

By hypothesis there is a negative path Q between x and y. By parity, $w(Q) \le -2$. Q passes through t since otherwise $Q + xt + ty$ would form a negative circuit. Moreover Q traverses the edge xt. For otherwise the weight of segment $Q[t, x]$ is at least 1, therefore the weight of $Q[t, y]$ is at most -3, and then $Q[t, y] + ty$ would form a negative circuit.

We also see that circuit $C = Q[t, y] + ty$ must have weight 0. Now C and P have solely node t in common. Indeed, let $z \in P \cap C - t$ be the first node of P (starting at t) distinct from t. By (*) $w(P[t, z]) < 0$. Hence the weight of both segments of C between z and t must be positive contradicting $w(C) = 0$. But now the paths P and $Q[t, y]$ together form a path of weight smaller than m, a contradiction.   □

Theorem 8.5′ immediately implies the Berge-Tutte formula for the maximum cardinality of a matching. It also has the following pretty corollary.

Corollary 8.8 (Frank, Sebő and Tardos 1984). *A graph can be made Eulerian by doubling (parallelly) at most* k *edges if and only if* $\Sigma q(V_i) \le 2k$ *for all partitions* $\{V_1, \ldots, V_m\}$ *of* V *where* $q(X)$ *denotes the number of components* C *of* $G - X$ *with* $V(C)$ *odd.*

It is interesting to observe that function q also played a role in Theorem 7.3.

Exploiting further the idea introduced in Lemma 8.7, A. Sebő (1987) found the following refinement of Theorem 8.5'.

Theorem 8.9 (Sebő 1987). *Let G be a partite graph and w a ± 1 weighting on the edges such that there is no negative circuit. Let s be a specified node of G and let $\lambda(v)$ denote the minimum w-weight of a path from s to v. Then for any integer i and for a component D of the subgraph induced by $V_i := \{v : \lambda(v) \leq i\}$ the cut $\nabla(D)$ contains exactly one negative edge if $s \notin D$ and no negative edge if $s \in D$.*

This theorem implies Theorem 8.1': the set of cuts of form $\nabla(D)$, where D is a component of the subgraph induced by V_i and $s \notin D$, provides the desired packing. It is also easily seen that Theorem 8.5' follows from Theorem 8.9.

Let F denote the set of edges e for which $w(e)$ is -1. Let T consist of those nodes of G that have an odd number of incident edges from F. For two nodes s, v let T' be the symmetric difference of T and $\{s, v\}$. It is not hard to see that $\lambda(v)$ is the difference of the minimum cardinality of a T'-join and the minimum cardinality of a T-join. Thus, by the remark after Theorem 8.1w, $\lambda(v)$ can be computed.

A striking consequence of Sebő's theorem is that, once the values $\lambda(v)$ have been computed, the packing of cuts in Theorem 8.1' is also immediately available.

If G is planar in Theorem 8.9, a planar-dual form of the theorem can be stated. In a certain sense we obtain this way a canonized form of Theorem 3.6. Namely, suppose that $G + H$ is planar and Eulerian and $G + H$ has a fixed embedding into the plane. Assume that the cut criterion holds for G and H. For any face C of $G + H$ let $\lambda(C) := \min(|E \cap P| - |F \cap P| : P$ a dual path from the unbounded face to a C). For each integer i let S_i denote the union of faces C of $G + H$ for which $\lambda(C) \leq i$. Then each non-empty S_i uniquely partitions into connected regions of the plane (connected in planar sense, that is, its boundary is a circuit of $G + H$). Call such a region an *island* if it is bounded.

Corollary 8.10 (Sebő 1989). *A circuit of $G + H$ bounding an island contains exactly one edge of H. Moreover, these circuits are edge-disjoint and each demand edge is contained in one of them.*

We conclude this section by mentioning a recent theorem by A.M.H. Gerards (1988). By an odd K_4 we mean a subdivided K_4 such that each face is odd. By a prism we mean the graph on six nodes consisting of two disjoint triangles and three disjoint edges connecting the two triangles. By an odd prism we mean a subdivided prism so that the two subdivided triangles are odd, while the two four-gons are even (see Figure 8.2).

Theorem 8.11 (Gerards 1988). *Let $G = (V, E)$ be a graph and $T \subseteq V$ a subset of nodes of even cardinality. If G contains neither odd K_4 nor odd prism, then the maximum number of disjoint T-cuts is equal to the minimum cardinality of a T-join.*

Note that both bipartite graphs and series-parallel graphs satisfy the assumptions therefore the theorem can be considered as a common generalization of two earlier results of Seymour.

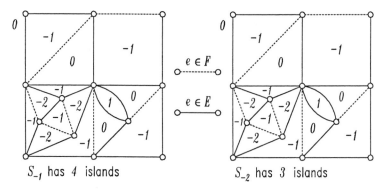

S_{-1} has 4 islands S_{-2} has 3 islands

Fig. 8.1

odd K_4 odd prism

Fig. 8.2

Almost bipartite graphs (graphs with a node covering all odd circuits) also satisfy the assumption. Applying first Theorem 8.11 for planar almost bipartite graphs and then taking the planar dual one can easily derive the following extension of Theorem 3.6.

Corollary 8.12. *Suppose that* $G+H$ *is planar and the degree* $d_{G+H}(v)$ *of every node* v *not on the infinite face of* $G+H$ *is even. Then the cut criterion is necessary and sufficient for the solvability of the edge-disjoint paths problem.*

9. Packing Cuts and Circuits

There is another fundamental theorem concerning packing of cuts. Let $G = (V, E)$ be a directed graph. For a subset X of nodes, if there is no edge leaving X, the (non-empty) set of edges entering X is called a *directed cut*.

Theorem 9.1 (Lucchesi and Younger 1978). *The maximum number of disjoint directed cuts is equal to the minimum number of edges covering all the cuts.*

For a short proof, see (Lovász 1976a). (Frank 1981) includes an algorithmic proof. By planar dualization one obtains:

Theorem 9.1′. *In a planar directed graph the maximum number of edge-disjoint directed circuits is equal to the minimum number of edges covering all the directed circuits.*

An analogous min-max relation holds for minimum directed cuts. By a *minumum directed cut* we mean a directed cut of least cardinality. The following result is a special case of a theorem of Edmonds and Giles (1977).

Theorem 9.2. *In a directed graph the maximum number of disjoint minimum directed cuts is equal to the minimum number of edges covering all the minimum directed cuts.*

This result can also be dualized. For example one gets: In a planar directed graph with no oppositely directed edges the maximum number of edge-disjoint directed triangles is equal to the minimum number of edges covering all directed triangles.

Sometimes the theorem of Lucchesi and Younger can be used for undirected graphs. For example, given a planar Eulerian graph G, what is the maximum number of circuits into which the edge set of G can be partitioned? D. Younger observed (as was communicated to me by W. Pulleyblank) that if we orient the edges of G in such a way that each face is surrounded by a directed circuit (we assume that G is 2-connected), then the maximum number of edge-disjoint directed circuits in the orientation of G is the same as the maximum number of edge-disjoint circuits in the undirected graph. (This is a useful exercise).

In Section 3 we briefly indicated how to derive Theorems 3.2-3.6 from their fractional forms. In the next few paragraphs we exhibit an approach, related to packing of cuts, by which the cut criterion can be proved to be sufficient for the existence of a multiflow, at least in some special cases.

Let us recall Theorem 2.0: a multiflow problem has a solution if and only if the distance criterion holds. Therefore if we want to show that in a certain case already the cut criterion is sufficient, we have to show that the cut criterion implies the distance criterion. One way to do so is, roughly, to point out that the vector w in Theorem 2.0 can be expressed as a non-negative linear combination of cuts.

Let $G = (V, E)$ and $H = (V, F)$ be graphs and w a non-negative rational weight function on E. Let $dist_w(u, v)$ denote the minimum w-weight of a path in G connecting u and v.

Theorem 9.3. *Suppose that either*

(a) *(Schrijver 1990) $G = (V, E)$ is planar, C_1 and C_2 are two specified faces of G and $H = (V, E)$ is the union of two complete graphs on $V(C_1)$ and $V(C_2)$, or*

(b) *$G + H$ is planar, or*

(c) $H = (V, F)$ is either K_4 or C_5 (Karzanov 1985a) or a double-star (Seymour 1978).

(A) Then there exists a fractional packing of cuts, that is an assigment on nonnegative variables $x(B)$ to cuts B such that for each edge $uv \in F$ $dist_w(u,v) = \Sigma(x(B) : B$ a cut, $uv \in B)$ and for each edge $uv \in E$ $w(uv) \geq \Sigma(x(B) : B$ a cut, $uv \in B)$.

(B) Moreover, if w is integer-valued such that every circuit of G has even w-weight, then x can be chosen integer-valued.

Proof of the fractional versions of Theorems 3.3, 3.6 and 3.8. We use part (A) of Theorem 9.3. The statement corresponding to cases (a),(b) and (c) will imply the fractional versions of Theorems 3.3, 3.6 and 3.8, respectively. Indeed, assume that there is a w violating the distance criterion. Let x be the variables in Theorem 9.3 assigned to the cuts. We have
$\Sigma(dist_w(u,v) : uv \in F) = \Sigma(\Sigma x(B) : B$ a cut and $uv \in B) : uv \in F) = \Sigma(x(B)|F \cap B| : B$ a cut$) \leq \Sigma(x(B)|E \cap B| : B$ a cut$) = \Sigma(\Sigma x(B) : B$ a cut and $uv \in B) : uv \in E) \leq \Sigma(w(uv) : uv \in E)$, contradicting the assumption that w violates the distance criterion. Here the first inequality follows from the cut criterion.   □

Notice that the above derivation works in the other direction as well, that is the fractional versions of Theorems 3.3, 3.6 and 3.8 imply part (A) of Theorem 9.3.

Remark. In this application we used only part (A) of Theorem 9.3. Part (B) should be considered interesting for its own sake. Actually, Schrijver, Karzanov and Seymour proved part B of cases (a) and (c), respectively, and observed that part (B) immediately implies part (A). (A relatively simple proof of Karzanov's theorem can be found in (Schrijver 1988a). Karzanov (1986b) gave a constructive proof of part B in case (a) that provides a strongly polynomial algorithm). The story of case (b) is different. We are going to show that part (B) of case (b) is equivalent to the following theorem of P. Seymour.

Theorem 9.4 (Seymour 1979). *Let $G' = (V', \overline{E})$ be a planar graph and $w' : \overline{E} \to \mathbb{Z}_+$ such that $w(B)$ is even for every cut B of G. There are non-negative integer variables $x(C)$ assigned to the circuits C of G' such that $w'(e) = \Sigma(x(C) : C$ a circuit and $e \in C)$ holds for every edge e if and only if $w'(e) \leq w'(B - e)$ holds for every cut B and edge $e \in B$.*

(The proof of this theorem is rather difficult.) By planar dualization we obtain

Theorem 9.4'. *Let $\overline{G} = (V, \overline{E})$ be a planar graph and $w' : \overline{E} \to \mathbb{Z}_+$ such that $w'(C)$ is even for every circuit C of $\overline{G}$. There are non-negative integer variables $x(B)$ assigned to the cuts B of $\overline{G}$ such that $w'(e) = \Sigma(x(B) : B$ a cut and $e \in B)$ holds for every edge e if and only if (*) $w'(e) \leq w'(C - e)$ holds for every cut C and edge $e \in C$.*

Proof of part (B) of Theorem 9.3b. Replace each edge e of G by a path of two edges e' and e'' (that is subdivide each edge by a new node) and let $w'(e) := \lfloor (w(e)/2) \rfloor$, $w'(e'') = \lceil (w(e)/2 \rceil$. Since $w(e) = w'(e') + w'(e'')$ this operation does not affect the w-distances of original nodes. For $uv \in F$ let $w'(uv) := dist_w(u,v)$ and let $\overline{G} = (V,\overline{E})$ be a graph where $\overline{E} = F \cup E' \cup E''$.
(Here E' and E'' denote the corresponding copies of E).

Since every circuit of G has even w-weight, every circuit of $\overline{G}$ has even w'-weight. An easy argument shows that the w'-distance of u and $v(uv \in F)$ in $\overline{G}$ is $dist_w(u,v)$. Therefore the hypotheses and (∗) of Theorem 8.4′ hold and then there is an x as described in the theorem. Since every cut of $\overline{G}$ which is not a star of a new node determines a cut of G, by leaving out these stars we obtain from x the desired solution to part (B) of Theorem 9.3b.   □

Proof of Theorem 9.4′ from Theorem 9.3. Let E and F be two copies of $\overline{E}$ (that is, to each edge $e \in \overline{E}$ there corresponds an edge in E and an edge in F that are parallel). Apply part B of Theorem 9.3b and let x be the integer vector provided by the theorem. Since (∗) implies that $dist_{w'}(u,v) = w'(e)$ for every edge $e = uv \in \overline{E}$, x will do for Theorem 9.4′ as well.   □

As far as part (A) of Theorem 9.3b is concerned, it follows from part (B) Theorem 9.3b but there is a more general result here. An equivalent reformulation of part (A) of Theorem 9.3b is the following.

Theorem 9.5′. *Let $G = (V,E)$ and $H = (V,F)$ be graphs for which $G + H$ is planar and let w and w' be two non-negative rational weight functions on E and on F, respectively. Then there exists a fractional packing of cuts, that is an assignment of non-negative variables $x(B)$ to cuts B such that for each edge $uv \in F$ $w'(f) \le \Sigma(x(B) : B$ a cut, $f \in B)$ and for each edge $uv \in E$ $w(uv) \ge \Sigma(x(B) : B$ a cut, $uv \in B)$ if and only if $w'(f) \le w(C - f)$ holds (equivalently, $w'(f) \le dist_w(u,v)$) for each circuit C of $G + H$ containing exactly one edge $f = uv$ from F.*

Now the promised generalization states that if we take the planar dual form of Theorem 9.5′, then planarity can be left out from the premises.

Theorem 9.5 (Seymour 1979). *Let $G = (V,E)$ and $H = (V,F)$ be two graphs and let w and w' be two non-negative rational weight functions on E and on F, respectively. Then there exists a fractional packing of circuits, that is an assignment of non-negative variables $x(C)$ to circuits C such that for each edge $uv \in F$ $w'(f) \le \Sigma(x(C) : C$ a circuit, $uv \in C)$ if and only if $w'(f) \le w(B - f)$ holds for each cut B of $G + H$ containing exactly one edge $f = uv$ from F.*

Proof. By Farkas' lemma if the desired x does not exist, then there is a vector $y : E \to \mathbb{R}$ with $y(e) \ge 0$ if $e \in E$, $y(e) \le 0$ if $e \in F$ such that $y(C) \ge 0$ for every circuit of $G + H$ and (∗) $\Sigma(y(e)w(e) : e \in E) + \Sigma(y(f)w'(f) : f \in F) < 0$.

Apply Theorem 8.3″ and let z be the vector in the theorem. We have $\Sigma(-y(f)w'(f) : f \in F) \le \Sigma(w'(f)\Sigma(z(B) : e \in B,$ B a cut containing solely f from F) : $e \in F) \le \Sigma(w(B - f)\Sigma(z(B) : B$ a cut containing solely f from

F) : $f \in F$) = $\Sigma(w(e)\Sigma(z(B)$: $e \in B$, B a cut containing one element of F) : $e \in E$) $\leq \Sigma(y(e)w(e))$: $e \in E$), contradicting (*).   □

(Note that the relation between Theorems 9.5 and 8.1″ is the same as the relation between the fractional versions of Theorems 3.3, 3.6 and 3.8 and part A of Theorem 9.3a, b and c, respectively.)

The problem of Theorem 9.5 can be interpreted so that one wants to find a fractional packing of circuits of $G + H$ such that certain edges satisfy an upper bound condition (edges in E) while other edges satisfy a lower bound condition (edges in F). We can impose both lower and upper bounds for every edge:

Theorem 9.6 (Seymour 1979). *Let* $\overline{G} = (V, \overline{E})$ *be an undirected graph endowed with two functions* $f : \overline{E} \to \mathbb{R}_+$, $g : \overline{E} \to \mathbb{R}_+ \cup \{\infty\}$ *for which* $f \leq g$. *There are non-negative variables* $x(C)$ *assigned to the circuits* C *of* G *for which* $f(e) \leq \Sigma(x(C) : C$ *a circuit and* $e \in C) \leq g(e)$ *holds for every edge* e *if and only if* $\Sigma(f(e) : e$ *enters* $X) \leq \Sigma(g(e) : e$ *leaves* $X)$ *holds for every subset* X *of nodes. Moreover, if* f *and* g *are integer-valued,* x *can be chosen integer-valued.*

An important difference between the directed and the undirected case is that the special case $f \equiv g$ is trivial for directed graphs while this is the crucial part in Seymour's proof of the undirected case.

Another essential difference is that for directed graphs one has the integrality result which is not so for undirected graphs. The Petersen graph shows that the integral packing of circuits does not necessarily exist: define f and g to be 2 on the edges of a specified perfect matching of the Petersen graph and 1 otherwise. In this view we should even more appreciate Theorem 9.4. (We note that even for planar graphs there is no known characterization for the existence of packing circuits if lower and upper bounds are imposed on the edges).

Let us conclude this section by presenting a generalization of Theorem 9.4. Let $G = (V, E)$ be an Eulerian graph. At every node $v \in V$ a partition $\mathcal{P}(v)$ of the edges incident to v is specified. A member of $\mathcal{P}(v)$ is called a *forbidden part* and a subset of a forbidden part is called a *forbidden set* if it has at least two elements. Let $\mathcal{P} := \cup(\mathcal{P}(v) : v \in V)$ denote the set of forbidden parts.

A circuit of G is called *good* if it includes no forbidden sets. If a cut S contains more than $|S|/2$ elements from a forbidden part P, then S is called *bad* (with respect to $\mathcal{P}$).

Theorem 9.7 (Fleischner and Frank 1988). *The edge set of a planar Eulerian graph can be partitioned into good circuits if and only if there are no bad cuts.*

This theorem immediately implies Theorem 9.4: replace each edge e by $w(e)$ parallel edges and let the forbidden parts consist of the sets of parallel edges. Another special case of the theorem is an earlier result of H. Fleischner (1980) when each forbidden part has at most two elements.

References

Abos, I., Frank, A., Tardos, É. (1982): Algorithms for edge-disjoint paths in a recti-linear grid and their application in layout design. Proc. 7th Colloq. on Microwave Communication, Akadémiai Kiadó, Budapest

Barahona, F. (1987): Planar multicommodity flows, max-cut and the Chinese postman problem. Report No. 87454-OR, Institut für Operations Research, Universität Bonn

Barahona, F., Cunningham, W. (1988): On dual integrality in matching problems. Report No. 88521-OR, Institut für Operations Research, Universität Bonn

Cherkasskij, B.V. (1977): Solution of a problem on multicommodity flows in a network. Ehkon. Mat. Metody **13(1)**, 143-151, in Russian

Dinits, E.A, Karzanov, A.V. (1979): On the problem of existing two integral flows of value 1. In: Combinatorial methods for flow problems (Inst. for System Studies, Moscow, issue 3), pp. 127-137, in Russian

Edmonds, J. (1965): Maximum matching and a polyhedron with (0,1)-vertices. J. Res. Nat. Bur. Standards **69B**, 125–130

Edmonds, J. (1973): Edge-disjoint branchings. In: Rustin, R. (ed.): Combinatorial algorithms. Algorithmics Press, New York, NY, pp. 91–96 (Courant Comput. Sci. Symp., Vol. 9)

Edmonds, J., Giles, R. (1977): A min-max relation for submodular functions on graphs. In: Hammer, P.L., Johnson, E.L., Korte, B., Nemhauser, G.L. (eds.): Studies in integer programming. North-Holland, Amsterdam, pp. 185–204 (Ann. Discrete Math., Vol. 1)

Edmonds, J., Johnson, E.L. (1973): Matching, Euler tours and the chinese postman. Math. Program. **5**, 88–124

Even, S., Itai, A., Shamir, A. (1976): On the complexity of timetable and multicommodity flow problems. SIAM J. Comput. **5**, 691–703

Fleischner, H. (1980): Eulersche Linien und Kreisüberdeckungen, die vorgegebene Durchgänge in den Kanten vermeiden. J. Comb. Theory, Ser. B **29**, 145–167

Fleischner, H., Frank, A. (1988): On circuit decomposition of planar Eulerian graphs. Submitted to J. Comb. Theory, Ser. B, to appear

Ford jr., L.R., Fulkerson, D.R. (1962): Flows in networks. Princeton University Press, Princeton, NJ

Fortune, S., Hopcroft, J.E., Wyllie, J. (1980): The directed subgraph homeomorphism problem. Theor. Comput. Sci. **10**, 111–121

Frank, A. (1981): How to make a digraph strongly connected. Combinatorica **1**, 145–153

Frank, A. (1982): Disjoint paths in a rectilinear grid. Combinatorica **2**, 361–371

Frank, A. (1985): Edge-disjoint paths in planar graphs. J. Comb. Theory, Ser. B **39**, 164–178

Frank, A. (1987): Graph connectivity and network flows. In: Graham, R.L., Grötschel, M., Lovász, L. (eds.): Handbook of combinatorics. (To appear)

Frank, A. (1988): Packing paths in planar graphs. Combinatorica, to appear

Frank, A. (1989): On connectivity properties of Eulerian digraphs. Ann. Discrete Math. **41**, 179-194

Frank, A., Sebő, A., Tardos, É. (1984): Covering directed and odd cuts. Math. Program. Study **22**, 99–112

Frank, A., Tardos, É. (1984): On routing around a rectangle. Proc. ETAN Network Theory Symp. 1984, Yugoslavia

Gallai, T. (1961): Maximum-Minimum-Sätze und verallgemeinerte Faktoren von Graphen. Acta Math. Akad. Sci. Hung. **12**, 131–163

Gallai, T. (1962): Graphen mit triangulierbaren ungeraden Vielecken. Közlemenyek, Magy. Tud. Akad., Mat. Kut. Intez. 7

Garey, M.R., Johnson, D.S. (1979): Computers and intractability: A guide to the theory of NP-completeness. W.H. Freeman, San Francisco, CA

Gerards, A.M.H. (1988): Graphs and polyhedra - Binary spaces and cutting planes. Ph.D.thesis, Tilburg University, Tilburg 1988

Guan, M.G. (1962): Graphic programming using odd or even points. Chin. J. Math. **1**, 273–277

Hirata, T., Kubota, K., Saito, O. (1984): A sufficient condition for a graph to be weakly k-linked. J. Comb. Theory, Ser. B **36**, 85–94

van Hoesel, C., Schrijver, A. (1990): Edge-disjoint homotopic paths in a planar graph with one hole. J. Comb. Theory, Ser. B **48**, 77–91

Hoffman, A. (1960): Some recent applications of the theory of linear inequalities to extremal combinatorial analysis. Proc. Symp. Appl. Math. *10*

Hu, T.C. (1963): Multicommodity network flows. Oper. Res. **11**, 344–360

Itai, A., Zehavi, A. (1984): Bounds on path connectivity. Discrete Math. **51**, 25–34

Jung, H.A. (1970): Eine Verallgemeinerung des n-fachen Zusammenhangs für Graphen. Math. Ann. **187**, 95–103

Karp, R.M. (1972): Reducibility among combinatorial problems. In: Miller, R.E., Thatcher, J.W. (eds.): Complexity of computer computations. Plenum Press, New York, NY, pp. 85–103 (IBM Res. Symp. Ser., Vol. 4)

Karp, R.M. (1975): On the computational complexity of combinatorial problems. Networks **5**, 45–68

Karzanov, A.V. (1985a): Metrics and undirected cuts. Math. Program. **32**, 183–198

Karzanov, A.V. (1985b): On multicommodity flow problems with integer-valued optimal solutions. Dokl. Akad. Nauk SSSR *280* (In Russian; English translation: Sov. Math. Dokl. *32*, 151–154)

Karzanov, A.V. (1986): An algorithm for maximum packing of odd-terminus cuts and its applications. In: A.S. Alekseev (Ed.): Studies in applied graph theory, Nauka, Novosibirsk, in Russian

Karzanov, A.V. (1987): Half-integral five-terminus flows. Discrete Appl. Math. **18**, 263-278

Karzanov, A.V. (1989a): Paths and metrics in a planar graph with three distinguished faces I. and II. J. Comb. Theory, submitted

Karzanov, A.V. (1989b): Packing of cuts realizing distances between certain vertices in a planar graph. Discrete Math., to appear

Karzanov, A.V., Lomonosov, M.V. (1978): Multiflows in undirected graphs. In: The problems of social and economic systems: Operations research model. The Institute for Systems Studies, Moscow (Math. Programming, Vol. 1, in Russian)

Kaufmann, M., Mehlhorn, K. (1990): This volume

Korach, E. (1982): Packing of T-cuts, and other aspects of dual integrality. Ph.D thesis, University of Waterloo Korach, E., Penn, M. (1985): Tight integral gap in the chinese postman problem. Preprint

Kramer, M.E., van Leeuwen, J. (1982): Wire routing is NP-complete. Technical Report RUU-CS-82-4, Department of Computer Science, Rijksuniversiteit te Utrecht

Kramer, M.E., van Leeuwen, J. (1984): The complexity of wire routing and finding the minimum area layouts for arbitrary VLSI circuits. In: Preparata, F.P. (ed.): VLSI theory. JAI Press, London, pp. 129–146 (Adv. Comput. Res., Vol. 2)

Lai, T.-H., Sprague, A. (1987): On the routability of a convex grid. J. Algorithms **8**, 372–384

LaPaugh, A.S. (1980): A polynomial-time algorithm for optimal routing around a rectangle. Proc. 21st IEEE FOCS, pp. 282–293

Larman, D.G., Mani, P. (1970): On the existence of certain configurations within graphs and the 1-skeletons of polytopes. Proc. Lond. Math. Soc., III. Ser. **20**, 144–160

Lawler, E.L. (1976): Combinatorial optimization: Networks and matroids. Holt, Rinehart and Winston, New York, NY

Lins, S. (1981): A minimax theorem on circuits in projective graphs. J. Comb. Theory, Ser. B **30**, 253–262

Lomonosov, M.V. (1979): Multiflow feasibility depending on cuts. Graph Theory Newsl. **9**, 4

Lomonosov, M.V. (1983): On the planar integer two-flow problem. Combinatorica **3**, 207–219

Lomonosov, M.V. (1985): Combinatorial approaches to multiflow problems. Discrete Appl. Math. **11**, 1–94

Lovász, L. (1976a): On two minimax theorems in graph theory. J. Comb. Theory, Ser. B **21**, 96–103

Lovász, L. (1976b): On some connectivity properties of Eulerian graphs. Acta Math. Akad. Sci. Hung. **28**, 129–138

Lovász, L., Plummer, M.D. (1986): Matching theory. North-Holland, Amsterdam (Ann. Discrete Math., Vol. 29)

Lucchesi, C.L., Younger, D.H. (1978): A minmax relation for directed graphs. J. Lond. Math. Soc., II. Ser. **17**, 369–374

Mader, W. (1978a): A reduction method for edge-connectivity in graphs. In: Bollobás, B. (ed.): Advances in graph theory. North-Holland, Amsterdam, pp. 145–164 (Ann. Discrete Math., Vol. 3)

Mader, W. (1978b): Über die Maximalzahl kantendisjunkter A-Wege. Arch. Math. **30**, 325–336

Mader, W. (1978c): Über die Maximalzahl kreuzungsfreier H-Wege. Arch. Math. **31**, 387–402

Mader, W. (1979): Connectivity and edge-connectivity in finite graphs. In: Bollobas, B. (ed.): Surveys in combinatorics. Cambridge University Press, Cambridge, pp. 293–309 (Lond. Math. Soc. Lect. Note Ser., Vol. 38)

Mader, W. (1985): Paths in graphs, reducing the edge-connectivity only by two. Graphs Comb. **1**, 81–89

Menger, K. (1927): Zur allgemeinen Kurventheorie. Fundam. Math. **10**, 96–115

Middendorf, M., Pfeiffer, F. (1989): On the complexity of the disjoint path problem. Report No. 89585-OR, Institut für Operations Research, Universität Bonn

Okamura, H. (1983): Multicommodity flows in graphs. Discrete Appl. Math. **6**, 55–62

Okamura, H. (1984): Multicommodity flows in graphs II. Jap. J. Math., New Ser. **10**, 99–116

Okamura, H. (1988): Weakly k-linked graphs. Preprint

Okamura, H., Seymour, P.D. (1981): Multicommodity flows in planar graphs. J. Comb. Theory, Ser. B **31**, 75–81

Papernov, B.A. (1976): Feasibility of multicommodity flows. In: Friedman, A.A. (ed.): Studies in discrete optimization. Nauka, Moscow, pp. 230–261 (In Russian)

Preparata, F.P., Lipski jr., W. (1984): Optimal three-layer channel routing. IEEE Trans. Comput. **33**

Richards, D. (1981): Complexity of single-layer routing. Manuscript (Unpublished)

Robertson, N., Seymour, P.D. (1986a): Graph minors VI: Disjoint paths across a disc. J. Comb. Theory, Ser. B **41**, 115–138

Robertson, N., Seymour, P.D. (1986b): Graph minors XIII: The disjoint paths problem. Preprint (J. Comb. Theory, Ser. B, to appear)

Rothschild, B., Whinston, A. (1966a): On two-commodity network flows. Oper. Res. **14**, 377–387

Rothschild, B., Whinston, A. (1966b): Feasibility of two-commodity network flows. Oper. Res. **14**, 1121–1129

Schrijver, A. (1988a): Short proofs on multicommodity flows and cuts. J. Comb. Theory, Ser. B, to appear

Schrijver, A. (1988b): The Klein bottle and multicommodity flows. Report OS-R8810, Mathematical Centre, Amsterdam (Combinatorica, to appear)

Schrijver, A. (1989): Distances and cuts in planar graphs. J. Comb. Theory, Ser. B **46**, 46–57

Sebő, A. (1987a): Undirected distances and the postman-structure of graphs. J. Comb. Theory, Ser. B, to appear

Sebő, A. (1987b): A quick proof of Seymour's theorem on T-joins. Discrete Math. **64**, 101–103

Sebő, A. (1988): Integer plane multicommodity flows with a bounded number of demands. Report No. 88534-OR, Institut für Operations Research, Universität Bonn

Sebő, A. (1989): Packing odd cuts and multicommodity flows. Report No. 89576-OR, Institut für Operations Research, Universität Bonn

Seymour, P. (1977): The matroids with the max-flow min-cut property. J. Comb. Theory, Ser. B **23**, 189-222

Seymour, P.D. (1978a): A two-commodity cut theorem. Discrete Math. **23**, 177–181

Seymour, P.D. (1978b): Sums of circuits. In: Bondy, J.A., Murty, U.S.R. (eds.): Graph theory and related topics. Academic Press, New York, NY, pp. 341–355

Seymour, P.D. (1979): A short proof of the two-commodity flow theorem. J. Comb. Theory, Ser. B **26**, 370–371

Seymour, P.D. (1980a): Four-terminus flows. Networks **10**, 79–86

Seymour, P.D. (1980b): Disjoint paths in graphs. Discrete Math. **29**, 293–309

Seymour, P.D. (1981a): On odd cuts and planar multicommodity flows. Proc. Lond. Math. Soc., III. Ser. **42**, 178–192

Seymour, P.D. (1981b): Matroids and multicommodity flows. Eur. J. Comb. **2**, 257–290

Shiloach, Y. (1979): Edge-disjoint branchings in directed multigraphs. Inf. Process. Lett. **8**, 24–27

Shiloach, Y. (1980): A polynomial solution to the undirected two paths problem. J. Assoc. Comput. Mach. **27**, 445–456

Strange, K.E., Toft, B. (1983): An introduction to the subgraph homeomorphism problem. In: Fiedler, M. (ed.): Graphs and other combinatorial topics. B.G. Teubner, Leipzig, pp. 296–301 (Teubner-Texte Math., Vol. 59)

Tardos, G. (1984). (Oral communication)

Thomassen, C. (1980): 2-linked graphs. Eur. J. Comb. **1**, 371–378

Thomassen, C. (1985): The 2-linkage problem for acyclic digraphs. Discrete Math. **55**, 73–87

Zubor, Z. (1989): Oral communication

Network Flow Algorithms

Andrew V. Goldberg, Éva Tardos and Robert E. Tarjan

0. Introduction

Network flow problems are central problems in operations research, computer science, and engineering and they arise in many real world applications. Starting with early work in linear programming and spurred by the classic book of Ford and Fulkerson [26], the study of such problems has led to continuing improvements in the efficiency of network flow algorithms. In spite of the long history of this study, many substantial results have been obtained within the last several years. In this survey we examine some of these recent developments and the ideas behind them.

We discuss the classical network flow problems, the *maximum flow problem* and the *minimum-cost circulation problem*, and a less standard problem, the *generalized flow problem*, sometimes called the problem of *flows with losses and gains*. The survey contains six chapters in addition to this introduction. Chapter 1 develops the terminology needed to discuss network flow problems. Chapter 2 discusses the maximum flow problem. Chapters 3, 4 and 5 discuss different aspects of the minimum-cost circulation problem, and Chapter 6 discusses the generalized flow problem. In the remainder of this introduction, we mention some of the history of network flow research, comment on some of the results to be presented in detail in later sections, and mention some results not covered in this survey.

We are interested in algorithms whose running time is small as a function of the size of the network and the numbers involved (e.g. capacities, costs, or gains). As a measure of the network size, we use n to denote the number of vertices and m to denote the number of arcs. As measures of the number sizes, we use U to denote the maximum arc capacity, C to denote the maximum arc cost, and B (in the case of the generalized flow problem) to denote the maximum numerator or denominator of the arc capacities and gains. In bounds using U, C, or B, we make the assumption that the capacities and costs are integers, and that the gains in the generalized flow problem are rational numbers.

We are most interested in polynomial-time algorithms. We make the following distinctions. An algorithm is *pseudopolynomial* if its running time has a bound that is polynomial in n, m, and the appropriate subset of U, C, and B. An algorithm is *polynomial* if its running time has a bound that is polynomial in n, m, and the appropriate subset of $\log U, \log C$, and $\log B$ [1]. An algorithm is *strongly polynomial* if its running time has a bound that is polynomial in n and m [2]. When comparing polynomial algorithms to strongly polynomial ones we shall use the *similarity assumption* that $\log U$, $\log C$ and $\log B$ are $\Theta(\log n)$ [33]. We shall also be interested in strongly polynomial algorithms, however.

The network flow problems discussed in this survey are special cases of linear programming, and thus they can be solved by general-purpose linear programming algorithms. However, the combinatorial structure of these problems makes it possible to obtain more efficient algorithms.

We shall not discuss in detail algorithms that are based on general linear programming methods. We should mention, however, that the first algorithm designed for network flow problems was the *network simplex method* of Dantzig [20]. It is a variant of the linear programming simplex method designed to take advantage of the combinatorial structure of network flow problems. Variants of the simplex method that avoid cycling give an exponential bound on the complexity of all the network flow problems. (Cunningham [19] gives an elegant anti-cycling strategy for the network simplex method based on graph-theoretic properties of the minimum-cost circulation problem). Recently, Goldfarb and Hao [52] have designed a variant of the primal network simplex method for the maximum flow problem that runs in strongly polynomial time (see Section 2.1). Orlin [82] designed a variant of the dual network simplex method for the minimum-cost circulation problem that runs in strongly polynomial time (see Chapter 5): For a long time, the network simplex method has been the method of choice in practice, in particular for the minimum-cost circulation problem (see e.g. [54]); for large instances of hard problems, the new scaling algorithms are probably better, however.

The first pseudopolynomial algorithm for the maximum flow problem is the *augmenting path* algorithm of Ford and Fulkerson [27, 26]. Dinic [21] and Edmonds and Karp [22] independently obtained polynomial versions of the augmenting path algorithm. Since then, several more-efficient algorithms have been developed. Chapter 2 presents the *push/relabel method*, recently proposed by Goldberg [40] and Goldberg and Tarjan [45], along with some of its more efficient variants.

The first pseudopolynomial algorithm for the minimum-cost circulation problem is the *out-of-kilter* method, which was developed independently by Yakovleva [105], Minty [77], and Fulkerson [32]. The first polynomial algorithm for the minimum-cost circulation problem is due to Edmonds and Karp [22]. To develop this algorithm Edmonds and Karp introduced the technique of *scaling*, which has proved to be a useful tool in the design and analysis of algorithms for a variety of combinatorial optimization problems. Chapter 3 and Section 5.2 are devoted to scaling algorithms for the minimum-cost circulation problem.

The maximum flow algorithms of Dinic [21] and Edmonds and Karp [22] are strongly polynomial, but the minimum-cost circulation algorithm of Edmonds

---

[1] All logarithms in this paper without an explicit base are base two.

[2] For a more formal definition of polynomial and strongly polynomial algorithms, see [55].

and Karp [22] is not. The first strongly polynomial algorithm for the minimum-cost circulation problem was designed by Tardos [96]. Chapter 4 and Section 5.3 are devoted to recent strongly polynomial algorithms for the minimum-cost circulation problem.

The first augmenting path algorithms for the generalized flow problem were developed independently by Jewell [62, 63] and Onaga [79]. Many pseudopolynomial minimum-cost circulation algorithms have been adapted for the generalized flow problem (see [102] for a survey). The first polynomial-time algorithm for the generalized flow problem was the ellipsoid method [70]. Kapoor and Vaidya [65] have shown how to speed up Karmarkar [66] — or Renegar [89] — type interior-point algorithms on network flow problems by taking advantage of the special structure of the matrices used in the linear programming formulations of these problems. Vaidya's algorithm [103] is the fastest currently known algorithm for the generalized flow problem. The first polynomial algorithms for the generalized flow problem that are not based on general-purpose linear programming methods are due to Goldberg, Plotkin, and Tardos [43]. These algorithms are discussed in Chapter 6. The existence of a strongly polynomial algorithm for the generalized flow problem is an interesting open question.

Important special cases of network flow problems that will not be covered in this survey are the *bipartite matching* problem and its weighted version, the *assignment* problem. These problems can be stated as maximum flow and minimum-cost circulation problems, respectively, on networks with unit capacities and a special structure (see e.g. [24, 99]). Some of the efficient algorithms for the more general problems have evolved from efficient algorithms developed earlier for these simpler problems.

König's [72] proof of a good characterization of the maximum size of a matching in a bipartite graph gives an $O(nm)$–time algorithm for finding a maximum matching. The Ford-Fulkerson maximum flow algorithm can be viewed as an extension of this algorithm. Hopcroft and Karp [58] gave an $O(\sqrt{n}m)$ algorithm for the bipartite matching problem. Even and Tarjan observed [25] that Dinic's maximum flow algorithm, when applied to the bipartite matching problem, behaves similarly to the Hopcroft-Karp algorithm and runs in $O(\sqrt{n}m)$ time as well. A variation of the Goldberg-Tarjan maximum flow algorithm (which can be viewed as a generalization of Dinic's algorithm) can be easily shown to lead to the same bound [5, 84]. In spite of recent progress on related problems, the $O(\sqrt{n}m)$ bound has not been improved.

The first algorithm for the assignment problem is the *Hungarian method* of Kuhn [73]. The out-of-kilter algorithm is an extension of this algorithm to the minimum-cost circulation problem. The Hungarian method solves the assignment problem in $O(n)$ shortest path computations. Edmonds and Karp [22] and Tomizawa [101] have observed that the dual variables can be maintained so that these shortest path computations are on graphs with non-negative arc costs. Combined with the shortest path algorithm of [29], this observation gives an $O(n(m + n \log n))$ bound for the problem. Gabow [33] used scaling to obtain an $O(n^{3/4} m \log C)$ algorithm for the problem. Extending ideas of the Hopcroft-Karp bipartite matching algorithm and those of the Goldberg-Tarjan minimum-cost

Table 1. Fastest currently known algorithms for network flow problems

Problem	Date	Discoverer	Running Time	References
Bipartite Matching	1973	Hopcroft and Karp	$O(\sqrt{n}m)$	[58]
Assignment	1955 1987	Kuhn Gabow and Tarjan	$O(n(m + n \log n))$ $O(\sqrt{n}m \log(nC))$	[73] [35]
Maximum Flow	1986 1988	Goldberg and Tarjan Ahuja, Orlin, and Tarjan	$O(nm \log(n^2/m))$ $O(nm \log(\frac{n}{m}\sqrt{\log U} + 2))$	[45] [4]
Minimum-Cost Circulation	1972 1987 1987 1988	Edmonds and Karp Goldberg and Tarjan Orlin Ahuja, Goldberg, Orlin, and Tarjan	$O(m \log U(m + n \log n))$ $O(nm \log(n^2/m) \log(nC))$ $O(m \log n(m + n \log n))$ $O(nm \log \log U \log(nC))$	[22] [49] [81] [1]
Generalized Flow	1989	Vaidya	$O(n^2 m^{1.5} \log(nB))$	[103]

circulation algorithm (discussed in Section 3), Gabow and Tarjan obtained an $O(\sqrt{n}m \log(nC))$ algorithm for the assignment problem.

A more recent pseudopolynomial algorithm for the assignment problem is the auction algorithm of Bertsekas [9] (first published in [10]). This algorithm contains some of the elements of the push/relabel algorithms discussed in Sections 2 and 3.

Some versions of the network simplex method have been shown to solve the assignment problem in polynomial time. In particular, Orlin [83] shows that a natural version of the primal simplex method runs in polynomial time, and Balinski [6] gives a *signature method* that is a dual simplex algorithm for the assignment problem and runs in strongly polynomial time.

For discussion of parallel algorithms for the bipartite matching and assignment problems, see [10, 34, 44, 68, 78].

In this survey we would like to highlight the main ideas involved in designing highly efficient algorithms for network flow problems, rather than to discuss in detail the fastest algorithms. For an introduction to network flow algorithms see [26, 74, 87, 99]. However, for easy reference, we summarize the running times of the fastest currently known algorithms in Table 1. For each problem we list all of the bounds that are best for some values of the parameters, but we only list the first algorithm achieving the same bound. Some of the bounds stated in the table depend on the $O(m + n \log n)$ implementation of Dijkstra's shortest path algorithm [29].

1. Preliminaries

In this chapter we define the problems addressed in this survey and review fundamental facts about these problems. These problems are the maximum flow problem, the minimum-cost circulation problem, the transshipment problem, and the generalized flow problem.

1.1 Flows and Residual Graphs

A *network* is a directed graph $G = (V, E)$ with a non-negative *capacity function* $u : E \to \mathbf{R}_\infty$.[3] We assume that G has no multiple arcs, i.e., $E \subset V \times V$. If there is an arc from a vertex v to a vertex w, this arc is unique by the assumption, and we will denote it by (v, w). This assumption is for notational convenience only. We also assume, without loss of generality, that the input graph G is symmetric: $(v, w) \in E \iff (w, v) \in E$. A *flow network* is a network with two distinguished vertices, the *source* s and the *sink* t.

A *pseudoflow* is a function $f : E \to \mathbf{R}$ that satisfies the following constraints:

(1.1) $f(v, w) \le u(v, w)$ $\forall (v, w) \in E$ (capacity constraint),

(1.2) $f(v, w) = -f(w, v)$ $\forall (v, w) \in E$ (flow antisymmetry constraint).

Remark. To gain intuition, it is often useful to think only about the non-negative components of a pseudoflow (or of a generalized pseudoflow, defined below). The antisymmetry constraints reflect the fact that a flow of value x from v to w can be thought of as a flow of value $(-x)$ from w to v. The negative flow values are introduced only for notational convenience. Note, for example, that one does not have to distinguish between lower and upper capacity bounds: the capacity of the arc (v, w) represents a lower bound on the flow value on the opposite arc.

Given a pseudoflow f, we define the *excess* function $e_f : V \to \mathbf{R}$ by $e_f(v) = \sum_{u \in V} f(u, v)$, the net flow into v. We will say that a vertex v has *excess* if $e_f(v)$ is positive, and has *deficit* if it is negative. For a vertex v, we define the *conservation constraint* by

(1.3) $e_f(v) = 0$ (flow conservation constraint).

Given a pseudoflow f, the *residual capacity* function $u_f : E \to \mathbf{R}$ is defined by $u_f(v, w) = u(v, w) - f(v, w)$. The *residual graph* with respect to a pseudoflow f is given by $G_f = (V, E_f)$, where $E_f = \{(v, w) \in E \mid u_f(v, w) > 0\}$.

1.2 The Maximum Flow Problem

To introduce the maximum flow problem, we need the following definitions in addition to the definitions of the previous section. Consider a flow network (G, u, s, t). A *preflow* is a pseudoflow f such that the excess function is non-negative for all vertices other than s and t. A *flow* f on G is a pseudoflow

---

[3] $\mathbf{R}_\infty = \mathbf{R} \cup \{\infty\}$.

satisfying the conservation constraints for all vertices except s and t. The *value* $|f|$ of a flow f is the net flow into the sink $e_f(t)$. A *maximum flow* is a flow of maximum value (also called *an optimal flow*). The maximum flow problem is that of finding a maximum flow in a given flow network.

Given a flow f, we define an *augmenting path* to be a source-to-sink path in the residual graph. The following theorem, due to Ford and Fulkerson, gives an optimality criterion for maximum flows.

1.2.1 Theorem [26]. *A flow is optimal if and only if its residual graph contains no augmenting path.*

1.3 The Minimum-Cost Circulation Problem

A *circulation* is a pseudoflow with zero excess at every vertex. A *cost function* is a real-valued function on arcs $c : E \to \mathbf{R}$. Without loss of generality, we assume that costs are antisymmetric:

(1.4) $c(v, w) = -c(w, v) \quad \forall(v, w) \in E$ (cost antisymmetry constraint).

The cost of a circulation f is given by

$$c(f) = \sum_{(v,w)\in E : f(v,w)\geq 0} f(v, w)c(v, w).$$

The *minimum-cost circulation* problem is that of finding a minimum-cost (*optimal*) circulation in an input network (G, u, c).

We have assumed that the capacities are non-negative. This assumption is no loss of generality. Given a circulation problem with some negative capacities one can find a feasible circulation f using one invocation of a maximum flow algorithm. (See [87], problem 11(e), p. 215.) The non-negative residual capacities $u(v, w) - f(v, w)$ then define an equivalent problem.

Next we state two criteria for the optimality of a circulation. Define the cost of a cycle Γ to be the sum of the costs of the arcs along the cycle and denote it by $c(\Gamma)$. The following theorem states the first optimality criterion.

1.3.1 Theorem   [15]. *A circulation is optimal if and only if its residual graph contains no negative-cost cycle.*

To state the second criterion, we need the notions of a price function and a reduced cost function. A *price function* is a vertex labeling $p : V \to \mathbf{R}$. The *reduced cost function* with respect to a price function p is defined by $c_p(v, w) = c(v, w) + p(v) - p(w)$. These notions, which originate in the theory of linear programming, are crucial for many minimum-cost flow algorithms. As linear programming dual variables, vertex prices have a natural economic interpretation as the current market prices of the commodity. We can interpret the reduced cost $c_p(v, w)$ as the cost of buying a unit of commodity at v, transporting it to w, and then selling it. Due to the conservation constraints, the reduced costs define an equivalent problem.

1.3.2 Theorem [26]. *A circulation f is optimal for the minimum-cost circulation problem (G, u, c) if and only if it is optimal for the problem (G, u, c_p) for every price function p.*

The second optimality criterion is as follows.

1.3.3 Theorem [26]. *A circulation f is optimal if and only if there is a price function p such that, for each arc (v, w),*

(1.5) $c_p(v, w) < 0 \Rightarrow f(v, w) = u(v, w)$ (complementary slackness constraint).

A minimum-cost circulation problem may have no optimal solution. The following theorem characterizes when this is the case.

1.3.4 Theorem. *There exists a minimum-cost circulation if and only if the input network contains no negative-cost cycle consisting of arcs with infinite capacity.*

Note that the maximum flow problem is a special case of the minimum-cost circulation problem (see e.g. [26]). To see this, consider a flow network and add a pair of new arcs (s, t) and (t, s) with $u(s, t) = 0$ and $u(t, s) = \infty$. Define the costs of all original arcs to be zero, and define $c(s, t) = -c(t, s) = 1$. A minimum-cost circulation in the resulting network restricted to the original network is a maximum flow in the original network.

1.4 The Transshipment Problem

In this section we define the (uncapacitated) *transshipment problem*. Although this problem is equivalent to the minimum-cost circulation problem, some algorithms are more natural in the context of the transshipment problem. For a more extensive discussion of this problem and the closely related transportation problem, see [74].

In the transshipment problem, all arc capacities are either zero or infinity. In addition, the input to the problem contains a demand function $d : V \to \mathbf{R}$ such that $\sum_{v \in V} d(v) = 0$. For the transshipment problem, the notion of the excess at a vertex is defined as follows:

(1.6) $$e_f(v) = \sum_{w \in V} f(w, v) - d(v).$$

A pseudoflow f is *feasible* if the conservation constraints (1.3) hold for all vertices. The *transshipment problem* is that of finding a minimum-cost feasible pseudoflow in an input network. In the special case of integer demands, we shall use D to denote the maximum absolute value of a demand.

Theorems analogous to Theorem 1.3.1, 1.3.3 and 1.3.2. hold for the transshipment problem, and the analog of Theorem 1.3.4 holds for a transshipment problem that has a feasible pseudoflow.

We make two simplifying assumptions about the transshipment problem. First, we assume that, for the zero flow, all residual arcs have non-negative cost. This assumption can be validated by first checking whether the condition of Theorem 1.3.4 is satisfied (using a shortest path computation). In the case of a transshipment problem, the residual graph of the zero flow consists of the arcs with infinite capacity. If this graph has no negative-cost cycles, then define $p(v)$ to be the cost of a minimum-cost path from a fixed vertex s to the vertex v. The costs $c_p(v, w)$ define an equivalent problem that satisfies the assumption. Second, we assume that the residual graph of the zero flow is strongly connected. (There is a path between each pair of vertices.) This condition can be imposed by adding two new vertices x and y and adding an appropriately high-cost arc from x to y with $u(x, y) = \infty$ and $u(y, x) = 0$, and arcs between the new vertices and every other vertex in both directions such that $u(v, x) = u(y, v) = \infty$ and $u(x, v) = u(v, y) = 0$ for every vertex v. If the original problem has a feasible solution, then in every minimum-cost solution of the transformed problem all of the dummy arcs introduced to make the graph strongly connected have zero flow.

Next we show that the transshipment problem is equivalent to the minimum-cost circulation problem (see e.g. [26]). Given an instance of the transshipment problem, we construct an equivalent instance of the minimum-cost circulation problem as follows. Add two new vertices, x and y, and arcs (y, x) and (x, y) with $u(x, y) = 0$ and $u(y, x) = \sum_{v:d(v)>0} d(v)$. Define the cost of (y, x) to be small enough so that any simple cycle containing (y, x) has negative cost; for example, define $c(y, x) = -c(x, y) = -nC$. For every vertex $v \in V$ with $d(v) > 0$, add arcs (x, v) and (v, x) and define $u(x, v) = d(v)$, $u(v, x) = 0$, and $c(x, v) = c(v, x) = 0$. For every vertex $v \in V$ with $d(v) < 0$, add arcs (v, y) and (y, v) and define $u(v, y) = -d(v)$, $u(y, v) = 0$, and $c(v, y) = c(y, v) = 0$. Consider an optimal solution f to the minimum-cost circulation problem in the resulting network. The transshipment problem is feasible if and only if f saturates all new arcs, and in this case the restriction of f to the original arcs gives an optimal solution to the transshipment problem.

Next we reduce the minimum-cost circulation problem to the transshipment problem. The reduction uses the technique of *edge-splitting*. Consider a minimum-cost circulation problem (G, u, c). First we make sure that for every edge $\{v, w\}$ of the network, either $u(v, w)$ or $u(w, v)$ is infinite. For every edge $\{v, w\}$ that does not satisfy the above condition, we introduce a new vertex $x_{(v,w)}$, and replace the arcs (v, w) and (w, v) by the arcs $(v, x_{(v,w)})$, $(x_{(v,w)}, w)$, $(w, x_{(v,w)})$, and $(x_{(v,w)}, v)$. Define the costs and capacities of the new arcs as follows:

$$u(v, x_{(v,w)}) = u(v, w), \quad c(v, x_{(v,w)}) = c(v, w),$$
$$u(x_{(v,w)}, w) = \infty, \quad c(x_{(v,w)}, w) = 0,$$
$$u(w, x_{(v,w)}) = u(w, v), \quad c(v, x_{(v,w)}) = 0,$$
$$u(x_{(v,w)}, v) = \infty, \quad c(x_{(v,w)}, v) = c(w, v).$$

This defines an equivalent minimum-cost circulation problem in which the capacity of every edge is unbounded in at least one direction.

To complete the transformation, we need to force the additional restriction that every finite capacity is zero. Initialize the demand function d to be zero on all vertices. Let $\{v, w\}$ be an edge such that $u(w, v) = \infty$ and $u(v, w)$ is finite but not zero. Replacing $u(v, w)$ by zero, and $d(v)$ and $d(w)$ by $d(v) + u(v, w)$ and $d(w) - u(v, w)$, respectively, defines an equivalent problem with $u(v, w) = 0$.

Remark. This transformation increases the number of vertices: from a network with n vertices and m edges, we obtain a network with up to $n + m$ vertices and $2m$ edges. However, the resulting network has a special structure. It is important for the analyses in Chapter 5 to notice that the newly introduced vertices can be eliminated for shortest path computations. Consequently, this blowup in the number of vertices will not affect the running time of shortest path computations in the residual graph.

The *capacitated transshipment problem* generalizes both the minimum-cost circulation problem and the (uncapacitated) transshipment problem discussed above. It is the same as the uncapacitated transshipment problem without the assumption that all arc capacities are infinite or zero. The reductions above can be extended to show that the capacitated transshipment problem is equivalent to the minimum-cost circulation problem. Whereas the previous simpler formulations are better suited for designing algorithms, the more general form can be useful in applications.

1.5 The Generalized Flow Problem

In this section we define generalized pseudoflows, generalized flows, and the generalized flow problem, and compare these concepts with their counterparts discussed in the previous sections. For alternative formulations of the problem, see e.g. [43, 74]. (This problem is also known as the problem of flows with losses and gains.)

The generalized flow problem is given by a network with a distinguished vertex, the *source* s, and a *gain function* $\gamma : E \rightarrow \mathbf{R}^{+4}$ on the arcs. We assume (without loss of generality) that the gain function is antisymmetric:

$$(1.7) \qquad \gamma(v, w) = \frac{1}{\gamma(w, v)} \quad \forall (v, w) \in E \quad \text{(gain antisymmetry constraints)}.$$

In the case of ordinary flows, if $f(v, w)$ units of flow are shipped from v to w, $f(v, w)$ units arrive at w. In the case of generalized flows, if $g(v, w)$ units of flow are shipped from v to w, $\gamma(v, w)g(v, w)$ units arrive at w. A *generalized pseudoflow* is a function $g : E \rightarrow \mathbf{R}$ that satisfies the capacity constraints (1.1) and the generalized antisymmetry constraints:

$$g(v, w) = -\gamma(w, v)g(w, v) \ \forall (v, w) \in E$$
$$(1.8) \qquad \text{(generalized antisymmetry constraints)}.$$

The gain of a path (cycle) is the product of the gains of the arcs on the path (cycle). For a given generalized pseudoflow g, the residual capacity and the residual graph $G_g = (V, E_g)$ are defined in the same way as for pseudoflows. The *excess* $e_g(v)$ of a generalized pseudoflow g at a vertex v is defined by

---

[4] $\mathbf{R}^+$ denotes the set of positive reals.

$$(1.9) \qquad\qquad e_g(v) = - \sum_{(v,w)\in E} g(v,w).$$

We will say that a vertex v has *excess* if $e_g(v)$ is positive and *deficit* if it is negative.

A *generalized flow* is a generalized pseudoflow that satisfies the conservation constraints (1.3) at all vertices except at the source. The *value* of a generalized pseudoflow g is the excess $e_g(s)$. The generalized flow problem is that of finding a generalized flow of maximum value in a given network. In our discussion of this problem, we shall assume that the capacities are integers, and that each gain is a rational number given as the ratio of two integers. Let B denote the largest integer used in the specification of the capacities and gains.

A *flow-generating cycle* is a cycle whose gain is greater than 1, and a *flow-absorbing cycle* is a cycle whose gain is less than 1. Observe that if one unit of flow leaves a vertex v and travels along a flow-generating cycle, more than one unit of flow arrives at v. Thus we can *augment* the flow along this cycle; that is, we can increase the excess at any vertex of the cycle while preserving the excesses at the other vertices, by increasing the flow along the arcs in the cycle (and correspondingly decreasing the flow on the opposite arcs to preserve the generalized antisymmetry constraints).

The generalized flow problem has the following interpretation in financial analysis. The commodity being moved is money, nodes correspond to different currencies and securities, arcs correspond to possible transactions, and gain factors represent the prices or the exchange rates (see Section 6.1). From the investor's point of view, a residual flow-generating cycle is an opportunity to make a profit. It is possible to take advantage of this opportunity, however, only if there is a way to transfer the profit to the investor's bank account (the source vertex). This motivates the following definition. A *generalized augmenting path (GAP)* is a residual flow-generating cycle and a (possibly trivial) residual path from a vertex on the cycle to the source. Given a generalized flow and a GAP in the residual graph, we can augment the flow along the GAP, increasing the value of the current flow. The role of GAP's in the generalized flow problem is similar to the role of negative-cost cycles in the minimum-cost circulation problem – both can be used to augment the flow and improve the value of the current solution. Onaga [80] proved that the non-existence of GAP's in the residual graph characterizes the optimality of a generalized flow.

1.5.1 Theorem [80]. *A generalized flow is optimal if and only if its residual graph contains no GAP.*

Using the linear programming dual of the problem, it is possible to give an alternate optimality criterion, similar to the criterion in Theorem 1.3.3 for the minimum-cost circulation problem. A *price function* p is a labeling of the vertices by real numbers, such that $p(s) = 1$. As in the case of the minimum-cost circulation problem, vertex prices can be interpreted as market prices of the commodity at vertices. If a unit of flow is purchased at v and shipped to w, then $\gamma(v,w)$ units arrive at w. Buying the unit at v costs $p(v)$, and selling $\gamma(v,w)$ units at w returns $p(w)\gamma(v, w)$. Thus the reduced cost of (v, w) is defined as

$$c_p(v, w) = p(v) - p(w)\gamma(v, w).$$

Linear programming duality theory provides the following optimality criterion.

1.5.2 Theorem [62]. *A generalized flow is optimal if and only if there exists a price function p such that the complementary slackness conditions (1.5) hold for each arc* $(v, w) \in E$.

One generalization of the problem, namely the *generalized flow problem with costs*, is worth mentioning here. As the name suggests, in this problem each arc (v, w) has a cost $c(v, w)$ in addition to its gain. The goal is to find a generalized flow that maximizes $e_g(s) - c(g)$, where $c(g) = \sum_{(v,w):g(v,w)>0} c(v, w)g(v, w)$. The introduction of costs enriches the combinatorial structure of the problem and allows the modeling of more complex problems, in particular economic processes. For example, a positive cost flow-generating cycle with a path leading to a negative cost flow-absorbing cycle may be used as a producer-consumer model. The generalized flow problem with costs has not been studied as much as the other problems discussed in this survey, and the only polynomial-time algorithms known for this problem are based on general linear programming methods [65, 103].

1.6 The Restricted Problem

Next we introduce a special variant of the generalized flow problem, and show that this variant is equivalent to the original problem.

Consider for a moment the following variation of the generalized flow problem: given a flow network with a source $s \in V$ and a sink $t \in V$, find a generalized pseudoflow with maximum excess at t and zero excess at each vertex other than s and t. Onaga [79] suggested the study of the special case of this problem in which the residual graph of the zero flow has no flow-generating cycles. We shall consider the corresponding special case of the generalized flow problem in which the residual graph of the zero flow has the property that all flow-generating cycles pass through the source. (If there are no flow-generating cycles, the zero flow is the optimal.) We shall also assume that the residual graph of the zero flow is strongly connected. A generalized flow network in which the residual graph of the zero flow is strongly connected and which has no flow-generating cycles not passing through the source is called a *restricted network*. The *restricted problem* is the generalized flow problem on a restricted network. The restricted problem has a simpler combinatorial structure that leads to simpler algorithms. Moreover, it turns out that the restricted problem is equivalent to the generalized flow problem. All of the algorithms that we review solve the restricted problem. In the rest of this section we shall review basic properties of the restricted problem, and we outline the reduction. In Chapter 6 we will use the term "generalized flow problem" to mean the restricted problem.

One of the nice facts about the restricted problem is that the optimality condition given by Theorem 1.5.2 simplifies in this case, and becomes very similar to Theorem 1.3.1. This characterization, which is also due to Onaga [79], can be deduced from Theorem 1.5.2 with the use of the following lemma. The lemma can be proved by induction on the number of arcs with positive flow.

1.6.1 Lemma. *Let g be a generalized pseudoflow in a restricted network. If the excess at every vertex is non-negative, then for every vertex v there exists a path from v to s in the residual graph G_g.*

1.6.2 Theorem [80]. *A generalized flow g in a restricted network is optimal if and only if the residual graph of g contains no flow-generating cycles.*

Note that the condition in the theorem can be formulated equivalently as "if and only if the residual graph of g has no negative-cost cycles, where the cost function is given by $c = -\log \gamma$."

1.6.3 Theorem [43]. *The generalized flow problem can be reduced to the restricted problem in $O(nm)$ time.*

Proof sketch: Given an instance of the generalized flow problem, reduce it to an instance in which all the flow-generating cycles pass through s, as follows. Let h be the generalized pseudoflow that saturates all arcs of gain greater than 1 and is zero on all arcs of gain 1. Define a new generalized flow network containing each arc (v, w) of the original network, with new capacity $u(v, w) - h(v, w)$. For each vertex $v \in V$ such that $e_h(v) > 0$, add an arc (s, v) of capacity $u(s, v) = e_h(v)/\alpha$, and with gain $\gamma(s, v) = \alpha$, where $\alpha = B^n$. Also add a reverse arc (v, s) with $u(v, s) = 0$. For each vertex $v \in V$ with $e_h(v) < 0$, add an arc (v, s) of capacity $u(v, s) = e_h(v)$ and with gain $\gamma(v, s) = \alpha$; also add a reverse arc (s, v) with $u(v, s) = 0$. Let $\hat{g}$ be an optimal generalized flow in the new network. Consider the generalized pseudoflow $\hat{g} + h$ restricted to the arcs of the original network. Using the Decomposition Theorem 1.7.3 below, one can show that the value of this generalized pseudoflow is equal to the value of the optimal generalized flow, and an optimal generalized flow can be constructed from this generalized pseudoflow in $O(nm)$ time. Intuitively, the new arcs ensure that vertices having excesses with respect to h are supplied with an adequate amount of "almost free" flow, and vertices having deficits with respect to h can deliver the corresponding amount of flow very efficiently to the source.

Having eliminated flow-generating cycles not containing the source, we can impose the strong connectivity condition by deleting all of the graph except the strongly connected component containing the source. This does not affect the value of an optimum solution by Theorem 1.5.2.   □

Note that this transformation increases the size of the numbers in the problem description. However, the polynomial-time algorithms described later depend only on T $(\leq B^n)$, the maximum product of gains along a simple path, and L $(\leq B^{2m})$, the product of denominators of arc gains; these parameters do not increase significantly.

1.7 Decomposition Theorems

A useful property of flows and generalized flows is the fact that they can be decomposed into a small number of "primitive elements". These elements depend on the problem under consideration. In this section we review decomposition theorems for circulations, pseudoflows, and generalized pseudoflows.

In the case of circulations, the primitive elements of the decomposition are flows along simple cycles.

1.7.1 Theorem [26]. *For every circulation f, there exists a collection of $k \leq m$ circulations $g_1, \ldots, g_k$ such that for every i, the graph induced by the set of arcs on which g_i is positive is a simple cycle consisting of arcs (v, w) with $f(v, w) > 0$, and*

$$(1.10) \qquad f(v, w) = \sum_i g_i.$$

Such a decomposition can be found in $O(nm)$ time.

For pseudoflows (or flows), the primitive elements of the decomposition are flows along simple cycles and flows along simple paths.

1.7.2 Theorem [26]. *For every pseudoflow f, there exists a collection of $k \leq m$ pseudoflows $g_1, \ldots, g_k$, such that for every i, the graph induced by the set of arcs on which g_i is positive is either a simple cycle or a simple path from a vertex with deficit to a vertex with excess; it consists of arcs (v, w) with $f(v, w) > 0$, and (1.10) is satisfied. Such a decomposition can be found in $O(nm)$ time.*

The five primitive elements of the decomposition for generalized pseudoflows g are defined as follows. The set of arcs on which an element of the decomposition of g is positive is a subset of the arcs on which g is positive. Let $T(g)$ denote the set of vertices with excesses and let $S(g)$ denote the set of vertices with deficits. The five types of primitive elements are classified according to the graph induced by the set of arcs with positive flow. *Type I*: A path from a vertex in $S(g)$ to a vertex in $T(g)$. Such a flow creates a deficit and an excess at the two ends of the path. *Type II*: A flow-generating cycle and a path leading from this cycle to a vertex in $T(g)$. Such a flow creates excess at the end of the path. (If the path ends at the source, then this corresponds to a GAP.) *Type III*: A flow-absorbing cycle and a path from a vertex in $S(g)$ to this cycle. Such a flow creates a deficit at the end of the path. *Type IV*: A cycle with unit gain. Such a flow does not create any excesses or deficits. *Type V*: A pair of cycles connected by a path, where one of the cycles generates flow and the other one absorbs it. Such a flow does not create any excesses or deficits.

1.7.3 Theorem [43, 53]. *For a generalized pseudoflow g there exist $k \leq m$ primitive pseudoflows $g_1, \ldots, g_k$ such that, for each i, g_i is of one of the five types described above, and (1.10) is satisfied. Such a decomposition can be found in $O(nm)$ time.*

Remark. In all three of these theorems the time to find the claimed decomposition can be reduced to $O(m \log n)$ by using the dynamic tree data structure that will be discussed in Section 2.5.

2. The Maximum Flow Problem

2.1 Introduction

The maximum flow problem has been studied for over thirty years. The classical methods for solving this problem are the Ford-Fulkerson *augmenting path method* [27, 26], the closely related *blocking flow method* of Dinic [21], and an appropriate variant of the *network simplex method* (see e.g. [61]). A *push-relabel method* for solving maximum flow problems has been recently proposed by Goldberg [40] and fully developed by Goldberg and Tarjan [45, 46]. Recently, parallel and distributed algorithms for the maximum flow problem have been studied as well.

Many maximum flow algorithms use scaling techniques and data structures to achieve better running time bounds. The scaling techniques used by maximum flow algorithms are *capacity scaling*, introduced by Edmonds and Karp [22] in the context of the minimum-cost circulation problem, and the closely related *excess scaling* of Ahuja and Orlin [3]. The dynamic tree data structure of Sleator and Tarjan [94, 95] makes it possible to speed up many maximum flow algorithms.

The technical part of this survey deals with the push-relabel method, which gives better sequential and parallel complexity bounds than previously known methods, and seems to outperform them in practice. In addition to describing the basic method, we show how to use excess scaling and dynamic trees to obtain faster implementations of the method. In this section we discuss the previous approaches and their relationship to the push-relabel method.

The augmenting path algorithm of Ford and Fulkerson is based on the fact that a flow is maximum if and only if there is no *augmenting path*. The algorithm repeatedly finds an augmenting path and augments along it, until no augmenting path exists. Although this simple generic method need not terminate if the network capacities are reals, and it can run in exponential time if the capacities are integers represented in binary [26], it can be made efficient by restricting the choice of augmenting paths. Edmonds and Karp [22] have shown that two strategies for selecting the next augmenting path give polynomial time bounds. The first such strategy is to select, at each iteration, a shortest augmenting path, where the length of a path is the number of arcs on the path. The second strategy is to select, at each iteration, a fattest path, where the fatness of a path is the minimum of the residual capacities of the arcs on the path.

Independently, Dinic [21] suggested finding augmenting paths in phases, handling all augmenting paths of a given shortest length in one phase by finding a *blocking flow* in a layered network. The shortest augmenting path length increases from phase to phase, so that the blocking flow method terminates in $n - 1$ phases. Dinic's discovery motivated a number of algorithms for the blocking flow problem. Dinic proposed an $O(nm)$ blocking flow algorithm. Soon thereafter, Karzanov [69] proposed an $O(n^2)$ algorithm, which achieves the best known bound for dense graphs. Karzanov also introduced the concept of a preflow; this concept is used by many maximum flow algorithms. Simpler algorithms achieving an $O(n^2)$ bound are described in [75, 97]. A sequence of algorithms achieving better and better running times on non-dense graphs has been proposed [17, 33, 36, 37, 49, 92, 95]. The algorithm of [33] uses capacity scaling; the algorithms of [95] and [49] use dynamic trees. The fastest currently-known sequential algorithm for the blocking flow problem, due to Goldberg and Tarjan, runs in $O(m \log(n^2/m))$ time [49].

Cherkasskij [18] proposed a method based on finding a *blocking preflow* in a layered network instead of a blocking flow. He gave an $O(n^2)$ algorithm for finding a blocking preflow based on Karzanov's algorithm, and proved an $O(n^3)$ bound on the total running time. The resulting algorithm is simpler than Karzanov's and seems to be better in practice. Conceptually, this algorithm is a step from the blocking flow method to the push/relabel method.

The push-relabel method [40, 41, 45, 49] replaces the layered network of the blocking flow method by a distance labeling of the vertices, which gives an estimate of the distance to the sink in the residual graph. The algorithm maintains a preflow and a distance labeling, and uses two simple operations, *pushing* and *relabeling*, to update the preflow and the labeling, repeating them until a maximum flow is found. The *pushing* operation is implicit in Karzanov's algorithm. The construction of a layered network at each iteration of Dinic's method can be viewed as a sequence of *relabeling* operations. Unlike the blocking flow method, the push-relabel method solves the maximum flow problem directly, without reducing the problem to a sequence of blocking flow subproblems. As a result, this method is more general and flexible than the blocking flow method, and leads to improved sequential and parallel algorithms. Ahuja and Orlin [3] suggested one way of recasting algorithms based on Dinic's approach into the push-relabel framework.

Goldberg's *FIFO*[5] algorithm [40] runs in $O(n^3)$ time. Goldberg and Tarjan [45, 46] showed how to use the dynamic tree data structure to improve the running time of this algorithm to $O(nm \log(n^2/m))$. They also gave a *largest-label* variant of the algorithm, for which they obtained the same time bounds – $O(n^3)$ without dynamic trees and $O(nm \log(n^2/m))$ with such trees. Cheriyan and Maheshwari [16] showed that (without dynamic trees) the FIFO algorithm runs in $\Omega(n^3)$ time, whereas the largest-label algorithm runs in $O(n^2\sqrt{m})$ time. Ahuja and Orlin [3] described an $O(nm + n^2 \log U)$-time version of the push-relabel method based on *excess scaling*. Ahuja, Orlin, and Tarjan [4] gave a modification of the Ahuja-Orlin algorithm that runs in $O(nm + n^2\sqrt{\log U})$ time without the use of dynamic trees and in $O(nm \log(\frac{n}{m}\sqrt{\log U} + 2))$ time with them.

The primal network simplex method (see e.g. [61] is another classical method for solving the maximum flow problem. Cunningham [19] gave a simple pivoting rule that avoids cycling due to degeneracy. Of course, if a simplex algorithm does not cycle, it must terminate in exponential time. Until recently, no polynomial time pivoting rule was known for the primal network simplex method. Goldfarb and Hao [52] have given such a rule. The resulting algorithm does $O(nm)$ pivots, which correspond to $O(nm)$ flow augmentations; it runs in $O(n^2m)$ time. Interestingly, the blocking flow method also makes $O(nm)$ flow augmentations. Unlike

---

[5] FIFO is an abbreviation of *first-in, first-out*.

the blocking flow method, the Goldfarb-Hao algorithm does not necessarily augment along shortest augmenting paths. In their analysis, Goldfarb and Hao use a variant of distance labeling and a variant of the *relabeling* operation mentioned above. Dynamic trees can be used to obtain an $O(nm \log n)$-time implementation of their algorithm [42].

The best currently known sequential bounds for the maximal flow problem are $O(nm \log(n^2/m))$ and $O(nm \log(\frac{n}{m}\sqrt{\log U} + 2))$. Note that, although the running times of the known algorithms come very close to an $O(mn)$ bound, the existence of a maximum flow algorithm that meets this bound remains open.

With the increasing interest in parallel computing, parallel and distributed algorithms for the maximum flow problem have received a great deal of attention. The first parallel algorithm for the problem, due to Shiloach and Vishkin [93], runs in $O(n^2 \log n)$ time on an n-processor PRAM [28] and uses $O(n)$ memory per processor. In a synchronous distributed model of computation, this algorithm runs in $O(n^2)$ time using $O(n^3)$ messages and $O(n)$ memory per processor. The algorithm of Goldberg [40, 41, 46] uses less memory than that of Shiloach and Vishkin: $O(m)$ memory for the PRAM implementation and $O(\Delta)$ memory per processor for the distributed implementation (where Δ is the processor degree in the network). The time, processor, and message bounds of this algorithm are the same as those of the Shiloach-Vishkin algorithm. Ahuja and Orlin [3] developed a PRAM implementation of their excess-scaling algorithm. The resource bounds are $\lceil m/n \rceil$ processors, $O(m)$ memory, and $O(n^2 \log n \log U)$ time. Cheriyan and Maheshwari [16] proposed a synchronous distributed implementation of the largest-label algorithm that runs in $O(n^2)$ time using $O(\Delta)$ memory per processor and $O(n^2 \sqrt{m})$ messages.

For a long time, the primal network simplex method was the method of choice in practice. A study of Goldfarb and Grigoriadis [51] suggested that the algorithm of Dinic [21] performs better than the network simplex method and better than the later algorithms based on the blocking flow method. Recent studies of Ahuja and Orlin (personal communication) and Grigoriadis (personal communication) show superiority of various versions of the push-relabel method to Dinic's algorithm. An experimental study of Goldberg [41] shows that a substantial speedup can be achieved by implementing the FIFO algorithm on a highly parallel computer.

More efficient algorithms have been developed for the special case of planar networks. Ford and Fulkerson [27] have observed that the the maximum flow problem on a planar network is related to a shortest path problem on the planar dual of the network. The algorithms in [8, 27, 30, 56, 57, 60, 64, 76, 88] make clever use of this observation.

2.2 A Generic Algorithm

In this section we describe the generic push-relabel algorithm [41, 45, 46]. First, however, we need the following definition. For a given preflow f, a *distance labeling* is a function d from the vertices to the non-negative integers such that

$d(t) = 0$, $d(s) = n$, and $d(v) \leq d(w) + 1$ for all residual arcs (v, w). The intuition behind this definition is as follows. Define a *distance graph* G_f^* as follows. Add an arc (s, t) to G_f. Define the length of all residual arcs to be equal to one and the length of the arc (s, t) to be n. Then d is a "locally consistent" estimate on the distance to the sink in the distance graph. (In fact, it is easy to show that d is a lower bound on the distance to the sink.) We denote by $d_{G_f^*}(v, w)$ the distance from vertex v to vertex w in the distance graph. The generic algorithm maintains a preflow f and a distance labeling d for f, and updates f and d using *push* and *relabel* operations. To describe these operations, we need the following definitions. We say that a vertex v is *active* if $v \notin \{s, t\}$ and $e_f(v) > 0$. Note that a preflow f is a flow if and only if there are no active vertices. An arc (v, w) is *admissible* if $(v, w) \in E_f$ and $d(v) = d(w) + 1$.

The algorithm begins with the preflow f that is equal to the arc capacity on each arc leaving the source and zero on all arcs not incident to the source, and with some initial labeling d. The algorithm then repetitively performs, in any order, the *update operations, push* and *relabel*, described in Figure 2.2. When there are no active vertices, the algorithm terminates. A summary of the algorithm appears in Figure 2.1.

```
procedure generic (V, E, u);
    [initialization]
    ∀(v, w) ∈ E do begin
        f(v, w) ← 0;
        if v = s then f(s, w) ← u(s, w);
        if w = s then f(v, s) ← −u(s, v);
    end;
    ∀w ∈ V do begin
        e_f(w) ← Σ_(v,w)∈E f(v, w);
        if w = s then d(w) = n else d(w) = 0;
    end;
    [loop]
    while ∃ an active vertex do
        select an update operation and apply it;
    return(f);
end.
```

Fig. 2.1   The generic maximum flow algorithm

The update operations modify the preflow f and the labeling d. A *push* from v to w increases $f(v, w)$ and $e_f(w)$ by up to $\delta = \min\{e_f(v), u_f(v, w)\}$, and decreases $f(w, v)$ and $e_f(v)$ by the same amount. The push is *saturating* if $u_f(v, w) = 0$ after the push and *nonsaturating* otherwise. A *relabeling* of v sets the label of v equal to the largest value allowed by the valid labeling constraints.

push(v, w).
Applicability: v is active **and** (v, w) is admissible.
Action: send $\delta \in (0, \min(e_f(v), u_f(v, w))]$ units of flow from v to w.

relabel(v).
Applicability: either s or t is reachable from v in G_f **and**
                $\forall w \in V \ u_f(v, w) = 0$ or $d(w) \geq d(v)$.
Action: replace $d(v)$ by $\min_{(v,w) \in E_f}\{d(w)\} + 1$.

Fig. 2.2 The update operations. The *pushing* operation updates the preflow, and the *relabeling* operation updates the distance labeling. Except for the excess scaling algorithm, all algorithms discussed in this section push the maximum possible amount δ when doing a push

There is one part of the algorithm we have not yet specified: the choice of an initial labeling d. The simplest choice is $d(s) = n$ and $d(v) = 0$ for $v \in V - \{s\}$. A more accurate choice (indeed, the most accurate possible choice) is $d(v) = d_{G_f^*}(v, t)$ for $v \in V$, where f is the initial preflow. The latter labeling can be computed in $O(m)$ time using backwards breadth-first searches from the sink and from the source in the residual graph. The resource bounds we shall derive for the algorithm are correct for any valid initial labeling. To simplify the proofs, we assume that the algorithm starts with the simple labeling. In practice, it is preferable to start with the most accurate values of the distance labels, and to update the distance labels periodically by using backward breadth-first search.

Remark. By giving priority to relabeling operations, it is possible to maintain the following invariant: Just before a push, d gives the exact distance to t in the distance graph. Furthermore, it is possible to implement the relabeling operations so that the total work done to maintain the distance labels is $O(nm)$ (see e.g. [46]). Since the running time bounds derived in this section are $\Omega(nm)$, one can assume that the relabeling is done in this way. In practice, however, maintaining exact distances is expensive; a better solution is to maintain a valid distance labeling and periodically update it to the exact labeling.

Next we turn our attention to the correctness and termination of the algorithm. Our proof of correctness is based on Theorem 1.2.1. The following lemma is important in the analysis of the algorithm.

2.2.1 Lemma. *If f is a preflow and v is a vertex with positive excess, then the source s is reachable from v in the residual graph G_f.*

Using this lemma and induction on the number of update operations, it can be shown that one of the two update operations must be applicable to an active vertex, and that the operations maintain a valid distance labeling and preflow.

2.2.2 Theorem. *Suppose that the algorithm terminates. Then the preflow f is a maximum flow.*

Proof. When the algorithm terminates, all vertices in $V - \{s,t\}$ must have zero excess, because there are no active vertices. Therefore f must be a flow. We show that if f is a preflow and d is a valid labeling for f, then the sink t is not reachable from the source s in the residual graph G_f. Then Theorem 1.2.1 implies that the algorithm terminates with a maximum flow.

Assume by way of contradiction that there is an augmenting path $s = v_0, v_1, \ldots, v_l = t$. Then $l < n$ and $(v_i, v_{i+1}) \in E_f$ for $0 \leq i < l$. Since d is a valid labeling, we have $d(v_i) \leq d(v_{i+1}) + 1$ for $0 \leq i < l$. Therefore, we have $d(s) \leq d(t) + l < n$, since $d(t) = 0$, which contradicts $d(s) = n$. $\qquad\square$

The key to the running time analysis of the algorithm is the following lemma, which shows that distance labels cannot increase too much.

2.2.3 Lemma. *At any time during the execution of the algorithm, for any vertex $v \in V$, $d(v) \leq 2n - 1$.*

Proof. The lemma is trivial for $v = s$ and $v = t$. Suppose $v \in V - \{s,t\}$. Since the algorithm changes vertex labels only by means of the *relabeling* operation, it is enough to prove the lemma for a vertex v such that s or t is reachable from v in G_f. Thus there is a simple path from v to s or t in G_f. Let $v = v_0, v_1, \ldots, v_l$ be such a path. The length l of the path is at most $n - 1$. Since d is a valid labeling and $(v_i, v_{i+1}) \in E_f$, we have $d(v_i) \leq d(v_{i+1}) + 1$. Therefore, since $d(v_l)$ is either n or 0, we have $d(v) = d(v_0) \leq d(v_l) + l \leq n + (n-1) = 2n - 1$. $\qquad\square$

Lemma 2.2.3 limits the number of relabeling operations, and allows us to amortize the work done by the algorithm over increases in vertex labels. The next two lemmas bound the number of relabelings and the number of saturating pushes.

2.2.4 Lemma. *The number of relabeling operations is at most $2n - 1$ per vertex and at most $(2n - 1)(n - 2) < 2n^2$ overall.*

2.2.5 Lemma. *The number of saturating pushes is at most nm.*

Proof. For an arc $(v, w) \in E$, consider the saturating pushes from v to w. After one such push, $u_f(v, w) = 0$, and another such push cannot occur until $d(w)$ increases by at least 2, a push from w to v occurs, and $d(v)$ increases by at least 2. If we charge each saturating push from v to w except the first to the preceding label increase of v, we obtain an upper bound of n on the number of such pushes. $\qquad\square$

The most interesting part of the analysis is obtaining a bound on the number of nonsaturating pushes. For this we use amortized analysis and in particular the *potential function* technique (see e.g. [98]).

2.2.6 Lemma. *The number of nonsaturating pushing operations is at most $2n^2 m$.*

Proof. We define the *potential* Φ of the current preflow f and labeling d by the formula $\Phi = \sum_{\{v | v \text{ is active}\}} d(v)$. We have $0 \leq \Phi \leq 2n^2$ by Lemma 2.2.3. Each nonsaturating push, say from a vertex v to a vertex w, decreases Φ by at least one, since $d(w) = d(v) - 1$ and the push makes v inactive. It follows that the total number of nonsaturating pushes over the entire algorithm is at most the sum of the increases in Φ during the course of the algorithm, since $\Phi = 0$ both at the beginning and at the end of the computation. Increasing the label of a vertex v by an amount k increases Φ by k. The total of such increases over the algorithm is at most $2n^2$. A saturating push can increase Φ by at most $2n - 2$. The total of such increases over the entire algorithm is at most $(2n - 2)nm$. Summing gives a bound of at most $2n^2 + (2n - 2)nm \leq 2n^2 m$ on the number of nonsaturating pushes. □

2.2.7 Theorem [46]. *The generic algorithm terminates after $O(n^2 m)$ update operations.*

Proof. Immediate from Lemmas 2.2.4, 2.2.5, and 2.2.6. □

The running time of the generic algorithm depends upon the order in which update operations are applied and on implementation details. In the next sections we explore these issues. First we give a simple implementation of the generic algorithm in which the time required for the nonsaturating pushes dominates the overall running time. Sections 2.3 and 2.4 specify orders of the update operations that decrease the number of nonsaturating pushes and permit $O(n^3)$ and $O(mn + n^2 \log U)$-time implementations. Section 2.5 explores an orthogonal approach. It shows how to use sophisticated data structures to reduce the time per nonsaturating push rather than the number of such pushes.

2.3 Efficient Implementation

Our first step toward an efficient implementation is a way of combining the update operations locally. We need some data structures to represent the network and the preflow. We call an unordered pair $\{v, w\}$ such that $(v, w) \in E$ an *edge* of G. We associate the three values $u(v, w)$, $u(w, v)$, and $f(v, w)(= -f(w, v))$ with each edge $\{v, w\}$. Each vertex v has a list of the incident edges $\{v, w\}$, in fixed but arbitrary order. Thus each edge $\{v, w\}$ appears in exactly two lists, the one for v and the one for w. Each vertex v has a *current edge* $\{v, w\}$, which is the current candidate for a pushing operation from v. Initially, the current edge of v is the first edge on the edge list of v. The main loop of the implementation consists of repeating the *discharge* operation described in Figure 2.3 until there are no active vertices. (We shall discuss the maintenance of active vertices later.) The *discharge* operation is applicable to an active vertex v. This operation iteratively attempts to push the excess at v through the current edge $\{v, w\}$ of v if a pushing operation is applicable to this edge. If not, the operation replaces $\{v, w\}$ as the current edge of v by the next edge on the edge list of v; or, if $\{v, w\}$ is the last edge on this list, it makes the first edge on the list the current one and relabels v. The operation stops when the excess at v is reduced to zero or v is relabeled.

discharge(v).
Applicability: *v* is active.
Action: let $\{v, w\}$ be the current edge of *v*;
    time-to-relabel ← **false;**
    repeat
     if (v, w) is admissible **then** *push(v, w)*
     else
      if $\{v, w\}$ is not the last edge on the edge list of *v* **then**
       replace $\{v, w\}$ as the current edge of *v* by the
       next edge on the list
      else begin
       make the first edge on the edge list of *v* the current edge;
       time-to-relabel ← **true;**
      end;
    until $e_f(v) = 0$ **or** *time-to-relabel*;
    if *time-to-relabel* **then** *relabel(v)*;

Fig. 2.3 The discharge operation

The following lemma shows that *discharge* does relabeling correctly; the proof of the lemma is straightforward.

2.3.1 Lemma. *The discharge operation does a relabeling only when the relabeling operation is applicable.*

2.3.2 Lemma. *The version of the generic push/relabel algorithm based on discharging runs in $O(nm)$ time plus the total time needed to do the nonsaturating pushes and to maintain the set of active vertices.*

Any representation of the set of active vertices that allows insertion, deletion, and access to some active vertex in $O(1)$ time results in an $O(n^2 m)$ running time for the discharge-based algorithm, by Lemmas 2.2.6 and 2.3.2. (Pushes can be implemented in $O(1)$ time per push.)

By processing active vertices in a more restricted order, we obtain improved performance. Two natural orders were suggested in [45, 46]. One, the *FIFO algorithm,* is to maintain the set of active vertices as a queue, always selecting for discharging the front vertex on the queue and adding newly active vertices to the rear of the queue. The other, the *largest-label algorithm,* is to always select for discharging a vertex with the largest label. The FIFO algorithm runs in $O(n^3)$ time [45, 46] and the largest-label algorithm runs in $O(n^2\sqrt{m})$ time [16]. We shall derive an $O(n^3)$ time bound for both algorithms, after first describing in a little more detail how to implement largest-label selection.

The implementation maintains an array of sets B_i, $0 \le i \le 2n - 1$, and an index *b* into the array. Set B_i consists of all active vertices with label *i*, represented as a doubly-linked list, so that insertion and deletion take $O(1)$ time. The index *b* is the largest label of an active vertex. During the initialization, when the arcs

```
procedure process-vertex;
    remove a vertex v from B_b;
    old-label ← d(v);
    discharge(v);
    add each vertex w made active by the discharge to B_{d(w)};
    if d(v) ≠ old-label then begin
        b ← d(v);
        add v to B_b;
    end
    else if B_b = ∅ then b ← b − 1;
end.
```

Fig. 2.4 The *process-vertex* procedure

going out of the source are saturated, the resulting active vertices are placed in B_0, and b is set to 0. At each iteration, the algorithm removes a vertex from B_b, processes it using the *discharge* operation, and updates b. The algorithm terminates when b becomes negative, *i.e.*, when there are no active vertices. This processing of vertices, which implements the *while* loop of the generic algorithm, is described in Figure 2.4.

To understand why the *process-vertex* procedure correctly maintains b, note that *discharge(v)* either relabels v or gets rid of all excess at v, but not both. In the former case, v is the active vertex with the largest distance label, so b must be increased to $d(v)$. In the latter case, the excess at v has been moved to vertices with distance labels of $b - 1$, so if B_b is empty, then b must be decreased by one. The total time spent updating b during the course of the algorithm is $O(n^2)$.

The bottleneck in both the FIFO method and the largest-label method is the number of nonsaturating pushes. We shall obtain an $O(n^3)$ bound on the number of such pushes by dividing the computation into *phases*, defined somewhat differently for each method. For the FIFO method, phase 1 consists of the discharge operations applied to the vertices added to the queue by the initialization of f; phase $i + 1$, for $i \leq 1$, consists of the discharge operations applied to the vertices added to the queue during phase i. For the largest-label method, a phase consists of a maximal interval of time during which b remains constant.

2.3.3 Lemma. *The number of phases during the running of either the FIFO or the largest-label algorithm is at most $4n^2$.*

Proof. Define the potential Φ of the current f and d by $\Phi = \max_{\{v | v \text{ is active}\}} d(v)$, with the maximum taken to be zero if there are no active vertices. (In the case of the largest-label algorithm, $\Phi = b$ except on termination.) There can be only $2n^2$ phases that do one or more relabelings. A phase that does no relabeling decreases Φ by at least one. The initial and final values of Φ are zero. Thus the number of phases that do no relabeling is at most the sum of the increases in Φ during the computation. The only increases in Φ are due to label increases;

an increase of a label by k can cause Φ to increase by up to k. Thus the sum of the increases in Φ over the computation is at most $2n^2$, and so is the number of phases that do no relabeling. $\square$

2.3.4 Theorem [46]. *Both the FIFO and the largest-label algorithm run in $O(n^3)$ time.*

Proof. For either algorithm, there is at most one nonsaturating push per vertex per phase. Thus by Lemma 2.3.3 the total number of nonsaturating pushes is $O(n^3)$, as is the running time by Lemma 2.3.2. $\square$

Cheriyan and Maheshwari [16], by means of an elegant balancing argument, were able to improve the bound on the number of nonsaturating pushes in the largest-label algorithm to $O(n^2\sqrt{m})$, giving the following result:

2.3.5 Theorem [16]. *The largest-label algorithm runs in $O(n^2\sqrt{m})$ time.*

2.4 Excess Scaling

A different approach to active vertex selection leads to running time bounds dependent on the size U of the largest capacity as well as on the graph size. This approach, *excess scaling*, was introduced by Ahuja and Orlin [3] and developed further by Ahuja, Orlin, and Tarjan [4]. We shall describe in detail a slight revision of the original excess-scaling algorithm, which has a running time of $O(nm + n^2\log U)$.

For the termination of the excess-scaling method, all arc capacities must be integral; hence we assume throughout this section that this is the case. The method preserves integrality of the flow throughout the computation. It depends on a parameter Δ that is an upper bound on the maximum excess of an active vertex. Initially $\Delta = 2^{\lceil \log U \rceil}$. The algorithm proceeds in phases; after each phase, Δ is halved. When $\Delta < 1$, all active vertex excesses must be zero, and the algorithm terminates. Thus the algorithm terminates after at most $\log_2 U + 1$ phases. To maintain the invariant that no active vertex excess exceeds Δ, the algorithm does not always push the maximum possible amount when doing a pushing operation. Specifically, when pushing from a vertex v to a vertex w, the algorithm moves an amount of flow δ given by $\delta = \min\{e_f(v), u_f(v,w), \Delta - e_f(w)\}$ if $w \neq t$, $\delta = \min\{e_f(v), u_f(v,w)\}$ if $w = t$. That is, δ is the maximum amount that can be pushed while maintaining the invariant.

The algorithm consists of initialization of the preflow, the distance labels, and Δ, followed by a sequence of *process-vertex* operations of the kind described in Section 2.3. Vertex selection for *process-vertex* operations is done by the *large excess, smallest label rule*: process an active vertex v with $e_f(v) > \Delta/2$; among all candidate vertices, choose one of smallest label. When every active vertex v has $e_f(v) \leq \Delta/2$, Δ is halved and a new phase begins; when there are no active vertices, the algorithm stops.

Since the excess-scaling algorithm is a version of the generic *process-vertex-*based algorithm described in Section 2.3, Lemma 2.3.2 applies. The following lemma bounds the number of nonsaturating pushes:

2.4.1 Lemma. *The number of nonsaturating pushes during the excess-scaling algorithm is $O(n^2 \log U)$.*

Proof. We define a potential Φ by $\Phi = \sum_{\{v | v \text{ is active}\}} e_f(v) \, d(v)/\Delta$. Since $e_f(v)/\Delta \leq$ 1 for any active vertex, $0 \leq \Phi \leq 2n^2$. Every pushing operation decreases Φ. Consider a nonsaturating push from a vertex v to a vertex w. The large excess, smallest label rule guarantees that before the push $e_f(v) > \Delta/2$ and either $e_f(w) \leq \Delta/2$ or $w = t$. Thus the push moves at least $\Delta/2$ units of flow, and hence decreases Φ by at least $1/2$. The initial and final values of Φ are zero, so the total number of nonsaturating pushes is at most twice the sum of the increases in Φ over the course of the algorithm. Increasing the label of a vertex by k can increase Φ by at most k. Thus relabelings account for a total increase in Φ of at most $2n^2$. A change in phase also increases Φ, by a factor of two, or at most n^2. Hence the sum of increases in Φ is at most $2n^2 + n^2 (\log_2 U + 1)$, and therefore the number of nonsaturating pushes is at most $4n^2 + 2n^2 (\log_2 U + 1)$.   □

The large excess, smallest label rule can be implemented by storing the active vertices with excess exceeding $\Delta/2$ in an array of sets, as was previously described for the largest-label rule. In the former case, the index b indicates the nonempty set of smallest index. Since b decreases only when a push occurs, and then by at most 1, the total time spent updating b is $O(nm + n^2 \log U)$. From Lemmas 2.3.2 and 2.4.1 we obtain the following result:

2.4.2 Theorem. *The excess-scaling algorithm runs in $O(nm + n^2 \log U)$ time.*

A more complicated version of excess scaling, devised by Ahuja, Orlin, and Tarjan [4], has a running time of $O(nm + n^2 \sqrt{\log U})$. This algorithm uses a hybrid vertex selection rule that combines a stack-based mechanism with the "wave" approach of Tarjan [97].

2.5 Use of Dynamic Trees

The two previous sections discussed ways of reducing the number of nonsaturating pushes by restricting the order of the update operations. An orthogonal approach is to reduce the *time* per nonsaturating push rather than the *number* of such pushes. The idea is to perform a succession of pushes along a single path in one operation, using a sophisticated data structure to make this possible. Observe that immediately after a nonsaturating push along an arc (v, w), (v, w) is still admissible, and we know its residual capacity. The *dynamic tree* data structure of Sleator and Tarjan [94, 95] provides an efficient way to maintain information about such arcs. We shall describe a dynamic tree version of the generic algorithm that has a running time of $O(nm \log n)$.

make-tree(v): Make vertex v into a one-vertex dynamic tree. Vertex v must be in no other tree.

find-root(v): Find and return the root of the tree containing vertex v.

find-value(v): Find and return the value of the tree arc connecting v to its parent. If v is a tree root, return infinity.

find-min(v): Find and return the ancestor w of v such that the tree arc connecting w to its parent has minimum value along the path from v to *find-root*(v). In case of a tie, choose the vertex w closest to the tree root. If v is a tree root, return v.

change-value(v, x): Add real number x to the value of every arc along the path from v to *find-root*(v).

link(v, w, x): Combine the trees containing v and w by making w the parent of v and giving the new tree arc joining v and w the value x. This operation does nothing if v and w are in the same tree or if v is not a tree root.

cut(v): Break the tree containing v into two trees by deleting the arc from v to its parent. This operation does nothing if v is a tree root.

Fig. 2.5 The dynamic tree operations

The dynamic tree data structure allows the maintenance of a collection of vertex-disjoint rooted trees, each arc of which has an associated real value. We regard a tree arc as directed toward the root, *i.e.*, from child to parent. We denote the parent of a vertex v by *parent*(v), and adopt the convention that every vertex is both an ancestor and a descendant of itself. The data structure supports the seven operations described in Figure 2.5. A sequence of l tree operations on trees of maximum size (number of vertices) k takes $O(l \log k)$ time.

In the dynamic tree algorithm, the arcs of the dynamic trees are a subset of the admissible arcs and every active vertex is a tree root. The value of an arc (v, w) is its residual capacity. Initially each vertex is made into a one-vertex dynamic tree using a *make-tree* operation.

The heart of the dynamic tree implementation is the *tree-push* operation described in Figure 2.6. This operation is applicable to an admissible arc (v, w) such that v is active. The operation adds (v, w) to the forest of dynamic trees, pushes as much flow as possible along the tree from v to the root of the tree containing w, and deletes from the forest each arc that is saturated by this flow change.

The dynamic tree algorithm is just the generic algorithm with the *push* operation replaced by the *tree-push* operation, with the initialization modified to make each vertex into a dynamic tree, and with a postprocessing step that extracts the correct flow value for each arc remaining in a dynamic tree. (This postprocessing takes one *find-value* operation per vertex.)

It is straightforward to show that the dynamic tree implementation is correct, by observing that it maintains the invariant that between tree-pushing operations

---

Tree-push (v, w)
Applicability: v is active and (v, w) is admissible.
Action: $link\ (v, w, u_f(v, w))$;
               $parent\ (v)\ \leftarrow w$;
               $\delta\ \leftarrow\ \min\{e_f(v),\ find\text{-}value(find\text{-}min\ (v))$;
               $change\text{-}value\ (v, -\delta)$;
               while $v \neq find\text{-}root(v)$ **and** $find\text{-}value(find\text{-}min(v)) = 0$ **do begin**
                    $z \leftarrow find\text{-}min(v)$;
                    $cut\ (z)$;
                    $f(z, parent(z)) \leftarrow u(z, parent(z))$;
                    $f(parent(z), z) \leftarrow -u(z, parent(z))$;
               end.

Fig. 2.6 The *tree-push* operation

every active vertex is a dynamic tree root and every dynamic tree arc is admissible. The following lemma bounds the number of *tree-push* operations:

2.5.1 Lemma. *The number of the* tree-push *operations done by the dynamic tree implementation is* $O(nm)$.

Proof. Each *tree-push* operation either saturates an arc (thus doing a saturating push) or decreases the number of active vertices by one. The number of active vertices can increase, by at most one, as a result of a *tree-push* operation that saturates an arc. By Lemma 2.2.5 there are at most nm saturating pushes. An upper bound of $2nm + n$ on the number of *tree-push* operations follows, since initially there are at most n active vertices. $\square$

If we implement the dynamic tree algorithm using the *discharge* operation of Section 2.3 with *push* replaced by *tree-push*, we obtain the following result (assuming active vertex selection takes $O(1)$ time):

2.5.2 Theorem. *The discharge-based implementation of the dynamic tree algorithm runs in* $O(nm \log n)$ *time.*

Proof. Each *tree-pushing* operation does $O(1)$ dynamic tree operations plus $O(1)$ per arc saturated. The theorem follows from Lemma 2.2.5, Lemma 2.3.2, and Lemma 2.5.1, since the maximum size of any dynamic tree is $O(n)$. $\square$

The dynamic tree data structure can be used in combination with the FIFO, largest-label, or excess-scaling method. The resulting time bounds are $O(nm \log(n^2/m))$ for the FIFO and largest-label methods [46] and $O(nm \log(\frac{n}{m}\sqrt{\log U}) + 2))$ for the fastest version of the excess-scaling method [4]. In each case, the dynamic tree method must be modified so that the trees are not too large, and the analysis needed to obtain the claimed bound is rather complicated.

3. The Minimum-Cost Circulation Problem: Cost Scaling

3.1 Introduction

Most polynomial-time algorithms for the minimum-cost circulation problem use the idea of scaling. This idea was introduced by Edmonds and Karp [22], who used it to develop the first polynomial-time algorithm for the problem. Scaling algorithms work by obtaining a sequence of feasible or almost-feasible solutions that are closer and closer to the optimum. The Edmonds-Karp algorithm scales capacities. Röck [90] was the first to propose an algorithm that scales costs. Later, Bland and Jensen [13] proposed a somewhat different cost-scaling algorithm, which is closer to the generalized cost scaling method discussed in this chapter.

The *cost-scaling* approach works by solving a sequence of problems, $P_0, P_1, \ldots,$ P_k, on the network with original capacities but approximate costs. The cost function c_i for P_i is obtained by taking the i most significant bits of the original cost function c. The first problem P_0 has zero costs, and therefore the zero circulation is optimal. An optimal solution to problem P_{i-1} can be used to obtain an optimal solution to problem P_i in at most n maximum flow computations [13, 90]. Note that for $k = \lceil \log_2 C \rceil$, $c_k = c$. Thus the algorithm terminates in $O(n \log C)$ maximum flow computations.

Goldberg and Tarjan [41, 48, 50] proposed a *generalized cost-scaling* approach. The idea of this method (which is described in detail below) is to view the maximum amount of violation of the complementary slackness conditions as an error parameter, and to improve this parameter by a factor of two at each iteration. Initially the error is at most C, and if the error is less than $1/n$, then the current solution is optimal. Thus the generalized cost-scaling method terminates in $O(\log(nC))$ iterations. The computation done at each iteration is similar to a maximum flow computation. The traditional cost-scaling method of Röck also improves the error from iteration to iteration, but it does so indirectly, by increasing the precision of the current costs and solving the resulting problem exactly. Keeping the original costs, as does the generalized cost-scaling approach, makes it possible to reduce the number of iterations required and to obtain strongly-polynomial running time bounds. Chapter 4 discusses a strongly-polynomial version of the generalized cost-scaling method. For further discussion of generalized versus traditional cost-scaling, see [49].

Time bounds for the cost-scaling algorithms mentioned above are as follows. The algorithms of Röck [90] and Bland and Jensen [13] run in $O(n \log(C) M(n, m, U))$ time, where $M(n, m, U)$ is the time required to compute a maximum flow on a network with n vertices, m arcs, and maximum arc capacity U. As we have seen in Chapter 2, $M = O(nm \log \min\{n^2/m, \frac{n}{m}\sqrt{\log U} + 2\})$. The fastest known implementation of the generalized cost-scaling method runs in $O(nm \log(n^2/m) \log(nC))$ time [48]. It is possible to combine cost scaling with capacity scaling. The first algorithm that combines the two scaling techniques is due to Gabow and Tarjan [35]. A different algorithm was proposed by Ahuja et al. [1]. The latter algorithm runs in $O(nm \log \log U \log(nC))$ time, which makes it the fastest known algorithm for the problem under the similarity assumption.

3.2 Approximate Optimality

A key notion is that of *approximate optimality*, obtained by relaxing the complementary slackness constraints in Theorem 1.3.3. For a constant $\varepsilon \geq 0$, a pseudoflow f is said to be *ε-optimal with respect to a price function p* if, for every arc (v, w), we have

(3.1) $f(v, w) < u(v, w) \Rightarrow c_p(v, w) \geq -\varepsilon$ (ε-optimality constraint).

A pseudoflow f is *ε-optimal* if f is ε-optimal with respect to some price function p.

An important property of ε-optimality is that if the arc costs are integers and ε is small enough, any ε-optimal circulation is minimum-cost. The following theorem, of Bertsekas [11] captures this fact.

3.2.1 Theorem [11]. *If all costs are integers and $\varepsilon < 1/n$, then an ε-optimal circulation f is minimum-cost.*

The ε-optimality constraints were first published by Tardos [96] in a paper describing the first strongly polynomial algorithm for the minimum-cost circulation problem. Bertsekas [11] proposed a pseudopolynomial algorithm based upon Theorem 3.2.1; his algorithm makes use of a fixed $\varepsilon < 1/n$. Goldberg and Tarjan [41, 49, 50] devised a successive approximation scheme that produces a sequence of circulations that are ε-optimal for smaller and smaller values of ε; when ε is small enough, the scheme terminates with an optimal circulation. We discuss this scheme below.

3.3 The Generalized Cost-Scaling Framework

Throughout the rest of this chapter, we assume that all arc costs are integral. We give here a high-level description of the generalized cost-scaling method (see Figure 3.1). The algorithm maintains an error parameter ε, a circulation f and a price function p, such that f is ε-optimal with respect to p. The algorithm starts with $\varepsilon = C$ (or alternatively $\varepsilon = 2^{\lceil \log_2 C \rceil}$), with $p(v) = 0$ for all $v \in V$, and with the zero circulation. Any circulation is C-optimal. The main loop of the algorithm repeatedly reduces the error parameter ε. When $\varepsilon < 1/n$, the current circulation is minimum-cost, and the algorithm terminates.

The task of the subroutine *refine* is to reduce the error in the optimality of the current circulation. The input to *refine* is an error parameter ε, a circulation f, and a price function p such that f is ε-optimal with respect to p. The output from *refine* is a reduced error parameter ε, a new circulation f, and a new price function p such that f is ε-optimal with respect to p. The implementations of *refine* described in this survey reduce the error parameter ε by a factor of two.

```
procedure min-cost(V, E, u, c);
    [initialization]
    ε ← C;
    ∀v,   p(v) ← 0;
    ∀(v, w) ∈ E,   f(v, w) ← 0;
    [loop]
    while ε ≥ 1/n do
        (ε, f, p) ← refine(ε, f, p);
    return(f);
end.
```

Fig. 3.1 The generalized cost-scaling method

The correctness of the algorithm is immediate from Theorem 3.2.1, assuming that *refine* is correct. The number of iterations of *refine* is $O(\log(nC))$. This gives us the following theorem:

3.3.1 Theorem [50]. *A minimum-cost circulation can be computed in the time required for $O(\log(nC))$ iterations of refine, if refine reduces ε by a factor of at least two.*

3.4 A Generic Refinement Algorithm

In this section we describe an implementation of *refine* that is a common generalization of the generic maximum flow algorithm of Section 2.2 and the auction algorithm for the assignment problem [9] (first published in [10]). We call this the *generic implementation*. This implementation, proposed by Goldberg and Tarjan [50], is essentially the same as the main loop of the minimum-cost circulation algorithm of Bertsekas [11], which is also a common generalization of the maximum flow and assignment algorithms. The ideas behind the auction algorithm can be used to give an alternative interpretation to the results of [41, 50] in terms of relaxation methods; see [12].

As we have mentioned in Section 3.3, the effect of *refine* is to reduce ε by a factor of two while maintaining the ε-optimality of the current flow *f* with respect to the current price function *p*. The generic *refine* subroutine is described on Figure 3.2. It begins by halving ε and saturating every arc with negative reduced cost. This converts the circulation *f* into an ε-optimal pseudoflow (indeed, into a 0-optimal pseudoflow). Then the subroutine converts the ε-optimal pseudoflow into an ε-optimal circulation by applying a sequence of the *update operations* *push* and *relabel*, each of which preserves ε-optimality.

The inner loop of the generic algorithm consists of repeatedly applying the two update operations, described in Figure 3.3, in any order, until no such operation applies. To define these operations, we need to redefine admissible arcs in the context of the minimum-cost circulation problem. Given a pseudoflow *f*

---

procedure *refine*(ε, f, p);
    [initialization]
    $\varepsilon \leftarrow \varepsilon/2$;
    $\forall (v, w) \in E$ **do if** $c_p(v, w) < 0$ **then begin**
                     $f(v, w) \leftarrow u(v, w); f(w, v) \leftarrow -u(v, w)$;
        end;
    [loop]
    while $\exists$ an update operation that applies **do**
        select such an operation and apply it;
    return(ε, f, p);
end.

Fig. 3.2 The generic *refine* subroutine

---

push(v, w).
Applicability: v is active **and** (v, w) is admissible.
Action: send $\delta = \min(e_f(v), u_f(v, w))$ units of flow from v to w.

relabel(v).
Applicability: v is active **and** $\forall w \in V$ ($u_f(v, w) = 0$ **or** $c_p(v, w) \geq 0$).
Action: replace $p(v)$ by $\max_{(v,w) \in E_f} \{p(w) - c(v, w) - \varepsilon\}$.

Fig. 3.3 The update operations for the generic refinement algorithm. Compare with Figure 2.2

and a price function p, we say that an arc (v, w) is *admissible* if (v, w) is a residual arc with negative reduced cost.

A *push* operation applies to an admissible arc (v, w) such that vertex v is active. It consists of pushing $\delta = \min\{e_f(v), u_f(v, w)\}$ units of flow from v to w, thereby decreasing $e_f(v)$ and $f(w, v)$ by δ and increasing $e_f(w)$ and $f(v, w)$ by δ. The push is *saturating* if $u_f(v, w) = 0$ after the push and *nonsaturating* otherwise.

A *relabel* operation applies to an active vertex v that has no exiting admissible arcs. It consists of decreasing $p(v)$ to the smallest value allowed by the ε-optimality constraints, namely $\max_{(v,w) \in E_f} \{-c(v, w) + p(w) - \varepsilon\}$.

If an ε-optimal pseudoflow f is not a circulation, then either a pushing or a relabeling operation is applicable. It is easy to show that any pushing operation preserves ε-optimality. The next lemma gives two important properties of the relabeling operation.

3.4.1 Lemma. *Suppose f is an ε-optimal pseudoflow with respect to a price function p and a vertex v is relabeled. Then the price of v decreases by at least ε and the pseudoflow f is ε-optimal with respect to the new price function p'.*

Proof. Before the relabeling, $c_p(v, w) \geq 0$ for all $(v, w) \in E_f$, i.e., $p(v) \geq p(w) - c(v, w)$ for all $(v, w) \in E_f$. Thus $p'(v) = \max_{(v,w)\in E_f} \{p(w) - c(v, w) - \varepsilon\} \leq p(v) - \varepsilon$.

To verify ε-optimality, observe that the only residual arcs whose reduced costs are affected by the relabeling are those of the form (v, w) or (w, v). Any arc of the form (w, v) has its reduced cost increased by the relabeling, preserving its ε-optimality constraint. Consider a residual arc (v, w). By the definition of p', $p'(v) \geq p(w) - c(v, w) - \varepsilon$. Thus $c_{p'}(v, w) = c(v, w) + p'(v) - p(w) \geq -\varepsilon$, which means that (v, w) satisfies its ε-optimality constraint. $\square$

Since the update operations preserve ε-optimality, and since some update operation applies if f is not a circulation, it follows that if *refine* terminates and returns (ε, f, p), then f is a circulation which is ε-optimal with respect to p. Thus *refine* is correct.

Next we analyze the number of update operations that can take place during an execution of *refine*. We begin with a definition. The *admissible graph* is the graph $G_A = (V, E_A)$ such that E_A is the set of admissible arcs. As *refine* executes, the admissible graph changes. An important invariant is that the admissible graph remains acyclic.

3.4.2 Lemma. *Immediately after a relabeling is applied to a vertex v, no admissible arcs enter v.*

Proof. Let (u, v) be a residual arc. Before the relabeling, $c_p(u, v) \geq -\varepsilon$ by ε-optimality. By Lemma 3.4.1, the relabeling decreases $p(v)$, and hence increases $c_p(u, v)$, by at least ε. Thus $c_p(u, v) \geq 0$ after the relabeling. $\square$

3.4.3 Corollary. *Throughout the running of* refine, *the admissible graph is acyclic.*

Proof. Initially the admissible graph contains no arcs and is thus acyclic. Pushes obviously preserve acyclicity. Lemma 3.4.2 implies that relabelings also preserve acyclicity. $\square$

Next we derive a crucial lemma, which generalizes Lemma 2.2.1.

3.4.4 Lemma. *Let f be a pseudoflow and f' a circulation. For any vertex v with $e_f(v) > 0$, there is a vertex w with $e_f(w) < 0$ and a sequence of distinct vertices $v = v_0, v_1, \ldots, v_{l-1}, v_l = w$ such that $(v_i, v_{i+1}) \in E_f$ and $(v_{i+1}, v_i) \in E_{f'}$ for $0 \leq i < l$.*

Proof. Let v be a vertex with $e_f(v) > 0$. Define $G_+ = (V, E_+)$, where $E_+ = \{(x, y) \mid f'(x, y) > f(x, y)\}$, and define $G_- = (V, E_-)$, where $E_- = \{(x, y) \mid f(x, y) > f'(x, y)\}$. Then $E_+ \subseteq E_f$, since $(x, y) \in E_+$ implies $f(x, y) < f'(x, y) \leq u(x, y)$. Similarly $E_- \subseteq E_{f'}$. Furthermore $(x, y) \in E_+$ if and only if $(y, x) \in E_-$ by antisymmetry. Thus to prove the lemma it suffices to show the existence in G_+ of a simple path $v = v_0, v_1 \ldots, v_l$ with $e_f(v_l) < 0$.

Let S be the set of all vertices reachable from v in G_+ and let $\overline{S} = V - S$. (Set $\overline{S}$ may be empty.) For every vertex pair $(x, y) \in S \times \overline{S}$, $f(x, y) \geq f'(x, y)$, for otherwise $y \in S$. We have

$$
\begin{aligned}
0 &= \sum_{(x,y) \in (S \times \bar{S}) \cap E} f'(x,y) & \text{since } f' \text{ is a circulation} \\
&\leq \sum_{(x,y) \in (S \times \bar{S}) \cap E} f(x,y) & \text{holds term-by-term} \\
&= \sum_{(x,y) \in (S \times \bar{S}) \cap E} f(x,y) + \\
&\quad \sum_{(x,y) \in (S \times S) \cap E} f(x,y) & \text{by antisymmetry} \\
&= \sum_{(x,y) \in (S \times V) \cap E} f(x,y) & \text{by definition of } \bar{S} \\
&= -\sum_{x \in S} e_f(x) & \text{by antisymmetry.}
\end{aligned}
$$

But $v \in S$. Since $e_f(v) > 0$, some vertex $w \in S$ must have $e_f(w) < 0$.   □

Using Lemma 3.4.4 we can bound the amount by which a vertex price can decrease during an invocation of *refine*.

3.4.5 Lemma. *The price of any vertex v decreases by at most $3n\varepsilon$ during an execution of refine.*

Proof. Let $f_{2\varepsilon}$ and $p_{2\varepsilon}$ be the circulation and price functions on entry to *refine*. Suppose a relabeling causes the price of a vertex v to decrease. Let f be the pseudoflow and p the price function just after the relabeling. Then $e_f(v) > 0$. Let $v = v_0, v_1, \ldots, v_l = w$ with $e_f(w) < 0$ be the vertex sequence satisfying Lemma 3.4.4 for f and $f' = f_{2\varepsilon}$.

The ε-optimality of f implies

$$
(3.2) \quad -l\varepsilon \leq \sum_{i=0}^{l-1} c_p(v_i, v_{i+1}) = p(v) - p(w) + \sum_{i=0}^{l-1} c(v_i, v_{i+1}).
$$

The 2ε-optimality of $f_{2\varepsilon}$ implies

$$
(3.3) \quad -2l\varepsilon \leq \sum_{i=0}^{l-1} c_{p_{2\varepsilon}}(v_{i+1}, v_i) = p_{2\varepsilon}(w) - p_{2\varepsilon}(v) + \sum_{i=0}^{l-1} c(v_{i+1}, v_i).
$$

But $\sum_{i=0}^{l-1} c(v_i, v_{i+1}) = -\sum_{i=0}^{l-1} c(v_{i+1}, v_i)$ by cost antisymmetry. Furthermore, $p(w) = p_{2\varepsilon}(w)$ since during *refine*, the initialization step is the only one that makes the excess of some vertices negative, and a vertex with negative excess has the same price as long as its excess remains negative. Adding inequalities (3.2) and (3.3) and rearranging terms thus gives

$$
p(v) \geq p_{2\varepsilon}(v) - 3l\varepsilon > p_{2\varepsilon}(v) - 3n\varepsilon.
$$ □

Now we count update operations. The following lemmas are analogous to Lemmas 2.2.4, 2.2.5 and 2.2.6.

3.4.6 Lemma. *The number of relabelings during an execution of refine is at most $3n$ per vertex and $3n(n-1)$ in total.*

3.4.7 Lemma. *The number of saturating pushes during an execution of refine is at most* $3nm$.

Proof. For an arc (v, w), consider the saturating pushes along this arc. Before the first such push can occur, vertex v must be relabeled. After such a push occurs, v must be relabeled before another such push can occur. But v is relabeled at most $3n$ times. Summing over all arcs gives the desired bound.   □

3.4.8 Lemma. *The number of nonsaturating pushes during one execution of refine is at most* $3n^2(m + n)$.

Proof. For each vertex v, let $\Phi(v)$ be the number of vertices reachable from v in the current admissible graph G_A. Let $\Phi = 0$ if there are no active vertices, and let $\Phi = \sum\{\Phi(v) | v \text{ is active}\}$ otherwise. Throughout the running of *refine*, $\Phi \geq 0$. Initially $\Phi \leq n$, since G_A has no arcs.

Consider the effect on Φ of update operations. A nonsaturating push decreases Φ by at least one, since G_A is always acyclic by Corollary 3.4.3. A saturating push can increase Φ by at most n, since at most one inactive vertex becomes active. If a vertex v is relabeled, Φ also can increase by at most n, since $\Phi(w)$ for $w \neq v$ can only decrease by Lemma 3.4.2. The total number of nonsaturating pushes is thus bounded by the initial value of Φ plus the total increase in Φ throughout the algorithm, *i.e.*, by $n + 3n^2(n - 1) + 3n^2m \leq 3n^2(m + n)$.   □

3.5 Efficient Implementation

As in the case of the generic maximum flow algorithm, we can obtain an especially efficient version of refine by choosing the order of the update operations carefully.

Local ordering of the basic operations is achieved using the data structures and the *discharge* operation of Section 2.3. The *discharge* operation is the same as the one in Figure 2.3, but uses the minimum-cost circulation versions of the update operations (see Figure 3.3). As in the maximum flow case, it is easy to show that *discharge* applies the relabeling operation correctly. The overall running time of *refine* is $O(nm)$ plus $O(1)$ per nonsaturating push plus the time needed for active vertex selection.

In the maximum flow case, one of the vertex selection methods we considered was the largest-label method. In the minimum-cost circulation case, a good method is *first-active* [49], which is a generalization of the largest-label method. The idea of the method is to process vertices in topological order with respect to the admissible graph.

The first-active method maintains a list L of all the vertices of G, in topological order with respect to the current admissible graph G_A, *i.e.*, if (v, w) is an arc of G_A, v appears before w in L. Initially L contains the vertices of G in any order. The method consists of repeating the following step until there are no active vertices: Find the first active vertex on L, say v, apply a *discharge* operation to v, and move v to the front of L if the *discharge* operation has relabeled v.

```
procedure first-active;
    let L be a list of all vertices;
    let v be the first vertex in L;
    while ∃ an active vertex do begin
        if v is active then begin
            discharge(v);
            if the discharge has relabeled v then
                move v to the front of L;
        end;
        else replace v by the vertex after v on L,
    end;
end.
```

Fig. 3.4 The *first-active* method

In order to implement this method, we maintain a *current vertex v* of L, which is the next candidate for discharging. Initially v is the first vertex on L. The implementation, described in Fig. 3.4, repeats the following step until there are no active vertices: If v is active, apply a discharge operation to it, and if this operation relabels v, move v to the front of L; otherwise (*i.e.*, v is inactive), replace v by the vertex currently after it on L. Because the reordering of L maintains a topological ordering with respect to G_A, no active vertex precedes v on L. This implies that the implementation is correct.

Define a *phase* as a period of time that begins with v equal to the first vertex on L and ends when the next relabeling is performed (or when the algorithm terminates).

3.5.1 Lemma. *The first-active method terminates after $O(n^2)$ phases.*

Proof. Each phase except the last ends with a relabeling operation. □

3.5.2 Theorem [50]. *The first-active implementation of refine runs in $O(n^3)$ time, giving an $O(n^3 \log(nC))$ bound for finding a minimum-cost circulation.*

Proof. Lemma 3.5.1 implies that there are $O(n^3)$ nonsaturating pushes (one per vertex per phase) during an execution of *refine*. The time spent manipulating L is $O(n)$ per phase, for a total of $O(n^3)$. All other operations require a total of $O(nm)$ time. □

A closely related strategy for selecting the next vertex to process is the *wave* method [41, 49, 50], which gives the same $O(n^3)$ running time bound for *refine*. (A similar pseudopolynomial algorithm, without the use of scaling and missing some of the implementation details, was developed independently in [11].) The only difference between the first-active method and the wave method is that the latter, after moving a vertex v to the front of L, replaces v by the vertex after the *old* position of v; if v is the last vertex on L, v is replaced by the first vertex on L.

As in the maximum flow case, the dynamic tree data structure can be used to obtain faster implementations of *refine*. A dynamic tree implementation of the generic version of *refine* analogous to the maximum flow algorithm discussed in Section 2.5 runs in $O(nm \log n)$ time [50]. A dynamic tree implementation of either the first-active method or the wave method runs in $O(nm \log(n^2/m))$ time [49]. In the latter implementation, a second data structure is needed to maintain the list L. The details are somewhat involved.

3.6 Refinement Using Blocking Flows

An alternative way to implement the refine subroutine is to generalize Dinic's approach to the maximum flow problem. Goldberg and Tarjan [49, 50] showed that refinement can be carried out by solving a sequence of $O(n)$ blocking flow problems on acyclic networks (*i.e.*, on networks for which the residual graph of the zero flow is acyclic); this extends Dinic's result, which reduces a maximum flow problem to $n-1$ blocking flow problems on layered networks. In this section we describe the Goldberg-Tarjan algorithm. At the end of this section, we make a few comments about blocking flow algorithms.

To describe the blocking flow version of refine we need some standard definitions. Consider a flow network (G, u, s, t). A flow f is *blocking* if any path from s to t *in the residual graph of zero flow* contains a saturated arc, *i.e.*, an arc (v, w) such that $u_f(v, w) = 0$. A maximum flow is blocking, but not conversely. A directed graph is *layered* if its vertices can be assigned integer layers in such a way that $layer(v) = layer(w) + 1$ for every arc (v, w). A layered graph is acyclic but not conversely.

An observation that is crucial to this section is as follows. Suppose we have a pseudoflow f and a price function p such that the vertices can be partitioned into two sets, S and $\overline{S}$, such that no admissible arc leads from a vertex in S to a vertex in $\overline{S}$; in other words, for every residual arc $(v, w) \in E_f$ such that $v \in S$ and $w \in \overline{S}$, we have $c_p(v, w) \geq 0$. Define p' to be equal to p on $\overline{S}$ and to $p - \varepsilon$ on S. It is easy to see that replacing p by p' does not create any new residual arc with reduced cost less than $-\varepsilon$. The blocking flow method augments by a blocking flow to create a partitioning of vertices as described above, and modifies the price function by replacing p by p'.

Figure 3.5 describes an implementation of *refine* that reduces ε by a factor of two by computing $O(n)$ blocking flows. This implementation reduces ε by a factor of two, saturates all admissible arcs, and then modifies the resulting pseudoflow (while maintaining ε-optimality with respect to the current price function) until it is a circulation. To modify the pseudoflow, the method first partitions the vertices of G into two sets S and $\overline{S}$, such that S contains all vertices reachable in the current admissible graph G_A from vertices of positive excess. Vertices in S have their prices decreased by ε. Next, an auxiliary network N is constructed by adding to G_A a source s, a sink t, an arc (s, v) of capacity $e_f(v)$ for each vertex v with $e_f(v) > 0$, and an arc (v, t) of capacity $-e_f(v)$ for each vertex with $e_f(v) < 0$. An arc $(v, w) \in E_A$ has capacity $u_f(v, w)$ in N. A blocking flow b on N is found.

---

procedure *refine*(ε, f, p);
  [initialization]
  $\varepsilon \leftarrow \varepsilon/2$;
  $\forall (v, w) \in E$ **do if** $c_p(v, w) < 0$ **then** $f(v, w) \leftarrow u(v, w)$;
  [loop]
  while f is not a circulation **do begin**
    $S \leftarrow \{v \in V | \exists u \in V$ such that $e_f(u) > 0$ and v is reachable from u in $G_A\}$;
    $\forall v \in S$, $p(v) \leftarrow p(v) - \varepsilon$;
    let N be the network formed from G_A by adding a source s, a sink t,
        an arc (s, v) of capacity $e_f(v)$ for each $v \in V$ with $e_f(v) > 0$, and
        an arc (v, t) of capacity $-e_f(v)$ for each $v \in V$ with $e_f(v) < 0$;
    find a blocking flow b on N;
    $\forall (v, w) \in E_A$, $f(v, w) \leftarrow f(v, w) + b(v, w)$;
  end;
  return(ε, f, p);
end.

Fig. 3.5 The blocking *refine* subroutine

Finally, the pseudoflow f is replaced by the pseudoflow $f'(v, w) = f(v, w) + b(v, w)$ for $(v, w) \in E$.

The correctness of the blocking flow method follows from the next lemma, which can be proved by induction on the number of iterations of the method.

3.6.1 Lemma. *The set S computed in the inner loop contains only vertices v with $e_f(v) \geq 0$. At the beginning of an iteration of the loop, f is an ε-optimal pseudoflow with respect to the price function p. Decreasing the prices of vertices in S by ε preserves the ε-optimality of f. The admissible graph remains acyclic throughout the algorithm.*

The bound on the number of iterations of the method follows from Lemma 3.4.5 and the fact that the prices of the vertices with deficit remain unchanged, while the prices of the vertices with excess decrease by ε during every iteration.

3.6.2 Lemma. *The number of iterations of the inner loop in the blocking flow implementation of refine is at most 3n.*

Since the running time of an iteration of the blocking flow method is dominated by the time needed for a blocking flow computation, we have the following theorem.

3.6.3 Theorem. *The blocking flow implementation of refine runs in $O(nB(n, m))$ time, giving an $O(nB(n, m)\log(nC))$ bound for finding a minimum-cost circulation, where $B(n, m)$ is the time needed to find a blocking flow in an acyclic network with n vertices and m arcs.*

The fastest known sequential algorithm for finding a blocking flow on an acyclic network is due to Goldberg and Tarjan [49] and runs in $O(m\log(n^2/m))$

time. Thus, by Theorem 3.6.3, we obtain an $O(nm \log(n^2/m) \log(nC))$ time bound for the minimum-cost circulation problem. This is the same as the fastest known implementation of the generic refinement method.

There is a crucial difference between Dinic's maximum flow algorithm and the blocking flow version of *refine*. Whereas the former finds blocking flows in layered networks, the latter must find blocking flows in acyclic networks, an apparently harder task. Although for sequential computation the acyclic case seems to be no harder than the layered case (the best known time bound is $O(m \log(n^2/m))$ for both), this is not true for parallel computation. The Shiloach-Vishkin PRAM blocking flow algorithm [93] for layered networks runs in $O(n^2 \log n)$ time using $O(n^2)$ memory and n processors. The fastest known PRAM algorithm for acyclic networks, due to Goldberg and Tarjan [47], runs in the same $O(n^2 \log n)$ time bound but uses $O(nm)$ memory and m processors.

4. Strongly Polynomial Algorithms Based on Cost-Scaling

4.1 Introduction

The question of whether the minimum-cost circulation problem has a strongly polynomial algorithm was posed in 1972 by Edmonds and Karp [22] and resolved only in 1984 by Tardos [96]. Her result led to the discovery of a number of strongly polynomial algorithms for the problem [31, 38, 81]. In this chapter we discuss several strongly polynomial algorithms based on cost scaling; in the next, we explore capacity-scaling algorithms, including strongly polynomial ones. Of the known strongly polynomial algorithms, the asymptotically fastest is the capacity-scaling algorithm of Orlin [81].

We begin by describing a modification of the generalized cost-scaling method that makes it strongly polynomial [49]. Then we describe the minimum-mean cycle-canceling algorithm of Goldberg and Tarjan [48]. This simple algorithm is a specialization of Klein's cycle-canceling algorithm [71]; it does not use scaling, but its analysis relies on ideas related to cost scaling.

4.2 Fitting Price Functions and Tight Error Parameters

In order to obtain strongly polynomial bounds on the generalized cost-scaling method, we need to take a closer look at the notion of ε-optimality defined in Section 3.2. The definition of ε-optimality motivates the following two problems:

1. Given a pseudoflow f and a constant $\varepsilon \geq 0$, find a price function p such that f is ε-optimal with respect to p, or show that there is no such price function (*i.e.*, that f is not ε-optimal).

2. Given a pseudoflow f, find the the smallest $\varepsilon \geq 0$ such that f is ε-optimal. For this ε, we say that f is ε-*tight*.

The problem of finding an optimal price function given an optimal circulation is the special case of Problem 1 with $\varepsilon = 0$. We shall see that the first problem can be reduced to a shortest path problem, and that the second problem requires the computation of a cycle of minimum average arc cost.

To address these problems, we need some results about shortest paths and shortest path trees (see e.g. [99]). Let G be a directed graph with a distinguished *source vertex* s from which every vertex is reachable and a cost $c(v, w)$ on every arc (v, w). For a spanning tree T rooted at s, the *tree cost function* $d : V \rightarrow R$ is defined recursively as follows: $d(s) = 0$, $d(v) = d(parent(v)) + c(parent(v), v)$ for $v \in V - \{s\}$, where $parent(v)$ is the parent of v in T. A spanning tree T rooted at s is a *shortest path tree* if and only if, for every vertex v, the path from s to v in T is a minimum-cost path from s to v in G, i.e., $d(v)$ is the cost of a minimum-cost path from s to v.

4.2.1 Lemma (see e.g. [99]). *Graph G contains a shortest path tree if and only if G does not contain a negative-cost cycle. A spanning tree T rooted at s is a shortest path tree if and only if $c(v, w) + d(v) \geq d(w)$ for every arc (v, w) in G.*

Consider Problem 1: given a pseudoflow f and a nonnegative ε, find a price function p with respect to which f is ε-optimal, or show that f is not ε-optimal. Define a new cost function $c^{(\varepsilon)} : E \rightarrow R$ by $c^{(\varepsilon)}(v, w) = c(v, w) + \varepsilon$. Extend the residual graph G_f by adding a single vertex s and arcs from it to all other vertices to form an *auxiliary graph* $G_{aux} = (V_{aux}, E_{aux}) = (V \cup \{s\}, E_f \cup (\{s\} \times V))$. Extend $c^{(\varepsilon)}$ to G_{aux} by defining $c^{(\varepsilon)}(s, v) = 0$ for every arc (s, v), where $v \in V$. Note that every vertex is reachable from s in G_{aux}.

4.2.2 Theorem. *Pseudoflow f is ε-optimal if and only if G_{aux} (or equivalently G_f) contains no cycle of negative $c^{(\varepsilon)}$-cost. If T is any shortest path tree of G_{aux} (rooted at s) with respect to the arc cost function $c^{(\varepsilon)}$, and d is the associated tree cost function, then f is ε-optimal with respect to the price function p defined by $p(v) = d(v)$ for all $v \in V$.*

Proof. Suppose f is ε-optimal. Any cycle in G_{aux} is a cycle in G_f, since vertex s has no incoming arcs. Let Γ be a cycle of length l in G_{aux}. Then $c(\Gamma) \geq -l\varepsilon$, which implies $c^{(\varepsilon)}(\Gamma) = c(\Gamma) + l\varepsilon \geq 0$. Therefore G_{aux} contains no cycle of negative $c^{(\varepsilon)}$-cost.

Suppose G_{aux} contains no cycle of negative $c^{(\varepsilon)}$-cost. Then, by Lemma 4.2.1, G_{aux} has some shortest path tree rooted at s. Let T be any such tree and let d be the tree cost function. By Lemma 4.2.1, $c^{(\varepsilon)}(v, w) + d(v) \geq d(w)$ for all $(v, w) \in E_f$, which is equivalent to $c(v, w) + d(v) - d(w) \geq -\varepsilon$ for all $(v, w) \in E_f$. But these are the ε-optimality constraints for the price function $p = d$. Thus f is ε-optimal with respect to p. $\square$

Using Theorem 4.2.2, we can solve Problem 1 by constructing the auxiliary graph G_{aux} and finding either a shortest path tree or a negative-cost cycle. Constructing G_{aux} takes $O(m)$ time. Finding a shortest path tree or a negative-cost cycle takes $O(nm)$ time using the Bellman-Ford shortest path algorithm (see e.g. [99]).

Let us turn to Problem 2: given a pseudoflow f, find the ε such that f is ε-tight. We need a definition. For a directed graph G with arc cost function c, the *minimum cycle mean* of G, denoted by $\mu(G, c)$, is the minimum, over all cycles Γ in G, of the mean cost of Γ, defined to be the total arc cost $c(\Gamma)$ of Γ divided by the number of arcs it contains. The connection between minimum cycle means and tight error parameters is given by the following theorem, which was discovered by Engel and Schneider [23] and later by Goldberg and Tarjan [50]:

4.2.3 Theorem [23]. *Suppose a pseudoflow f is not optimal. Then f is ε-tight for*
$$\varepsilon = -\mu(G_f, c).$$

Proof. Assume f is not optimal. Consider any cycle Γ in G_f. Let the length of Γ be l. For any ε, let $c^{(\varepsilon)}$ be the cost function defined above: $c^{(\varepsilon)}(v, w) = c(v, w) + \varepsilon$ for $(v, w) \in E_f$. Let ε be such that f is ε-tight, and let $\mu = \mu(G_f, c)$. By Theorem 4.2.2, $0 \le c^{(\varepsilon)}(\Gamma) = c(\Gamma) + l\varepsilon$, i.e., $c(\Gamma)/l \ge -\varepsilon$. Since this is true for any cycle Γ, $\mu \ge -\varepsilon$, i.e., $\varepsilon \ge -\mu$. Conversely, for any cycle Γ, $c(\Gamma)/l \ge \mu$, which implies $c^{(-\mu)}(\Gamma) \ge 0$. By Theorem 4.2.2, this implies $-\mu \ge \varepsilon$. $\qquad\square$

Karp [67] observed that the minimum mean cycle can be computed in $O(nm)$ time by extracting information from a single run of the Bellman-Ford shortest path algorithm. This gives the fastest known strongly polynomial algorithm for computing the minimum cycle mean. The fastest scaling algorithm is the $O(\sqrt{nm}\log(nC))$-time algorithm of Orlin and Ahuja [85]. Since we are interested here in strongly polynomial algorithms, we shall use Karp's bound of $O(nm)$ as an estimate of the time to compute a minimum cycle mean.

The following observation is helpful in the analysis to follow. Suppose f is an ε-tight pseudoflow and $\varepsilon > 0$. Let p be a price function such that f is ε-optimal with respect to p, and let Γ be a cycle in G_f with mean cost $-\varepsilon$. Since $-\varepsilon$ is a lower bound on the reduced cost of an arc in G_f, every arc of Γ must have reduced cost exactly $-\varepsilon$.

4.3 Fixed Arcs

The main observation that leads to strongly polynomial cost-scaling algorithms for the minimum-cost circulation problem is the following result of Tardos [96]: if the absolute value of the reduced cost of an arc is significantly greater then the current error parameter ε, then the value of any optimal circulation on this arc is the same as the value of the current circulation. The following theorem is a slight generalization of this result (to get the original result, take $\varepsilon' = 0$). This theorem can be used in two ways. The first is to drop the capacity constraint for an arc of large reduced cost. This approach is used in [96]. The second, discussed below, is to consider the arcs that have the same flow value in every ε-optimal circulation for the current value of the error parameter ε and to notice that the flow through these arcs will not change. This approach is used in [49, 48].

4.3.1 Theorem [96]. *Let $\varepsilon > 0$, $\varepsilon' \ge 0$ be constants. Suppose that a circulation f is ε-optimal with respect to a price function p, and that there is an arc $(v, w) \in E$*

such that $|c_p(v, w)| \geq n(\varepsilon + \varepsilon')$. Then, for any ε'-optimal circulation f', we have $f(v, w) = f'(v, w)$.

Proof. By antisymmetry, it is enough to prove the theorem for the case $c_p(v, w) \geq n(\varepsilon + \varepsilon')$. Let f' be a circulation such that $f'(v, w) \neq f(v, w)$. Since $c_p(v, w) > \varepsilon$, the flow f through the arc (v, w) must be as small as the capacity constraints allow, namely $-u(w, v)$, and therefore $f'(v, w) \neq f(v, w)$ implies $f'(v, w) > f(v, w)$. We show that f' is not ε'-optimal, and the theorem follows.

Consider $G_> = \{(x, y) \in E | f'(x, y) > f(x, y)\}$. Note that $G_>$ is a subgraph of G_f, and (v, w) is an arc of $G_>$. Since f and f' are circulations, $G_>$ must contain a simple cycle Γ that passes through (v, w). Let l be the length of Γ. Since all arcs of Γ are residual arcs, the cost of Γ is at least

$$c_p(v, w) - (l - 1)\varepsilon \geq n(\varepsilon + \varepsilon') - (n - 1)\varepsilon > n\varepsilon'.$$

Now consider the cycle $\overline{\Gamma}$ obtained by reversing the arcs on Γ. Note that $\overline{\Gamma}$ is a cycle in $G_< = \{(x, y) \in E | f'(x, y) < f(x, y)\}$ and is therefore a cycle in $G_{f'}$. By antisymmetry, the cost of $\overline{\Gamma}$ is less than $-n\varepsilon'$ and thus the mean cost of $\overline{\Gamma}$ is less than $-\varepsilon'$. But Theorem 4.2.3 implies that f' is not ε'-optimal.   □

To state an important corollary of Theorem 4.3.1, we need the following definition. We say that an arc $(v, w) \in E$ is ε-*fixed* if the flow through this arc is the same for all ε-optimal circulations.

4.3.2 Corollary [49]. *Let $\varepsilon > 0$, suppose f is ε-optimal with respect to a price function p, and suppose that (v, w) is an arc such that $|c_p(v, w)| \geq 2n\varepsilon$. Then (v, w) is ε-fixed.*

Define F_ε to be the set of ε-fixed arcs. Since the generalized cost-scaling method decreases ε, an arc that becomes ε-fixed stays ε-fixed. We show that when ε decreases by a factor of $2n$, a new arc becomes ε-fixed.

4.3.3 Lemma. *Assume $\varepsilon' \leq \frac{\varepsilon}{2n}$. Suppose that there exists an ε-tight circulation f. Then $F_{\varepsilon'}$ properly contains F_ε.*

Proof. Since every ε'-optimal circulation is ε-optimal, we have $F_\varepsilon \subseteq F_{\varepsilon'}$. To show that the containment is proper, we have to show that there is an ε'-fixed arc that is not ε-fixed.

Since the circulation f is ε-tight, there exists a price function p such that f is ε-optimal with respect to p, and there exists a simple cycle Γ in G_f every arc of which has reduced cost $-\varepsilon$. (See Section 4.2.) Since increasing f along Γ preserves ε-optimality, the arcs of Γ are not ε-fixed.

We show that at least one arc of Γ is ε'-fixed. Let f' be a circulation that is ε'-optimal with respect to some price function p'. Since the mean cost of Γ is $-\varepsilon$, there is an arc (v, w) of Γ with $c_{p'}(v, w) \leq -\varepsilon \leq -2n\varepsilon'$. By Corollary 4.3.2, the arc (v, w) is ε'-fixed.   □

In the next section we show how to use this lemma to get a strongly polynomial bound for a variation of the generalized cost-scaling method.

4.4 The Strongly Polynomial Framework

The minimum-cost circulation framework of Section 3.3 has the disadvantage that the number of iterations of *refine* depends on the magnitudes of the costs. If the costs are huge integers, the method need not run in time polynomial in n and m; if the costs are irrational, the method need not even terminate. In this section we show that a natural modification of the generalized cost-scaling approach produces strongly polynomial algorithms. The running time bounds we derived for algorithms based on the approach of Section 3.3 remain valid for the modified approach presented in this section. The main idea of this modification is to improve ε periodically by finding a price function that fits the current circulation better than the current price function. This idea can also be used to improve the practical performance of the method.

The changes needed to make the generalized cost-scaling approach strongly polynomial, suggested by Lemma 4.3.3, are to add an extra computation to the main loop of the algorithm and to change the termination condition. Before calling *refine* to reduce the error parameter ε, the new method computes the value λ and a price function p_λ such that the current circulation f is λ-tight with respect to p_λ. The strongly polynomial method is described on Figure 4.1. The value of λ and the price function p_λ in line (∗) are computed as described in Section 4.2. The algorithm terminates when the circulation f is optimal, *i.e.*, $\lambda = 0$.

```
procedure min-cost(V, E, u, c);
    [initialization]
    ε ← C;
    ∀v,  p(v) ← 0;
    ∀(v, w) ∈ E,  f(v, w) ← 0;
    [loop]
    while ε > 0 do begin
(*)     find λ and pλ such that f is λ-tight with respect to pλ;
        if λ > 0 then (ε, f, p) ← refine(λ, f, pλ)
        else return(f);
end.
```

Fig. 4.1 The strongly polynomial algorithm

The time to perform line (∗) is $O(nm)$. (See Section 4.2.) Since all the implementations of *refine* that we have considered have a time bound greater than $O(nm)$, the time per iteration in the new version of the algorithm exceeds the time per iteration in the original version by less than a constant factor. Since each iteration at least halves ε, the bound of $O(\log(nC))$ on the number of iterations derived in Chapter 3 remains valid, assuming that the costs are integral. For arbitrary real-valued costs, we shall derive a bound of $O(m \log n)$ on the number of iterations.

4.4.1 Theorem [48]. *The total number of iterations of the while loop in procedure* min-cost *is* $O(m \log n)$.

Proof. Consider a time during the execution of the algorithm. During the next $O(\log n)$ iterations, either the algorithm terminates, or the error parameter is reduced by a factor of $2n$. In the latter case, Lemma 4.3.3 implies that an arc becomes fixed. If all arcs become fixed, the algorithm terminates in one iteration of the loop. Therefore the total number of iterations is $O(m \log n)$.   □

The best strongly polynomial implementation of the generalized cost-scaling method [49], based on the dynamic tree implementation of *refine*, runs in $O(nm^2 \log(n^2/m) \log n)$ time.

4.5 Cycle-Canceling Algorithms

The ideas discussed in Sections 4.2 – 4.4 are quite powerful. In this section we use these ideas to show that a simple cycle-canceling algorithm of Klein [71] becomes strongly polynomial if a careful choice is made among possible cycles to cancel. Klein's algorithm consists of repeatedly finding a residual cycle of negative cost and sending as much flow as possible around the cycle. This algorithm can run for an exponential number of iterations if the capacities and costs are integers, and it need not terminate if the capacities are irrational [26]. Goldberg and Tarjan [48] showed that if a cycle with the minimum mean cost is canceled at each iteration, the algorithm becomes strongly polynomial. We call the resulting algorithm the *minimum-mean cycle-canceling* algorithm.

The minimum-mean cycle-canceling algorithm is closely related to the shortest augmenting path maximum flow algorithm of Edmonds and Karp [22]. The relationship is as follows. If a maximum flow problem is formulated as a minimum-cost circulation problem in a standard way, then Klein's cycle-canceling algorithm corresponds exactly to the Ford-Fulkerson maximum flow algorithm, and the minimum-mean cycle-canceling algorithm corresponds exactly to the Edmonds-Karp algorithm. The minimum-mean cycle-canceling algorithm can also be interpreted as a steepest descent method using the L_1 metric.

For a circulation f we define $\varepsilon(f)$ to be zero if f is optimal and to be the unique number $\varepsilon' > 0$ such that f is ε'-tight otherwise. We use $\varepsilon(f)$ as a measure of the quality of f. Let f be an arbitrary circulation, let $\varepsilon = \varepsilon(f)$, and let p be a price function with respect to which f is ε-optimal. Holding ε and p fixed, we study the effect on $\varepsilon(f)$ of a minimum-mean cycle cancellation that modifies f. Since all arcs on a minimum-mean cycle have negative reduced cost with respect to p, cancellation of such a cycle does not introduce a new residual arc with negative reduced cost, and hence $\varepsilon(f)$ does not increase.

4.5.1 Lemma. *A sequence of m minimum-mean cycle cancellations reduces $\varepsilon(f)$ to at most $(1 - 1/n)\varepsilon$, i.e., to at most $1 - 1/n$ times its original value.*

Proof. Let p a price function such that f is ε-tight with respect to p. Holding ε and p fixed, we study the effect on the admissible graph G_A (with respect to the circulation f and price function p) of a sequence of m minimum-mean cycle cancellations that modify f. Initially every arc $(v, w) \in E_A$ satisfies $c_p(v, w) \geq -\varepsilon$. Canceling a cycle all of whose arcs are in E_A adds only arcs of positive reduced cost to E_f and deletes at least one arc from E_A. We consider two cases.

Case 1: None of the cycles canceled contains an arc of nonnegative reduced cost. Then each cancellation reduces the size of E_A, and after m cancellations E_A is empty, which implies that f is optimal, i.e., $\varepsilon(f) = 0$. Thus the lemma is true in this case.

Case 2: Some cycle canceled contains an arc of nonnegative reduced cost. Let Γ be the first such cycle canceled. Every arc of Γ has a reduced cost of at least $-\varepsilon$, one arc of Γ has a nonnegative reduced cost, and the number of arcs in Γ is at most n. Therefore the mean cost of Γ is at least $-(1 - 1/n)\varepsilon$. Thus, just before the cancellation of Γ, $\varepsilon(f) \leq (1 - 1/n)\varepsilon$ by Theorem 4.2.3. Since $\varepsilon(f)$ never increases, the lemma is true in this case as well.   □

Lemma 4.5.1 is enough to derive a polynomial bound on the number of iterations, assuming that all arc costs are integers.

4.5.2 Theorem [48]. *If all arc costs are integers, then the minimum-mean cycle-canceling algorithm terminates after $O(nm \log(nC))$ iterations.*

Proof. The lemma follows from Lemmas 3.2.1 and 4.5.1 and the observation that the initial circulation is C-optimal.   □

To obtain a strongly polynomial bound, we use the ideas of Section 4.3. The proof of the next theorem uses the following inequality:

$$(1 - \frac{1}{n})^{n(\ln n + 1)} \leq \frac{1}{2n} \text{ for } n \geq 2.$$

4.5.3 Theorem [48]. *For arbitrary real-valued arc costs, the minimum-mean cycle-canceling algorithm terminates after $O(nm^2 \log n)$ cycle cancellations.*

Proof. Let $k = m(n\lceil \ln n + 1 \rceil)$. Divide the iterations into groups of k consecutive iterations. We claim that each group of iterations fixes the flow on a distinct arc (v, w), i.e., iterations after those in the group do not change $f(v, w)$. The theorem is immediate from the claim.

To prove the claim, consider any group of iterations. Let f be the flow before the first iteration of the group, f' the flow after the last iteration of the group, $\varepsilon = \varepsilon(f)$, $\varepsilon' = \varepsilon(f')$, and let p' be a price function for which f' satisfies the ε'-optimality constraints. Let Γ be the cycle canceled in the first iteration of the group. By Lemma 4.5.1, the choice of k implies that $\varepsilon' \leq \varepsilon(1 - \frac{1}{n})^{n\lceil \ln n + 1 \rceil} \leq$

$\frac{\varepsilon}{2n}$. Since the mean cost of Γ is $-\varepsilon$, some arc on Γ, say (v, w), must have $c_{p'}(v, w) \leq -\varepsilon \leq -2n\varepsilon'$. By Corollary 4.3.2, the flow on (v, w) will not be changed by iterations after those in the group. But $f(v, w)$ *is* changed by the first iteration in the group, which cancels Γ. Thus each group fixes the flow on a distinct arc. $\square$

4.5.4 Theorem [48]. *The minimum-mean cycle-canceling algorithm runs in* $O(n^2 m^3 \log n)$ *time on networks with arbitrary real-valued arc costs, and in* $O(n^2 m^2 \min\{\log (nC), m \log n\})$ *time on networks with integer arc costs.*

Proof. Immediate from Theorems 4.5.2 and 4.5.3. $\square$

Although the minimum-mean cycle-canceling algorithm seems to be of mostly theoretical interest, it has a variant that is quite efficient. This variant maintains a price function and, instead of canceling the minimum-mean-cost residual cycle, cancels residual cycles composed entirely of negative reduced cost arcs; if no such cycle exists, the algorithm updates the price function to improve the error parameter ε. An implementation of the algorithm using the dynamic tree data structure runs in $O(nm \log n \log(nC))$ time. See [48] for details.

The techniques used to analyze the minimum-mean cycle-canceling algorithm also provide an approach to making the primal network simplex algorithm for the minimum-cost circulation problem more efficient. Tarjan [100] has discovered a pivot rule that gives a bound of $O(n^{\frac{1}{2} \log n + O(1)})$ on the number of pivots and the total running time. This is the first known subexponential time bound. For the extended primal network simplex method in which cost-increasing as well as cost-decreasing pivots are allowed, he has obtained a strongly polynomial time bound. In contrast, the *dual* network simplex method is known to have strongly polynomial variants, as mentioned in Chapter 5.

Barahona and Tardos [7] have exhibited another cycle-canceling algorithm that runs in polynomial time. Their algorithm is based on an algorithm of Weintraub [104], which works as follows. Consider the improvement in the cost of a circulation obtained by canceling a negative cycle. Since a symmetric difference of two circulations can be decomposed into at most m cycles, canceling the cycle that gives the best improvement reduces the difference between the current and the optimal values of the cost by a factor of $(1 - 1/m)$. If the input data is integral, only a polynomial number of such improvements can be made until an optimal solution is obtained. Finding the cycle that gives the best improvement is NP-hard, however. Weintraub shows how to find a collection of cycles whose cancellation reduces the cost by at least as much as the best improvement achievable by canceling a single cycle. His method requires a superpolynomial number of applications of an algorithm for the assignment problem. Each such application yields a minimum-cost collection of vertex-disjoint cycles. Barahona and Tardos [7] have shown that the algorithm can be modified so that the required collection of cycles is found in at most m assignment computations. The resulting minimum-cost circulation algorithm runs in polynomial time.

5. Capacity-Scaling Algorithms

5.1 Introduction

In this chapter, we survey minimum-cost circulation algorithms based on capacity scaling. We concentrate on two algorithms: the first polynomial-time algorithm, that of Edmonds and Karp [22], who introduced the idea of scaling, and the algorithm of Orlin [81]. The latter is an extension of the Edmonds-Karp algorithm and is the fastest known strongly polynomial algorithm.

In this chapter we shall consider the (uncapacitated) transshipment problem, which is equivalent to the minimum-cost circulation problem. This simplifies the presentation considerably. We begin by describing a generic augmenting path algorithm, which we call the *minimum-cost augmentation* algorithm, that is the basis of most capacity-scaling algorithms. The algorithm is due to Jewel [62], Busacker and Gowen [14], and Iri [59]. The use of dual variables as described below was proposed independently by Edmonds and Karp [22] and Tomizawa [101]. The idea of the algorithm is to maintain a pseudoflow f that satisfies the complementary slackness constraints while repeatedly augmenting flow to gradually get rid of all excesses. Two observations justify this method: (1) augmenting flow along a minimum-cost path preserves the invariant that the current pseudoflow has minimum cost among those pseudoflows with the same excess function; (2) a shortest path computation suffices both to find a path along which to move flow and to find price changes to preserve complementary slackness.

The algorithm maintains a pseudoflow f and a price function p such that $c_p(v,w) \geq 0$ for every residual arc (v,w). The algorithm consists of repeating the . *augmentation step*, described in Figure 5.1, until every excess is zero. Then the current pseudoflow is optimal.

The correctness of this algorithm follows from the observation that the price transformation in the *augment* step preserves complementary slackness and makes the new reduced cost of any minimum-cost path from s to t equal to zero. One iteration of the *augment* step takes time proportional to that required by a single-source shortest path computation on a graph with arcs of non-negative cost. The fastest known strongly polynomial algorithm for this computation is Fredman and Tarjan's implementation [29] of Dijkstra's algorithm, which runs in

---

augment(s, t).
  Applicability: $e(s) > 0$ **and** $e(t) < 0$.
  Action: For every vertex v, compute $\pi(v)$, the minimum reduced cost of a residual path from s to v. For every vertex v, replace $p(v)$ by $p(v) + \pi(v)$. Move a positive amount of flow from s to t along a path of minimum reduced cost.

Fig. 5.1 The augmentation step

$O(n \log n + m)$ time. The problem can also be solved in $O(m \log \log C)$ time [86] or $O(n\sqrt{\log C} + m)$ time [2], where C is the maximum arc cost. Since we are mainly interested here in strongly polynomial algorithms, we shall use $O(n \log n + m)$ as our estimate of the time for each *augment* step.

Remark. The only reason to maintain prices in this algorithm is to simplify the minimum-cost path computations by guaranteeing that each such computation is done on a graph all of whose arc costs are non-negative. If prices are not maintained in this way, each shortest path computation takes $O(nm)$ time using the Bellman-Ford algorithm (see e.g. [99]).

5.2 The Edmonds-Karp Algorithm

The overall running time of the augmentation algorithm depends on the number of augmentation steps, which cannot be polynomially bounded without specifying their order more precisely. To impose an efficient order, Edmonds and Karp introduced the idea of capacity scaling, which Orlin [82] in his description of Edmonds-Karp algorithm reinterpreted as excess scaling. We shall present a modification of Orlin's version of the Edmonds-Karp algorithm.

The algorithm maintains a *scaling parameter* Δ such that the flow on every arc is an integer multiple of Δ. For a given pseudoflow f and value of the scaling parameter Δ, we denote the sets of vertices with large excesses and large deficits as follows:

$$(5.1) \quad \begin{aligned} S_f(\Delta) &= \{v \in V : e_f(v) > \Delta\}; \\ T_f(\Delta) &= \{v \in V : e_f(v) < -\Delta\}. \end{aligned}$$

The algorithm consists of a number of scaling phases. During a Δ-scaling phase the residual capacity of every arc is an integer multiple of Δ. The algorithm chooses vertices $s \in S_f(\Delta/2)$, $t \in T_f(\Delta/2)$, performs *augment*(s, t), which sends Δ units of flow along a minimum-cost path from s to t, and repeats such augmentations until either $S_f(\Delta/2)$ or $T_f(\Delta/2)$ is empty. Note that this can create a deficit in place of an excess, and vice versa. Vertices with the new excesses and deficits, however, are not in $S_f(\Delta/2) \bigcup T_f(\Delta/2)$. Then the algorithm halves Δ and begins a new scaling phase. The initial value of Δ is $2^{\lceil \log D \rceil}$. The algorithm terminates after the 1-scaling phase, assuming all supplies are integral. Figure 5.2 provides a detailed description of a phase of the algorithm.

The following lemma follows from the description of the algorithm.

5.2.1 Lemma. *During a Δ-scaling phase, the residual capacity of every arc is an integer multiple of Δ.*

5.2.2 Lemma. *Let f' be the pseudoflow at the end of a 2Δ-scaling phase. If $S_{f'}(\Delta) = \emptyset$, then there are at most $|S_{f'}(\Delta/2)|$ flow augmentations during the Δ-scaling phase. An analogous statement is true if $T_{f'}(\Delta) = \emptyset$ at the end of a 2Δ-scaling phase.*

while $S_f(\Delta/2) \neq \emptyset$ **and** $T_f(\Delta/2) \neq \emptyset$ **do begin**
    Choose $s \in S_f(\Delta/2)$ and $t \in T_f(\Delta/2)$. Perform *augment* (s,t), sending
    Δ units of flow along the augmenting path selected.
end;

Fig. 5.2 A phase of the Edmonds-Karp Algorithm.

Proof. By assumption, $S_{f'}(\Delta) = \emptyset$. Therefore, for every vertex $v \in V$, $e_f(v) < \Delta$ during the Δ-scaling phase. This implies that pushing Δ units of flow along a path from $s \in S_f(\Delta/2)$ to $t \in T_f(\Delta/2)$ removes s from $S_f(\Delta/2)$. Note that since $e_f(v) < -\Delta/2$ for $v \in T(\Delta/2)$, vertices with new excesses are not in $S_f(\Delta/2)$. $\square$

There are $\lceil \log D \rceil$ phases, each of which consists of up to n single-source shortest path computations on networks with non-negative arc costs. Thus we have the following results for networks with integral supplies:

5.2.3 Theorem [22]. *The Edmonds-Karp algorithm solves the transshipment problem in $O((n \log D)(n \log n + m))$ time, and the minimum-cost circulation problem in $O((m \log U)(n \log n + m))$ time.*

The excess-scaling idea can be combined with the idea of augmenting along approximately minimum-cost paths rather than exactly minimum-cost paths. Approximately minimum-cost paths can be defined using the ε-optimality notion discussed in Chapter 3. An appropriate combination of these ideas yields the *double scaling algorithm* of Ahuja, Goldberg, Orlin, and Tarjan [1]. An implementation of this algorithm based on dynamic trees has a time bound of $O(nm(\log \log U) \log(nC))$ for the minimum-cost circulation problem.

5.3 Strongly Polynomial Algorithms

After Tardos [96] discovered the first strongly polynomial algorithm, Fujishige [31] and independently Orlin [82] developed more efficient strongly polynomial algorithms. Orlin's algorithm [81], that is the main focus of this section, is an extension of these algorithms. These algorithms are based on the method of Section 5.2 and the following dual version of Theorem 4.3.1.

5.3.1 Theorem [31, 82]. *Let f be a pseudoflow, and let p be a price function such that the complementary slackness conditions are satisfied and $f(v,w) > \sum_{v:e_f(v)>0} |e_f(v)|$. Then every optimal price function p^* satisfies $c_{p^*}(v,w) = 0$.*

Proof. Let p^* be an optimal price function, and let f^* be a corresponding optimal pseudoflow. Assume by way of contradiction that $c_{p^*}(w,v) \neq 0$. Define the pseudoflow f' by $f'(x,y) = f(x,y) - f^*(x,y)$ (*i.e.*, we can obtain an optimal pseudoflow by augmenting f by f'). The Decomposition Theorem (Theorem 1.7.2) implies that f' can be decomposed into flows along a collection of cycles and a collection of paths from vertices with excess to vertices with deficits (both excesses and deficits are with respect to f). Furthermore, the Decomposition Theorem also implies that for any arc (x, y) that is not on one of the cycles, $|f'(x, y)| \leq \sum_{v:e_f(v)>0} e_f(v)$.

The arcs on the cycles of the decomposition are in E_f, and the arcs opposite to the ones on the cycles are in E_{f^*}. This implies that the cycles have zero cost, and that every arc on the cycles must have zero reduced cost with respect to both p and p^*. We have assumed that $c_{p^*}(v, w) \neq 0$; therefore (w, v) cannot be on such a cycle. Thus $f'(w, v) \leq \sum_{v:e_f(v)>0} e_f(v)$ and thus $f^*(v, w) > 0$. But $f^*(v, w) > 0$ and $c_{p^*}(v, w) \neq 0$ contradict the complementary slackness constraint.   □

The idea behind the strongly polynomial algorithms based on capacity-scaling is to contract the arcs satisfying the condition of Theorem 5.3.1. Let f and p be a pseudoflow and a price function, respectively, and let (v, w) be an arc satisfying the condition of Theorem 5.3.1. The reduced cost function c_p defines an equivalent problem with $c_p(v, w) = 0$. By the above theorem the optimal prices of the vertices v and w are the same. Define a transshipment problem on the network formed by contracting the arc (v, w), with cost function c_p, capacity function u, and demand function d such that the demand of the new vertex is $d(v) + d(w)$.

5.3.2 Theorem. *For any optimal price function p^* for the new problem, the price function $p'(r) = p(r) + p^*(r)$ for $r \in V$ (with $p^*(v) = p^*(w)$ defined to be the price of the new vertex) is optimal for the original problem.*

Proof. The optimal price function is the optimal solution of a linear program, the linear programming dual of the minimum-cost flow problem. The price function $p' = p + p^*$ is the optimal solution of same linear program with the additional constraint $c(v, w) + p'(v) - p'(w) = 0$. By Theorem 5.3.1, the two linear programs have the same set of optimal solutions.   □

We shall describe a simple strongly polynomial algorithm based on this idea. It is a variation of Fujishige's algorithm [31], though our description is simplified by using ideas from [81]. One iteration of the algorithm consists of running the Edmonds-Karp algorithm for the first $2 \log n$ scaling phases starting with $\Delta = \max_{v \in V} |d(v)|$ (we cannot find the power of two used by the Edmonds-Karp algorithm in strongly polynomial time), and then contracting all arcs that satisfy the condition of Theorem 5.3.1. The next iteration considers the contracted network. We shall prove that each iteration will contract at least one arc. The algorithm terminates when the current pseudoflow is a feasible solution. Theorem 5.3.2 guarantees that the price function found at this point is optimal. Given an optimal price function, the optimal flow can be found by a single maximum flow computation. (See Figure 5.3 for a more detailed description.)

The following theorem implies that at least one arc is contracted at each iteration of the inner loop.

Step 1 Run the Edmonds-Karp algorithm for the first $\lceil 2\log n \rceil$ scaling phases with cost function c_p, starting with $\Delta = \max_{v \in V}|d(v)|$. Let f' be the pseudoflow and let p' be the price function found by the algorithm.

Step 2 Contract every arc (v,w) with $|f(v,w)| > n\Delta$ and update the price function by setting $p(v) \leftarrow p(v) + p'(v)$ for all $v \in V$ (where all vertices $v \in V$ contracted into the same vertex v' of the current network have the same price $p'(v) = p'(v')$).

Fig. 5.3 The inner loop of the simple strongly polynomial algorithm.

5.3.3 Theorem [81]. *Let f and Δ be the pseudoflow and the scaling parameter at the end of an iteration of the algorithm of Figure 5.3. Then $\sum_{v \in V}|e_f(v)| < n\Delta$. Furthermore, if f is not feasible, then there exists an arc (v,w) with $|f(v,w)| \geq n\Delta$.*

Proof. After the Δ-scaling phase either $S_f(\Delta/2)$ or $T_f(\Delta/2)$ is empty. For a pseudoflow f, the excesses sum to zero. Therefore $\sum_{v \in V}|e_f(v)| < n\Delta$. Also, every excess is less than $n\Delta$. After $\lceil 2\log n \rceil$ scaling phases $\Delta \leq (\max_{w \in V}|d(w)|) \cdot n^{-2}$. For a vertex v whose demand has the maximum absolute value this implies that

$$(5.2) \qquad \left| \sum_{w \in V} f(v,w) \right| \geq |d(v)| - n\Delta \geq n(n-1)\Delta.$$

Consequently, at least one arc incident to v carries at least $n\Delta$ flow. $\square$

One iteration takes $O(n(n\log n + m)\log n)$ time. Each iteration contracts at least one arc, and therefore there are at most n iterations.

5.3.4 Theorem. *The algorithm of Figure 5.3 solves the transshipment problem in $O(n^2 \log n(n\log n + m))$ time, and the minimum-cost circulation problem in $O(m^2 \log n(n\log n + m))$ time.*

This simple algorithm wastes a lot of time by throwing away the current flow after each contraction. If there are several vertices with demand close to the maximum then it might make sense to run somewhat more than $2\log n$ scaling phases. The proof of Theorem 5.3.3 would then apply to more than one vertex, and more than one arc could be contracted in an iteration. This idea does not help much if there are not enough vertices with close-to-maximum demand. On the other hand, by Lemma 5.2.2 the number of shortest path computations during a phase can be bounded in terms of the number of vertices with relatively large demand (it is bounded either by $|S(\Delta/2)|$ or by $|T(\Delta/2)|$). One could try to find a point for stopping the scaling algorithm that balances the number of shortest path computations done and the number of arcs contracted.

Galil and Tardos [38] developed an $O(n^2(n\log + m)\log n)$-time minimum-cost circulation algorithm based on this idea. In fact, the balancing technique of Galil

Step 1 Let $\Delta = \max_{v \in V} |d(v)|$. **If** $\Delta = 0$ **then** STOP.
Step 2 while $S_f(\Delta/2) \neq \emptyset$ **and** $T_f(\Delta/2) \neq \emptyset$ **do begin**
        Choose $s \in S_f(\Delta/2)$ and $t \in T_f(\Delta/2)$. Perform *augment(s,t)*, sending
        Δ units of flow along the augmenting path selected. Contract every
        arc (v, w) with $f(v, w) > 4n\Delta$.
   end.
Step 3 If f is zero on all uncontracted arcs go to Step 1, otherwise let $\Delta = \Delta/2$
        and go to Step 2.

Fig. 5.4 Orlin's Algorithm

and Tardos [38], together with Orlin's [81] proof of Theorem 5.3.3, can be used to give an $O(n(n \log n + m) \log n)$-time algorithm for the transshipment problem. Instead of pursuing this approach further, however, we shall present a variant of a much simpler algorithm due to Orlin [81] that has same efficiency and does not rely on delicate balancing. This algorithm contracts arcs during the scaling phases, without starting the scaling from scratch after each contraction; the algorithm finds an optimal flow directly, without using an optimal price function. The algorithm is a modification of the Edmonds-Karp algorithm, but runs in $O(n \log n)$ iterations instead of the $O(n \log D)$ iterations of the Edmonds-Karp algorithm.

Orlin's algorithm runs the Edmonds-Karp scaling algorithm starting with $\Delta = \max_{v \in V} |d(v)|$ and contracts all arcs that carry at least $4n\Delta$ flow. The scaling algorithm continues as long as the current pseudoflow is non-zero on some uncontracted arc. When the pseudoflow is zero on every arc, the scaling is restarted by setting Δ to be the maximum absolute value of a demand. Roughly speaking, a vertex v is the start of at most $2 \log n$ shortest path computations during the algorithm. For a vertex v of the original graph, this follows from the fact that $v \notin S_f(\Delta) \cup T_f(\Delta)$ unless $|d(v)| \geq \Delta$, and therefore an arc incident to v will be contracted at most $2 \log n$ scaling phases after v first served as a starting vertex. (See the proof of Theorem 5.3.3.) This may not be true for the contracted vertices, however. A vertex created by a contraction may have very small demand relative to its current excess.

We say that a vertex v is *active* if $v \in S_f(\Delta/2) \cup T_f(\Delta/2)$. A vertex v is *activated* by a Δ-scaling phase if v is not active at the end of the 2Δ-scaling phase, and it becomes active during the Δ-scaling phase (*i.e.*, either it becomes active at the beginning of the phase or the vertex is created by contraction during the phase). We can prove the following lemma using a proof similar to that of Lemma 5.2.2.

5.3.5 Lemma. *The number of shortest path computations during a phase is bounded by the number of vertices activated by the phase.*

5.3.6 Theorem. *A vertex can be activated at most $\lceil 2 \log n \rceil + 1$ times before it is contracted.*

Proof. A vertex can be activated once due to contraction. When a vertex v is activated for the second time, it must already exist at the end of the previous scaling phase, when it was not active. Let Δ be the scaling parameter in the phase when v is activated for the second time. At the beginning of this phase, $\Delta/2 < |e_f(v)| \leq \Delta$. But $d(v) - e_f(v)$ is an integer multiple of 2Δ. This implies that $\Delta/2 \leq |e_f(v)| \leq |d(v)|$. After $O(\log n)$ more scaling phases the scaling parameter will be less than $|d(v)|/4n^2$. Using an argument analogous to the proof of Theorem 5.3.3 we can conclude that some arc incident to v is contracted.   □

The previous lemma and theorem bound the number of shortest path computations during the algorithm. All other work done is linear per scaling phase. At least one arc is contracted in each group of $O(\log n)$ scaling phases. Therefore, there are at most $O(n \log n)$ scaling phases.

5.3.7 Theorem [81]. *Orlin's algorithm solves the transshipment problem in $O(n \log \min\{n, D\}(n \log n + m))$ time, and the minimum-cost circulation problem in $O(m \log \min\{n, U\}(n \log n + m))$ time.*

Remark. In contrast to the simpler strongly polynomial algorithm discussed earlier, Orlin's algorithm constructs an optimal pseudoflow directly, without first constructing an optimal price function. To see this, consider the amount of flow moved during the Δ-scaling phase for some value Δ. Lemma 5.3.5 gives a $2n\Delta$ bound. Suppose an arc is contracted during the Δ-scaling phase. The overall amount of flow that is moved after this contraction can be bounded by $4n\Delta$. Therefore the pseudoflow constructed by the algorithm is feasible in the uncontracted network.

Orlin [82] has observed that capacity scaling ideas can be used to guide pivot selection in the dual network simplex algorithm for the transshipment problem. More recently Orlin, Plotkin and Tardos (personal communication, 1989) have obtained an $O(nm \log \min(n, D))$ bound on the number of pivots based on such a pivot selection strategy.

6. The Generalized Flow Problem

6.1 Introduction

The generalized flow problem models the following situation in financial analysis. An investor wants to take advantage of the discrepancies in the prices of securities on different stock exchanges and of currency conversion rates. His objective is to maximize his profit by trading on different exchanges and by converting currencies. The generalized flow problem, considered in this chapter, models the above situation, assuming that a bounded amount of money is available to the investor and that bounded amounts of securities can be traded without affecting the prices. Vertices of the network correspond to different currencies and securities, and arcs correspond to possible transactions.

The generalized flow problem was first considered by Jewell [62]. This problem is very similar to the minimum-cost circulation problem, and several of the early minimum-cost circulation algorithms have been adapted to this problem. The first simple combinatorial algorithms were developed by Jewell [62] and Onaga [80]. These algorithms are not even pseudopolynomial, however, and for real-valued data they need not terminate. Several variations suggested in the early 70's result in algorithms running in finite (but exponential) time. The paper by Truemper [102] contains an excellent summary of these results.

The generalized flow problem is a special case of linear programming, and therefore it can be solved in polynomial time using any general-purpose linear programming algorithm, such as the ellipsoid method [70] and the interior-point algorithms [66, 89]. Kapoor and Vaidya [65] developed techniques to use the structure of the matrices that arise in linear programming formulations of network flow problems to speed up interior-point algorithms. The first two polynomial-time *combinatorial* algorithms were developed by Goldberg, Plotkin, and Tardos [43]. We shall review one of their algorithms in detail and sketch the other one.

The current fastest algorithm for the generalized flow problem, due to Vaidya, is based on an interior-point method for linear programming and runs in $O(n^2 m^{1.5} \log(nB))$ time [103]. This algorithm combines the ideas from the paper of Kapoor and Vaidya [65] with the current fastest linear programming algorithm (which uses fast matrix multiplication). The two combinatorial algorithms due to Goldberg, Plotkin and Tardos [43] run in $O(mn^2(m + n \log n) \log n \log B)$ and $O(m^2 n^2 \log n \log^2 B)$ time, respectively. More recently Maggs (personal communication, 1989) improved the latter bound through better balancing to

$$O\left(n^2 m^2 \log^2 B \frac{\log n}{\log((n^2 \log n)/m) + \log \log B}\right).$$

6.2 Simple Combinatorial Algorithms

Jewell [62] and Onaga [79] suggested solving the generalized flow problem by using adaptations of augmenting path algorithms for the maximum flow and minimum-cost circulation problems. Theorem 6.3.4 is the basis of Onaga's augmenting path algorithm for the restricted problem. Starting from the zero flow, this algorithm iteratively augments the flow along a flow-generating cycle in the residual graph, thereby increasing the excess at the source. The structure of the restricted problem makes it possible to find the most efficient flow-generating cycle, i.e., the residual flow-generating cycle with highest-gain. The highest-gain cycle consists of an arc (r, s) and a highest-gain simple path from s to r for some vertex r. The highest-gain simple path to r is the shortest path, if the length of each arc (v, w) is defined as $l(v, w) = -\log \gamma(v, w)$. Such a shortest path exists since deleting the arcs entering the source from the residual graph of the initial zero flow yields a graph with no negative cycles. The main observation of Onaga is that if the augmentation is done using the most-efficient flow-generating cycle, then all flow-generating cycles in the residual graph of the resulting generalized flow pass through the source. Onaga's algorithm iteratively augments the generalized flow along the most efficient flow-generating cycle in the residual graph. This algorithm maintains the invariant that every vertex is reachable from s in the current residual graph, a property that can be verified by induction on the number of augmentations.

Consider the special case of a network with two distinguished vertices $s, t \in V$ and with all gains other than $\gamma(t, s)$ and $\gamma(s, t)$ equal to one. The generalized flow problem in this network is equivalent to the maximum flow problem, and Onaga's algorithm specializes to the Ford-Fulkerson maximum flow algorithm. Consequently, the algorithm does not run in finite time.

The next algorithm we describe uses a maximum flow computation as a subroutine. To describe this algorithm, we need to introduce a relabeling technique of Glover and Klingman [39]. This technique can be motivated as follows. Recall the financial analysis interpretation of the generalized flow problem, in which vertices correspond to different securities or currencies, and arcs correspond to possible transactions. Suppose one country decides to change the unit of currency. (For example, Great Britain could decide to introduce the penny as the basic currency unit, instead of the pound, or Italy could decide to erase a couple of 0's at the end of its bills.) This causes an appropriate update of the exchange rates. Some of the capacities change as well (a million $\mathscr{L}$ limit on the $\mathscr{L}$ – DM exchange would now read as a limit of 100 million pennies). It is easy to see that such a relabeling defines an equivalent problem. Such a relabeling can be used, for example, to normalize local units of currency to current market conditions.

To formally define the *relabeled problem*, let $\mu(v) \in \mathbf{R}^+$ denote the number of old units corresponding to each new unit at vertex $v \in V$. Given a function μ, we shall refer to $\mu(v)$ as the *label* of v.

Definition: For a function $\mu : V \longmapsto \mathbf{R}^+$ and a network $N = (V, E, \gamma, u)$, the *relabeled network* is $N_\mu = (V, E, \gamma_\mu, u_\mu)$, where the *relabeled capacities* and the *relabeled gains* are defined by

$$
\begin{aligned}
u_\mu(v, w) &= u(v, w)/\mu(v) \\
\gamma_\mu(v, w) &= \gamma(v, w)\mu(v)/\mu(w).
\end{aligned}
$$

For a generalized pseudoflow g and a labeling μ, the generalized pseudoflow corresponding to g in the relabeled network is $g_\mu(v, w) = g(v, w)/\mu(v)$, the *relabeled residual capacity* is defined by $u_{g,\mu}(v, w) = (u(v, w) - g(v, w))/\mu(v)$, and the *relabeled excess* by $e_{g,\mu}(v) = e_g(v)/\mu(v)$. The corresponding pseudoflows have the same residual graph.

Now we present a *canonical relabeling*. The residual graph of a generalized pseudoflow g can be canonically relabeled if every vertex $v \in V$ is reachable from the source and every flow-generating cycle in the residual graph contains the source. For a vertex $v \in V$, the canonical label $\mu(v)$ is defined to be the gain of a highest-gain simple path from s to v in the residual graph. That is, one new unit corresponds to the amount of flow that can reach the vertex v if one old unit of flow is pushed along a most-efficient simple path in the residual graph from s to v, ignoring capacity restrictions along the path. Observe that in a restricted network, the highest-gain paths from s to each other vertex can be found using any single-source shortest path algorithm.

Step 1 Find μ, a canonical labeling from the source.
If $\gamma_\mu(v, w) \leq 1$ on every arc of the residual graph of the current generalized flow g, then STOP (the current flow is optimal).

Step 2 Let $\alpha = \max_{(v,w) \in E_g} \gamma_\mu(v, w)$. Consider the network G' consisting of all arcs (v, w) with $\gamma_\mu(v, w) = 1$, and all arcs in $A = \{(v, s) | \gamma_\mu(v, s) = \alpha\}$ and their opposites.

Find a (standard) circulation f' in G' that maximizes $\sum_{(v,s) \in A} f'(v, s)$, the flow into s.

Step 3 Let g' be the generalized flow corresponding to f', i.e, $g'(v, w) = f'(v, w)$ if $v \neq s$ and $g'(s, v) = -\gamma_\mu(v, s) f'(v, s)$.
Update the current solution by setting $g(v, w) = g(v, w) + g'(v, w)\mu(v) \; \forall (v, w) \in E$.

Fig. 6.1 Inner loop of Truemper's algorithm

6.2.1 Theorem. *After a canonical relabeling, the following properties hold: Every arc $(v, w) \in E_g$ such that $w \neq s$ has $\gamma_\mu(v, w) \leq 1$; there exists a path from s to every other vertex r in the residual graph with $\gamma_\mu(v, w) = 1$ for all arcs (v, w) on the path; the most efficient flow-generating cycles each consist of an arc $(r, s) \in E_g$ and an (s, r)-path for some $r \in V$, such that $\gamma_\mu(v, w) = 1$ along the path and $\gamma_\mu(r, s) = \max(\gamma_\mu(v, w) : (v, w) \in E_g)$.*

Next we describe a simple finite algorithm, due to Truemper [102]. The algorithm, described in Figure 6.1, is a refinement of Onaga's algorithm in which augmentation along all of the maximum gain cycles is done simultaneously. The algorithm maintains a generalized pseudoflow g whose residual graph has every vertex reachable from the source and has all flow-generating cycles passing through the source. The algorithm first canonically relabels the residual graph. Now the highest-gain residual cycles have maximum relabeled gain on arcs entering the source and have a relabeled gain of one on all other arcs. Consider the subgraph induced by arcs with relabeled gain of one and arcs of maximum relabeled gain entering the source. A circulation that maximizes the sum of the flow on the latter arcs can be found by a maximum flow computation. This circulation gives an augmentation of the current generalized flow g. After such an augmentation, all gain cycles in the residual graph pass through the source, and the maximum gain of a flow-generating cycle in the residual graph is decreased. Therefore, this algorithm runs in finite time.

Truemper's algorithm is in some sense an analog of Jewell's [62] minimum-cost flow algorithm, which augments the flow along all of the cheapest augmenting paths at once using a maximum flow subroutine. Using the same network as Zadeh [106] used for the minimum-cost flow problem, Ruhe [91] gave an example on which Truemper's algorithm takes exponential time.

Goldberg, Plotkin, and Tardos [43] gave two algorithms, one that uses a minimum-cost flow computation as a subroutine, and one that builds on ideas from several maximum and minimum-cost flow algorithms. In the next section we describe the first algorithm in detail and very briefly outline the second one.

6.3 Polynomial-Time Combinatorial Algorithms

The main idea of the first polynomial-time algorithm of Goldberg-Plotkin-Tardos is best described by contrasting the algorithm with Truemper's. In each iteration, both algorithms solve a simpler flow problem, and interpret the result as an augmentation in the generalized flow network. Truemper's algorithm is slow because at each iteration it considers only arcs with unit relabeled gain and some of the arcs adjacent to the source, disregarding the rest of the graph completely. The Goldberg-Plotkin-Tardos algorithm, which we shall call *algorithm MCF*, considers all arcs. It assigns a cost $c(v, w) = -\log \gamma_\mu(v, w)$ to each arc (v, w) and solves the resulting minimum-cost circulation problem (disregarding gains).

The *interpretation* of a pseudoflow f is a generalized pseudoflow g, such that $g(v, w) = f(v, w)$ if $f(v, w) \geq 0$ and $g(v, w) = -\gamma_\mu(w, v)f(w, v)$ otherwise. In Truemper's algorithm, the interpretation of a feasible circulation on the subnetwork G' is a feasible generalized flow on the original network. In Algorithm MCF, however, the interpretation of a minimum-cost circulation is a generalized pseudoflow. Arcs of the flow that have a relabeled gain of less than 1 produce vertices with deficits in the interpretation. A connection between a pseudoflow f and its interpretation is given by the following lemma.

6.3.1 Lemma. *The residual graphs of a pseudoflow f and its interpretation g as a generalized pseudoflow are the same.*

The first iteration of the algorithm solves a minimum-cost circulation problem, and it creates excess at the source and deficits at various other vertices. Each subsequent iteration uses some of the excess at the source to balance the deficits created by the previous iterations by solving a capacitated transshipment problem. More precisely, Algorithm MCF, shown in Figure 6.2, maintains a generalized pseudoflow g in the original (non-relabeled) network, such that the excess at every vertex other than the source is non-positive. The algorithm proceeds in iterations. At each iteration it canonically relabels the residual graph, solves the corresponding capacitated transshipment problem in the relabeled network, and interprets the result as a generalized augmentation.

The most important property of a minimum-cost pseudoflow, for this application, is that its residual graph has no negative cycles. This implies (by Lemma 6.3.1) that the residual graphs of the generalized pseudoflows produced by the algorithm in Step 3 have no flow-generating cycles. The following lemma (analogous to Lemma 1.6.1) implies that this is enough to ensure that the new residual graph can be canonically relabeled.

6.3.2 Lemma. *Let g be a generalized pseudoflow in a restricted network. If the residual graph of G_g contains no flow-generating cycles, and the excess at every vertex other than s is non-positive, then every vertex is reachable from s in G_g.*

Step 1 Find μ, a canonical labeling of the residual graph of the generalized pseudoflow g. If $\gamma_\mu(v, w) \leq 1$ on every arc of the residual graph and $\forall v \in (V - \{s\}) : e_{g,\mu}(v) = 0$, then HALT (the current flow is optimal).

Step 2 Introduce costs $c(v, w) = -\log \gamma_\mu(v, w)$ on the arcs of the network. Solve the capacitated transshipment problem in the residual relabeled network of G with demands $d(v) = e_{g,\mu}(v)$ for $v \in V - \{s\}$.

Step 3 Let g' be the interpretation of f'. Update the current solution by setting $g(v, w) = g(v, w) + g'(v, w)\mu(v)$ $\forall (v, w) \in E$.

Fig. 6.2 Inner loop of Algorithm MCF

The relabeled gain factors are at most 1 (except for arcs entering the source in the first iteration); therefore the flow computation creates deficits, but no excesses at vertices other than the source. Using the decomposition of pseudoflows (Theorem 1.7.3), one can prove that in each relabeled network there exists a flow satisfying all the deficits. These properties are summarized in the following lemma.

6.3.3 Lemma. *For a generalized pseudoflow g that is constructed by Step 3, the following properties hold: The residual graph of g has no flow-generating cycle; canonical relabeling applies to the residual graph of g; all excesses, except the excess of the source, are non-positive; the pseudoflow required in Step 2 of the next iteration exists.*

The algorithm terminates once a generalized flow has been found. Throughout the computation the residual graph contains no flow-generating cycles; therefore (by Theorem 1.6.2) if the algorithm ever finds a generalized flow, then this flow is optimal.

Next we bound the number of iterations of the algorithm. Consider the generalized pseudoflow g existing at the beginning of an iteration. Because there are deficits but no flow-generating cycles in the residual graph, the current excess at the source is an overestimate of the value of the maximum possible excess. It is easy to see that, after a canonical relabeling, the sum of the deficits at all the vertices other than the source is a lower bound on the amount of the overestimate. We use this value, $Def(g, \mu) = \sum_{v \neq s}(-e_{g,\mu}(v))$, as a measure of the proximity of a generalized pseudoflow to an optimal generalized flow. It is not too hard to show that if $Def(g, \mu)$ is smaller than L^{-1} then the algorithm terminates in one more iteration. (Recall that L is the product of denominators of arc gains.)

6.3.4 Theorem [43]. *If $Def(g, \mu) < L^{-1}$ before Step 2 of an iteration, then the algorithm produces a generalized flow in Step 3.*

An important observation is that the labels μ are monotonically decreasing during the algorithm. The decrease in the labels is related to the price function associated with the minimum-cost pseudoflow computation. Let p' denote the optimal price function associated with the pseudoflow f' found in Step 2. Assume, without loss of generality, that $p'(s) = 0$.

6.3.5 Lemma. *For each vertex v, the canonical relabeling in Step 1 of the next iteration decreases the label $\mu(v)$ by at least a factor of $2^{-p'(v)}$.*

The key idea of the analysis is to distinguish two cases: Case 1, in which the pseudoflow f' is along "cheap" paths, (*i.e.*, $p'(v)$ is small, say $p'(v) < \log 1.5$, for every $v \in V$); and Case 2, in which there exists a vertex v such that $p'(v)$ is "large" ($\geq \log 1.5$). In the first case, $Def(f, \mu)$ decreases significantly; in the second case, Lemma 6.3.5 implies that at least one of the labels decreases significantly. The label of a vertex v is the gain of the current most efficient path from s to v. This limits the number of times Case 2 can occur. Theorem 6.3.4 can be used to guarantee that Case 1 cannot occur too often. The following lemma provides a tool for estimating the total deficit created when interpreting a minimum-cost pseudoflow as a generalized pseudoflow.

6.3.6 Lemma. *Let f' be a pseudoflow along a simple path P from s to some other vertex v that satisfies one unit of deficit at v. Let g' be the interpretation of f' as a generalized pseudoflow. Assume that all relabeled gains along the path p are at most 1, and denote them by $\gamma_1, \ldots, \gamma_k$. Then after augmenting by g', the unit of deficit at v is replaced by deficits on vertices along P that sum up to at most $(\prod_{1 \leq i \leq k} \gamma_i)^{-1} - 1$.*

Proof. The deficit created at the ith vertex of the path is $(1 - \gamma_i)$, for $i = 1, \ldots, k$. Using the assumption that the gain factors along the path are at most 1, the sum of the deficits can be bounded by

$$\sum_{i=1}^{k} (1 - \gamma_i) \leq \sum_{i=1}^{k} \frac{1 - \gamma_i}{\prod_{j=1}^{i} \gamma_j} = \frac{1}{\prod_{i=1}^{k} \gamma_i} - 1. \qquad \square$$

This lemma can be used to bound the value of $Def(g, \mu)$ after an application of Step 3. Let p' be an optimal price function associated with the pseudoflow f', such that $p'(s) = 0$. Let $\beta = \max_{v \in V} p'(v)$ be the maximum price. Using the pseudoflow decomposition theorem (Theorem 1.7.2), one can show that a minimum-cost pseudoflow f' can be decomposed into flows along paths from the source s to the other vertices and cycles in the residual graph of g, such that the cost of the cycles is zero and the cost of each path ending at a vertex v is at most $p'(v)$. To bound the amount of deficit created by the interpretation, we shall consider the interpretation of the flow on the cycles and the paths one-by-one. Since the cycles consists of zero-cost arcs only, the interpretation of the flow along these cycles does not create deficits. When interpreting flow along a path from s to v, the deficit at v is replaced by deficits that sum to at most $2^{p'(v)} - 1 \leq 2^\beta - 1$ times the deficit at v satisfied by this path. This proves the following lemma.

6.3.7 Lemma. *The value of Def(g, μ) after an application of Step 3 can be bounded by $2^\beta - 1$ times its value before the step. In particular, if $p'(v) < \log 1.5$ for every vertex v, then Def(g, μ) decreases by a factor of 2.*

The remaining difficulty is the fact that the function $Def(g, \mu)$ can increase, both when Case 2 applies in Step 3 and due to the relabeling in Step 1 (changing the currency unit from $\mathscr{L}$ to penny increases a 5 $\mathscr{L}$ deficit to 500 pennies). The increase in $Def(g, \mu)$ can be related to the decrease in the labels, however. The deficit at a vertex increases in Step 1 by a factor of α if and only if the label of this vertex decreases by the same factor. The increase in $Def(g, \mu)$ during Step 3 is bounded by 2^β, where $\beta = \max p'(v)$, by Lemma 6.3.7. By Lemma 6.3.5, this means that $\mu(v)$ for some vertex v decreases by at least β during Step 1 of the next iteration. Hence, in both cases, if $Def(g, \mu)$ increases by α during a step, then there exists a vertex v for which $\mu(v)$ decreases by at least α during the next execution of Step 1. Let $T \leq B^n$ denote an upper bound on the gain of any simple path. The label of a vertex v is the gain of a simple path from s to v in the residual graph. Therefore the labels are at least T^{-1} and at most T, and the label of a vertex cannot decrease by more than a factor of T^2 during the algorithm. This gives the following lemma.

6.3.8 Lemma. *The growth of the function Def(g, μ) throughout an execution of the algorithm is bounded by a factor of $T^{O(n)} = B^{O(n^2)}$.*

6.3.9 Theorem [43]. *The algorithm terminates in $O(n^2 \log B)$ iterations.*

Proof. The label of any vertex v is the gain of a simple path from s to v, and it is monotonically decreasing during the algorithm. Therefore, it cannot decrease by a constant factor more than $O(\log T)$ times for any given vertex, and Case 2 cannot occur more than $O(n \log T)$ times. When Case 1 applies, the value of $Def(g, \mu)$ decreases by a factor of 2. The value of $Def(g, \mu)$ is at most $O(nBT)$ after the first iteration; and, by Theorem 6.3.4, the algorithm terminates when this value decreases below L^{-1}. Lemma 6.3.8 limits the increase of $Def(g, \mu)$ to $T^{O(n)}$ during the algorithm. Hence, Case 1 cannot occur more than $O(\log(nBT) + n \log T + \log L) = O(n^2 \log B)$ times. □

To get a bound on the running time, one has to decide which algorithm to use for computing the minimum-cost pseudoflow in Step 2. The best choice is Orlin's $O(m(m + n \log n) \log n)$ time algorithm, discussed in Chapter 5.

6.3.10 Theorem [43]. *Algorithm MCF solves the generalized flow problem using at most $O(n^2 m(m + n \log n) \log n \log B)$ arithmetic operations on numbers whose size is bounded by $O(m \log B)$.*

We conclude this section with a brief discussion of the other algorithm of [43]. The algorithm is based on ideas from two flow algorithms: the cycle-canceling algorithm of Goldberg and Tarjan [48] described in Chapter 4, and the fat-path h maximum flow algorithm of Edmonds and Karp [22], which finds the maximum flow in a network by repeatedly augmenting along a highest-capacity residual path.

Each iteration of the algorithm starts with cycle-canceling. Canceling a flow-generating cycle in a generalized flow network creates an excess at some vertex of the cycle. If we cancel cycles other than the most efficient ones, the residual graph of the resulting generalized pseudoflow will have flow-generating cycles that do not contain the source. The algorithm cancels flow-generating cycles by an adaptation of the Goldberg-Tarjan algorithm. Using the dynamic tree data structure, this phase can be implemented to run in $O(n^2 m \log n \log B)$ time. The resulting excesses at vertices other than the source are moved to the source along augmenting paths in the second phase of the algorithm.

Consider a generalized pseudoflow that has non-negative excesses and whose residual graph has no flow-generating cycles. The key idea of the second phase is to search for augmenting paths from vertices with excess to the source that result in a significant increase in the excess of the source. The algorithm maintains a scaling parameter Δ such that the maximum excess at the source is at most $2m\Delta$ more than the current excess. It looks for an augmenting path that can increase the excess of the source by at least Δ. If the residual graph of the current generalized pseudoflow has no flow-generating cycles, then one can find a sequence of such paths such that, after augmenting the generalized flow along these paths, the maximum excess at the source is at most $m\Delta$ more than the current excess. This phase can be implemented in $O(m(m + n \log n))$ time by a sequence of $O(m)$ single-source shortest path computations on graphs with arcs of non-negative cost. Now Δ is divided by two and a new iteration is started. After $O(m \log B)$ iterations, when the excess at the source is very close to the optimum value, Truemper's algorithm can be applied to bring all of the remaining excesses to the source.

Maggs (personal communication, 1989) has observed that this algorithm can be improved through better balancing of the two phases. Dividing Δ by a factor of 2^α after each iteration decreases the number of iterations by a factor of α and increases the time required for the second phase by a factor of 2^α. The parameter α is chosen so that the time required for the two phases is balanced. The resulting algorithm runs in $O\left(n^2 m^2 \log^2 B \frac{\log n}{\log((n^2 \log n)/m) + \log \log B}\right)$ time.

Acknowledgements. We would like to thank Serge Plotkin for his helpful comments on an earlier draft, and for pointing out an error.

The research of the first author was partially supported by NSF Presidential Young Investigator Grant CCR-8858097, IBM Faculty Development Award, ONR Contract N00014-88-K-0166, and AT&T Bell Laboratories.

The survey was written while the second author was visiting the Department of Mathematics at M.I.T. Her research was partially supported by Air Force contract AFOSR-89-0271 and DARPA contract N00014-89-J-1988.

The research of the third author was partially supported by the National Science Foundation, Grant No. DCR-8605961, and the Office of Naval Research, Contract No. N00014-87-K-0467.

References

[1] Ahuja, R.K., Goldberg, A.V., Orlin, J.B., Tarjan, R.E. (1988): Finding minimum-cost flows by double scaling. Technical Report CS-TR-164-88, Department of Computer Science, Princeton University

[2] Ahuja, R.K., Mehlhorn, K., Orlin, J.B., Tarjan, R.E. (1987): Faster algorithms for the shortest path problem. Technical Report CS-TR-154-88, Department of Computer Science, Princeton University

[3] Ahuja, R.K., Orlin, J.B. (1987): A fast and simple algorithm for the maximum flow problem. Sloan Working Paper 1905-87, Sloan School of Management, M.I.T.

[4] Ahuja, R.K., Orlin, J.B., Tarjan, R.E. (1987): Improved time bounds for the maximum flow problem. Technical Report CS-TR-118-87, Department of Computer Science, Princeton University (SIAM J. Comput., to appear)

[5] Awerbuch, B. (1985). (Personal communication, Laboratory for Computer Science, M.I.T.)

[6] Balinski, M.L. (1985): Signature methods for the assignment problem. Oper. Res. *33*, 527–536

[7] Barahona, F., Tardos, É. (1989): Note on Weintraub's minimum-cost circulation algorithm. SIAM J. Comput. *18*, 579–583

[8] Berge, C. (1973): Graphs and hypergraphs. North-Holland, Amsterdam

[9] Bertsekas, D.P. (1979): A distributed algorithm for the assignment problem. Laboratory for Decision Systems, M.I.T. (Unpublished)

[10] Bertsekas, D.P. (1985): A distributed asynchronous relaxation algorithm for the assignment problem. Proc. 24th IEEE Conf. on Decision and Control, Ft. Lauderdale, FL

[11] Bertsekas, D.P. (1986): Distributed asynchronous relaxation methods for linear network flow problems. Technical Report LIDS-P-1986, Laboratory for Decision Systems, M.I.T.

[12] Bertsekas, D.P., Eckstein, J. (1988): Dual coordinate step methods for linear network flow problems. Math. Program. *42*, 203–243

[13] Bland, R.G., Jensen, D.L. (1985): On the computational behavior of a polynomial-time network flow algorithm. Technical Report 661, School of Operations Research and Industrial Engineering, Cornell University

[14] Busacker, R.G., Gowen, P.J. (1961): A procedure for determinimg a family of minimal-cost network flow patterns. Technical Report 15, O.R.O.

[15] Busacker, R.G., Saaty, T.L. (1965): Finite graphs and networks: an introduction with applications. McGraw-Hill, New York, NY

[16] Cheriyan, J., Maheshwari, S.N. (1987): Analysis of a preflow push algorithm for maximum network flow. Manuscript, Department of Computer Science and Engineering, Indian Institute of Technology, New Dehli (Unpublished)

[17] Cherkasskij, B.V. (1977): Algorithm for construction of maximal flow in networks with complexity of $O(V^2\sqrt{E})$ operations. Math. Methods Sol. Econ. Probl. 7, 112–125 (In Russian)

[18] Cherkasskij, B.V. (1979): A fast algorithm for computing maximum flow in a network. In: A.V. Karzanov (Ed.): Collected Papers, Issue 3: Combinatorial methods for flow problems, pp. 90-96. The Institute for Systems Studies, Moscow, (In Russian)

[19] Cunningham, W.H. (1976): A network simplex method. Math. Program. *11*, 105–116

[20] Dantzig, G.B. (1962): Linear programming and extensions. University Press, Princeton, NJ

[21] Dinic, E.A. (1970): Algorithm for solution of a problem of maximum flow in networks with power estimation. Sov. Math. Dokl. *11*, 1277–1280

[22] Edmonds, J., Karp, R.M. (1972): Theoretical improvements in algorithmic efficiency for network flow problems. J. Assoc. Comput. Mach. *19*, 248–264

[23] Engel, G.M., Schneider, H. (1975): Diagonal similarity and equivalence for matrices over groups with 0. Czech. Math. J. *25*, 389–403

[24] Even, S. (1979): Graph algorithms. Computer Science Press, Potomac, MD

[25] Even, S., Tarjan, R.E. (1975): Network flow and testing graph connectivity. SIAM J. Comput. *4*, 507–518

[26] Ford jr., L.R., Fulkerson, D.R. (1962): Flows in networks. Princeton University Press, Princeton, NJ

[27] Ford jr., L.R., Fulkerson, D.R. (1956): Maximal flow through a network. Can. J. Math. *8*, 399–404

[28] Fortune, S., Wyllie, J. (1978): Parallelism in random access machines. Proc. 10th ACM STOC, pp. 114–118

[29] Fredman, M.L., Tarjan, R.E. (1987): Fibonacci heaps and their uses in improved network optimization algorithms. J. Assoc. Comput. Mach. *34*, 596–615

[30] Fredrickson, G. (1987): Fast algorithms for shortest paths in planar graphs with applications. SIAM J. Comput. *16*, 1004–1022

[31] Fujishige, S. (1986): A capacity-rounding algorithm for the minimum-cost circulation problem: A dual framework of the Tardos algorithm. Math. Program. *35*, 298–308

[32] Fulkerson, D.R. (1961): An out-of-Kilter method for minimal cost flow problems. SIAM J. Appl. Math. *9*, 18–27

[33] Gabow, H.N. (1985): Scaling algorithms for network problems. J. Comput. Syst. Sci. *31*, 148–168

[34] Gabow, H.N., Tarjan, R.E. (1988): Almost-optimal speed-ups of algorithms for matching and related problems. Proc. 20th ACM STOC, pp. 514–527

[35] Gabow, H.N., Tarjan, R.E.: Faster scaling algorithms for network problems. SIAM J. Comput., to appear

[36] Galil, Z. (1980): An $O(V^{5/3}E^{2/3})$ algorithm for the maximal flow problem. Acta Inf. *14*, 221–242

[37] Galil, Z., Naamad, A. (1980): An $O(EV\log^2 V)$ algorithm for the maximal flow problem. J. Comput. Syst. Sci. *21*, 203–217

[38] Galil, Z., Tardos, É. (1988): An $O(n^2(m+n\log n)\log n)$ minimum cost flow algorithm. J. Assoc. Comput. Mach. *35*, 374–386

[39] Glover, F., Klingman, D. (1973): On the equivalence of some generalized network problems to pure network problems. Math. Program. *4*, 269–278

[40] Goldberg, A.V. (1985): A new max-flow algorithm. Technical Report MIT/LCS/TM-291, Laboratory for Computer Science, M.I.T.

[41] Goldberg, A.V. (1987): Efficient graph algorithms for sequential and parallel computers. Ph. D. Thesis, M.I.T. (Also available as Technical Report TR-374, Laboratory for Computer Science, M.I.T.)

[42] Goldberg, A.V., Grigoriadis, M.D., Tarjan, R.E. (1988): Efficiency of the network simplex algorithm for the maximum flow problem. Technical Report LCSR-TR-117, Laboratory for Computer Science Research, Department of Computer Science, Rutgers University

[43] Goldberg, A.V., Plotkin, S.A., Tardos, É. (1988): Combinatorial algorithms for the generalized circulation problem. Technical Report STAN-CS-88-1209, Stanford University (Also available as Technical Memorandum MIT/LCS/TM-358, Laboratory for Computer Science, M.I.T.)

[44] Goldberg, A.V., Plotkin, S.A., Vaidya, P.M. (1988): Sublinear-time parallel algorithms for matching and related problems. Proc. 29th IEEE FOCS, pp. 174–185

[45] Goldberg, A.V., Tarjan, R.E. (1986): A new approach to the maximum flow problem. Proc. 18th ACM STOC, pp. 136–146

[46] Goldberg, A.V., Tarjan, R.E. (1988): A new approach to the maximum flow problem. J. Assoc. Comput. Mach. 35, 921–940

[47] Goldberg, A.V., Tarjan, R.E. (1988): A parallel algorithm for finding a blocking flow in an acyclic network. Technical Report STAN-CS-88-1228, Department of Computer Science, Stanford University

[48] Goldberg, A.V., Tarjan, R.E. (1987): Finding minimum-cost circulations by canceling negative cycles. Technical Report MIT/LCS/TM-334, Laboratory for Computer Science, M.I.T. (Also available as Technical Report CS-TR 107-87, Department of Computer Science, Princeton University)

[49] Goldberg, A.V., Tarjan, R.E. (1987): Finding minimum-cost circulations by successive approximation. Technical Report MIT/LCS/TM-333, Laboratory for Computer Science, M.I.T. (Math. Oper. Res., to appear)

[50] Goldberg, A.V., Tarjan, R.E. (1987): Solving minimum-cost flow problems by successive approximation. Proc. 19th ACM STOC, pp. 7–18

[51] Goldfarb, D., Grigoriadis, M.D. (1988): A computational comparison of the Dinic and network simplex methods for maximum flow. Ann. Oper. Res.1383–123

[52] Goldfarb, D., Hao, J. (1988): A primal simplex algorithm that solves the maximum flow problem in at most nm pivots and $O(n^2m)$ time. Technical Report, Department of Industrial Engineering and Operations Research, Columbia University

[53] Gondran, M., Minoux, M. (1984): Graphs and algorithms. J. Wiley & Sons, New York, NY

[54] Grigoriadis, M.D. (1986): An efficient implementation of the network simplex method. Math. Program. Study 26, 83–111

[55] Grötschel, M., Lovász, L., Schrijver, A. (1988): Geometric algorithms and combinatorial optimization. Springer Verlag, Berlin, Heidelberg (Algorithms and Combinatorics, Vol. 2)

[56] Hassin, R. (1981): Maximum flows in (s,t) planar networks. Inf. Process. Lett. 13, 107–108

[57] Hassin, R., Johnson, D.B. (1985): An $O(n\log^2 n)$ algorithm for maximum flow in undirected planar network. SIAM J. Comput. 14, 612–624

[58] Hopcroft, J.E., Karp, R.M. (1973): An $n^{5/2}$ algorithm for maximum matching in bipartite graphs. SIAM J. Comput. 2, 225–231

[59] Iri, M. (1960): A new method of solving transportation-network problems. J. Oper. Res. Soc. Japan 3, 27–87

[60] Itai, A., Shiloach, Y. (1979): Maximum flow in planar network. SIAM J. Comput. 8, 135–150

[61] Jensen, P.A., Barnes, J.W. (1980): Network flow programming. J. Wiley & Sons, New York, NY

[62] Jewell, W.S. (1958): Optimal flow through networks. Technical Report 8, M.I.T.

[63] Jewell, W.S. (1962): Optimal flow through networks with gains. Oper. Res. 10, 476–499

[64] Johnson, D.B., Venkatesan, S. (1982): Using divide and conquer to find flows in directed planar networks in $O(n^{1.5}\log n)$ time. Proc. 20th Annual Allerton Conf. on Communication, Control and Computing, pp. 898–905

[65] Kapoor, S., Vaidya, P.M. (1986): Fast algorithms for convex quadratic programming and multicommodity flows. Proc. 18th ACM STOC, pp. 147–159

[66] Karmarkar, N. (1984): A new polynomial-time algorithm for linear programming. Combinatorica 4, 373–395

[67] Karp, R.M. (1978): A characterization of the minimum cycle mean in a digraph. Discrete Math. 23, 309–311

[68] Karp, R.M., Upfal, E., Wigderson, A. (1986): Constructing a maximum matching is in random NC. Combinatorica 6, 35–48

[69] Karzanov, A.V. (1974): Determining the maximal flow in a network by the method of preflows. Sov. Math. Dokl. 15, 434–437

[70] Khachiyan, L.G. (1980): Polynomial algorithms in linear programming. Zh. Vychisl. Mat. Mat. Fiz. 20, 53–72

[71] Klein, M. (1967): A primal method for minimal cost flows with applications to the assignment and transportation problems. Manage. Sci. 14, 205–220

[72] König, D. (1950): Theorie der endlichen und unendlichen Graphen. Chelsea Publishing Co., New York, NY

[73] Kuhn, H.W. (1955): The Hungarian method for the assignment problem. Nav. Res. Logist. 2, 83–97

[74] Lawler, E.L. (1976): Combinatorial optimization: Networks and matroids. Holt, Rinehart and Winston, New York, NY

[75] Malhotra, V.M., Kumar, M.P., Maheshwari, S.N. (1978): An $O(3V3^3)$ algorithm for finding maximum flows in networks. Inf. Process. Lett. 7, 277–278

[76] Miller, G.L., Naor, J. (1989): Flow in planar graphs with multiple sources and sinks. To appear in Proc. 21th IEEE FOCS

[77] Minty, G.J. (1960): Monotone networks. Proc. R. Soc. Lond., Ser. A 257, 194–212

[78] Mulmuley, K., Vazirani, U.V., Vazirani, V.V. (1987): Matching is as easy as matrix inversion. Combinatorica 7, 105–131

[79] Onaga, K. (1966): Dynamic programming of optimum flows in lossy communication nets. IEEE Trans. Circuit Theory 13, 308–327

[80] Onaga, K. (1967): Optimal flows in general communication networks. J. Franklin Inst. 283, 308–327

[81] Orlin, J.B. (1988): A faster strongly polynomial minimum cost flow algorithm. Proc. 20th ACM STOC, pp. 377–387

[82] Orlin, J.B. (1984): Genuinely polynomial simplex and non-simplex algorithms for the minimum cost flow problem. Technical Report No. 1615-84, Sloan School of Management, M.I.T.

[83] Orlin, J.B. (1985): On the simplex algorithm for networks and generalized networks. Math. Program. Study 24, 166–178

[84] Orlin, J.B., Ahuja, R.K. (1987): New distance-directed algorithms for maximum-flow and parametric maximum-flow problems. Sloan Working Paper 1908-87, Sloan School of Management, M.I.T.

[85] Orlin, J.B., Ahuja, R.K. (1988): New scaling algorithms for assignment and minimum cycle mean problems. Sloan Working Paper 2019-88, Sloan School of Management, M.I.T.

[86] Van Emde Boas, P., Kaas, R., Zijlstra, E. (1977): Design and implementation of an efficient priority queue. Math. Syst. Theory 10, 99–127

[87] Papadimitriou, C.H., Steiglitz, K. (1982): Combinatorial optimization: Algorithms and complexity. Prentice-Hall, Englewood Cliffs, NJ

[88] Reif, J.H. (1983): Minimum $s - t$ cut of a planar undirected network in $O(n \log^2 n)$ time. SIAM J. Comput. 12, 71–81

[89] Renegar, J. (1988): A polynomial time algorithm, based on Newton's method, for linear programming. Math. Program. *40*, 59–94

[90] Röck, H. (1980): Scaling techniques for minimal cost network flows. In: Pape, U. (ed.): Discrete structures and algorithms. Carl Hansen, Munich, pp. 181–191

[91] Ruhe, G. (1988): Parametric maximal flows in generalized networks - complexity and algorithms. Optimization *19*, ,235–251

[92] Shiloach, Y. (1978): An $O(nI \log^2 I)$ maximum-flow algorithm. Technical Report STAN-CS-78-802, Department of Computer Science, Stanford University

[93] Shiloach, Y., Vishkin, U. (1982): An $O(n^2 \log n)$ parallel max-flow algorithm. J. Algorithms *3*, 128–146

[94] Sleator, D.D. (1980): An $O(nm \log n)$ algorithm for maximum network flow. Technical Report STAN-CS-80-831, Department of Computer Science, Stanford University

[95] Sleator, D.D., Tarjan, R.E. (1983): A data structure for dynamic trees. J. Comput. Syst. Sci. *26*, 362–391

[96] Tardos, É. (1985): A strongly polynomial minimum cost circulation algorithm. Combinatorica *5*, 247–255

[97] Tarjan, R.E. (1984): A simple version of Karzanov's blocking flow algorithm. Oper. Res. Lett. *2*, 265–268

[98] Tarjan, R.E. (1985): Amortized computational complexity. SIAM J. Algebraic Discrete Methods *6*, 306–318

[99] Tarjan, R.E. (1983): Data structures and network algorithms. SIAM, Philadelphia, PA

[100] Tarjan, R.E. (1988): Efficiency of the primal network simplex algorihm for the minimum-cost circulation problem. Technical Report CS-TR-187-88, Department of Computer Science, Princeton University

[101] Tomizawa, N. (1972): On some techniques useful for solution of transportation networks problems. Networks *1*, 173–194

[102] Truemper, K. (1977): On max flows with gains and pure min-cost flows. SIAM J. Appl. Math. *32*, 450–456

[103] Vaidya, P.M. (1989): Speeding up linear programming using fast matrix multiplication. Technical Memorandum, AT&T Bell Laboratories, Murray Hill, NJ

[104] Weintraub, A. (1974): A primal algorithm to solve network flow problems with convex costs. Manage. Sci. *21*, 87–97

[105] Yakovleva, M.A. (1959): A problem on minimum transportation cost. In: Nemchinov, V.S. (ed.): Applications of mathematics in economic research. Izdat. Social'no-Ekon. Lit., Moscow, pp. 390–399

[106] Zadeh, N. (1973): A bad network problem for the simplex method and other minimum cost flow problems. Math. Program. *5*, 255–266

Routing Problems in Grid Graphs

Michael Kaufmann and Kurt Mehlhorn

1. Introduction

The routing problem lies at the heart of VLSI design. A routing problem is given by a routing region and a set of nets. In this paper the routing region will always be a grid graph, i.e., a finite subgraph of the integer grid.

Definition. The integer grid has vertices $x \in \mathbb{Z}^2$ and edges $\{x, y\}$ where $x = (x_1, x_2)$, $y = (y_1, y_2)$ and $|x_1 - y_1| + |x_2 - y_2| = 1$.

Grid graphs model the popular constraint that wires in a VLSI layout can only run horizontally and vertically in a natural way. A net or demand is specified by a pair of vertices (also called terminals of the net); a rough routing (also called global routing or homotopy) may also be specified. A solution to a routing problem called layout consists of a set of grid paths, one for each net, such that the following two conditions hold:

1) The paths are pairwise edge-disjoint.
2) For each net, the path for this net connects the two terminals of the net. In addition, if a rough routing for the net is specified, then the path must be homotopic to the rough routing; cf. Section 3 for a definition of homotopy.

A routing problem is naturally viewed as a multi-commodity flow problem. Each net represents the demand to send one unit of flow of a certain commodity from one terminal to the other terminal of the net; also each edge has capacity one. The additional constraint is that a commodity has to be sent along a single path and cannot be split up into pieces. If all nets have the same terminals then Menger's theorem provides us with a solution: The number of edge-disjoint paths is given by the capacity of a minimum cut. In other words, a set of demands can be satisfied, if there is no oversaturated cut, i.e., the cut condition holds. Okamura and Seymour extend Menger's Theorem to multi-commodity flow problems in planar graphs where all terminals lie on the boundary of the *same* face. They show that the cut condition together with an evenness condition implies solvability. This theorem together with an algorithmic version of it is discussed in Section 2; cf. also Frank's paper in this volume. In Section 3 we then turn to the homotopic routing problem in grid graphs. The terminals are now allowed to lie on the boundary of many faces; however global routings have to be specified. Again, the cut condition together with evenness implies routability. Section 4 is then devoted to a discussion of the evenness condition. Section 5 deals with special cases of the problem of Section 2. It is shown that for grid graphs without holes, e.g. rectangles or convex polygons, faster algorithms than for the general case can be obtained. Section 6 discusses the routing problem for grid graphs with a single hole. Terminals are allowed to lie on both non-trivial faces (the outer face and the hole) and no rough routing is given.

Mathematical models frequently suppress important features of real life problems. This is also true for the routing problem as discussed in Section 2 to 6; in particular, *layer assignment* is not dealt with and *multi-terminal nets* are not treated. They are the subject of Sections 7 and 8.

2. Edge-Disjoint Paths in Planar Graphs

Problem. Planar Edge-Disjoint Paths Problem (PED).

Input:

a) An embedded planar graph $G = (V, E)$.
b) A set $\mathcal{N}$ of nets where each $N \in \mathcal{N}$ is a pair of vertices on the boundary of the unbounded face of G.

Output: A family $\{p(N); N \in \mathcal{N}\}$ of paths such that

1) if $N = \{s, t\}$ then $p(N)$ is a path with endpoints s and t.
2) $p(N)$ and $p(N')$ are edge-disjoint for $N, N' \in \mathcal{N}, N \neq N'$.   □

Subsets $X \subseteq V$ are also called *cuts* in the sequel. For a subset $X \subseteq V$ we define the *capacity* $cap(X)$ of X as the number of edges having exactly one endpoint in X and the *density* $dens(X)$ of X as the number of nets having exactly one terminal in X, i.e.,

$$cap(X) = \{\{a, b\} \in E; |\{a, b\} \cap X| = 1\}$$
$$dens(X) = \{\{s, t\} \in \mathcal{N}; |\{s, t\} \cap X| = 1\}.$$

The *free capacity* of X is then given by $fcap(X) = cap(X) - dens(X)$. A cut X is *saturated* if $fcap(X) = 0$ and *oversaturated* if $fcap(X) < 0$. An edge-disjoint path problem $(V, E, \mathcal{N})$ is *even* if $fcap(X)$ is even for every cut X; it satisfies the *cut condition* if $fcap(X) \geq 0$ for all cuts X.

Theorem 1 (Okamura/Seymour). *Let $P = (V, E, \mathcal{N})$ be an even planar edge-disjoint path problem. Then P is solvable iff P satisfies the cut condition.*

For a proof of this result we refer the reader to the paper by A. Frank in this volume. The proof directly yields an algorithm which we now discuss. Let P be an even problem. The idea is to construct a sequence $P_0, P_1, \ldots$ of problems such that

1) $P_0 = P$;
2) P_{i+1} has one less edge than P_i;
3) P_i is even;

4) if P_i satifies the cut condition then P_{i+1} does;
5) if P_{i+1} is solvable then P_i is solvable.

The construction ends when the algorithm either detects a violation of the cut condition or reaches a problem $P_m = (V_m, E_m, \mathcal{N}_m)$ which has an empty set of nets and is therefore trivially solvable. The problem P_{i+1} is constructed from P_i as follows. Let $e_0, e_1, \ldots, e_{k-1}$ be the edges on the boundary of the outer face in clockwise order with $e_0 = \{a, b\}$. Let us call a cut X *simple* if there are precisely two edges e_j having exactly one endpoint in X and let us call a cut X a *cut through e_0* if $a \notin X$ and $b \in X$. We now distinguish cases.

Case a): There is a simple cut through e_0 with negative free capacity: Stop and declare the problem unsolvable.

Case b): There is no saturated simple cut through e_0: Then P_{i+1} is obtained by deleting the edge e_0 and adding the net $\{a, b\}$ to the set of nets.

Case c): There is a saturated simple cut through e_0: Let X be a saturated simple cut through e_0 of minimal cardinality and let $\{s, t\}$ be a net such that $s \in X, t \notin X$ and t is as close as possible to a in a counterclockwise traversal of the boundary of G. We delete the edge e_0 and replace the net $\{s, t\}$ by the two nets $\{a, t\}$ and $\{b, s\}$. If $a = t$ then $\{a, t\}$ is not added and if $b = s$ then $\{b, s\}$ is not added. Case c) is illustrated by Figure 2.

In the above case distinction it is assumed that the first case which applies is taken. For the proof of correctness of this algorithm, we again refer the reader to the paper by A. Frank. The problem of implementation is discussed by Matsumoto/Nishizeki/Saito [MNS85] and Becker/Mehlhorn [BM85].

Theorem 2 [BM85]. *Let* $P = (V, E, \mathcal{N})$ *be an even planar edge-disjoint path problem with* $n = |V|$.

a) *The solvability of P can be decided in time* $O(n^2)$. *Moreover, within the same time a solution can be constructed, if there is one.*
b) *If* (V, E) *is a grid graph then time* $O(n^{3/2})$ *suffices.*

Open Problem. Improve the running time.

For the following Sections it is useful to have a more "topological" definition of cuts and nets. Consider the dual of the planar graph $G = (V, E)$ and let $M = \{F_{ext}\}$; here F_{ext} denotes the unbounded or external face. A cut C (in the new sense) is any non-trivial simple path in the dual having both its endpoints in M. A cut C in the new sense induces a cut in the old sense as follows: Remove the edges intersected by C from G and let X be one of the two connected components obtained in this way. For two distinct cuts C_1 and C_2 starting with the dual of the same edge e_0 it is natural to define the ordering relation "C_1 *is left of* C_2" as follows. Traverse C_1 and C_2 starting at their common origin until

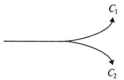

Fig. 1 C_1 is left of C_2

they separate. If at the point of separation the situation is as shown in Figure 1 then C_1 is left of C_2. With this definition, we can reformulate the selection of cut X in case c) as follows:

Cut Selection Rule. Let X be the leftmost saturated cut through e_0, i.e., X is saturated and X is left of any other saturated cut through e_0, cf. Figure 1.

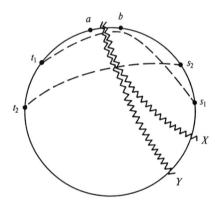

Fig. 2 X is left of Y, $\{a, t_1\}$ is right of $\{a, t_2\}$

Let us turn to the selection of the net N next. For $N = \{s, t\}, s \in X, t \notin X$ let us call the net $\{a, t\}$ the tail of the decomposition of N with respect to the edge $e_0 = \{a, b\}$. We may view the tail $\{a, t\}$ as a path from a to t running counterclockwise along the boundary of G and turning to the right at t. With this interpretation the selection of $\{s, t\}$ in part c) can be formulated as follows:

Net Selection Rule. Choose $N = \{s, t\}, s \in X, t \notin X$, such that the tail of the decomposition of N with respect to the edge e_0 is *rightmost*, cf. Figure 2.

3. Homotopic Edge-Disjoint Path Problems

We first state the problem and then define the concepts net, grid path and homotopy used in its definition.

Problem. Homotopic Routing Problem in Grid Graphs (HRP).

Input: A grid graph R and nets $q_1, \ldots, q_k$.

Output: Pairwise edge-disjoint grid paths $p_1, \ldots, p_k$ such that p_i is homotopic to q_i, $1 \le i \le k$, or an indication that no such paths exist. □

We call a bounded face F of R trivial if it has exactly four vertices on its boundary and nontrivial otherwise. We use M to denote the set of nontrivial bounded faces together with the unbounded face F_{ext} and $\mathcal{O}$ to denote the union of the interiors of the faces in M. A nontrivial face is also called a hole.

A path P is a continuous function $p : [0, 1] \to \mathbb{R}^2 - \mathcal{O}$. A path p is called a *net* if $\{p(0), p(1)\} \subseteq V \cap \partial\mathcal{O}$ where $\partial\mathcal{O}$ is the boundary of $\mathcal{O}$. Two paths p and q are homotopic, denoted $p \sim q$, if there is a continuous function $F : [0, 1] \times [0, 1] \to \mathbb{R}^2 - \mathcal{O}$ such that $F(0, x) = p(x)$ and $F(1, x) = q(x)$ for all $x \in [0, 1]$, and $F(t, 0) = p(0)$ and $F(t, 1) = p(1)$ for all $t \in [0, 1]$. A path p is called a *grid path* if $p(x)$ belongs to R for all x. Fig. 3 gives an example of an HRP.

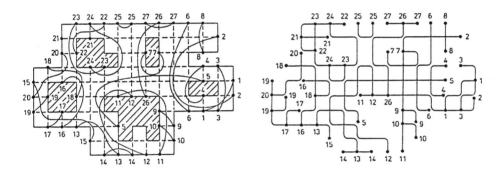

Fig. 3

Remark. In the previous Section all terminals had to lie on the infinite face. With $M = \{F_{ext}\}$ any two paths with the same endpoints are homotopic. It therefore sufficed to specify a net by its endpoints.

Theorem 3 [KM86]. *Let $P = (R, \mathcal{N})$ be an even bounded HRP. Here R is a grid graph and $\mathcal{N}$ a set of nets:*

$$P \text{ is solvable if and only if } fcap(X) \ge 0 \text{ for every cut } X.$$

As before, a cut is a simple path in the dual of R connecting two (not necessarily distinct) faces in M. The capacity $cap(C)$ of a cut C is the number of intersections with edges of R. If C is a cut and p is a path then $cross(p, C)$ is the number of intersections of p and C and $mincross(p, C) = \min\{cross(q, C); q \sim p\}$. Finally, the *density* $dens(C)$ of cut C is defined by

$$dens(C) = \sum_{p \in \mathcal{N}} mincross(p, C)$$

and the *free capacity* $fcap(C)$ is given by

$$fcap(C) = cap(C) - dens(C).$$

A cut C is *saturated* if $fcap(C) = 0$ and *oversaturated* if $fcap(C) < 0$.

An HRP is *even* if $fcap(C)$ is even for every cut C.

Let v be a vertex in R. We denote the degree of v by $deg(v)$ and the number of nets having v as an endpoint by $ter(v)$. An HRP is *bounded*, if $deg(v) + ter(v) \le 4$ for all vertices v.

Remark. The definition of density given above extends the definition in the previous Section. Clearly, in the situation discussed there, $mincross(p, C) = 1$ if the endpoints of p lie on different sides of the cut C and $mincross(p, C) = 0$ otherwise.

The algorithm to solve HRP's is very similar to the algorithm given in the previous Section. We assume for simplicity that P satisfies the cut condition. We again construct a sequence $P_0, P_1, \ldots$ of HRP's such that

1) $P_0 = P$;
2) P_{i+1} has one less edge than P_i;
3) P_i is even and bounded;
4) if P_i satisfies the cut condition then P_{i+1} does;
5) if P_{i+1} is solvable then P_i is solvable.

The construction stops when a trivial problem is reached. The problem P_{i+1} is obtained from P_i as follows. Again we have to distinguish several cases. In all cases we use juxtaposition to denote concatenation of paths, i.e., if p and q are paths with $p(1) = q(0)$ then $pq(\lambda) = p(2\lambda)$ for $0 \le \lambda \le 1/2$ and $pq(\lambda) = q(2\lambda - 1)$ for $1/2 \le \lambda \le 1$. We also use a more careful definition for the ordering relation "right-of" on nets. If N is a net let $can(N)$ be the shortest path homotopic to N. We think of $can(N)$ as slightly extended into the incident non-trivial faces at both its terminals, cf. Figure 4a. Consider now two nets N_1 and N_2 with the same starting point. Then, if $N_1 \ne N_2$, $can(N_i)$ is not a prefix of $can(N_{3-i})$ for $i = 1, 2$. We say that N_1 is right-of N_2 if $can(N_1)$ and $can(N_2)$ separate as shown in Figure 4b.

We are now ready for the algorithm. Again we distinguish cases:

Case a): There is a cut X with $cap(X) = 1$: Let N be the unique net with $mincross(N, X) = 1$, let e be the edge intersected by X and let N_1 and N_2 be such that $N \sim N_1 e N_2$ and $mincross(N_1, X) = mincross(N_2, X) = 0$. Delete e and replace N by the nets N_1 and N_2.

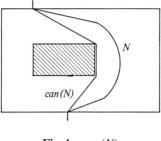

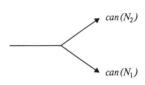

Fig. 4a *can*(*N*)   **Fig. 4b** The relation "right-of"

Case b): There is a vertex v with $deg(v) = ter(v)$. Let v be a vertex with $deg(v) = ter(v)$; let $e_i, 1 \le i \le 2$, be the edges incident to v and let $N_i, 1 \le i \le 2$, be the nets incident to v where the edges are numbered as shown in Figure 5 and N_1 is right-of N_2. Let $N_1 \sim e_1 N_1'$ where N_1' does not use edge e_1; remove edge e_1, reserve it for net N_1 and replace net N_1 by net N_1'.

Fig. 5 A vertex v with $deg(v) = 2$ and the edges incident to it. The face in $\mathcal{O}$ is shown hatched

Case c): No cut of capacity one exists and $ter(v) < deg(v)$ for all v. Let vertex a be the left upper corner of the routing region, i.e. there is no vertex with either larger y-coordinate or smaller x-coordinate and same y-coordinate. Let b be the lower neighbor of a and $e^* = \{a, b\}$. The edge e^* plays the role of the edge e_0 in Section 2.

A cut is called a *straight-line cut* if it consists of a sequence of horizontal and vertical straight-line segments. For this definition to make sense, we view the dual as a grid graph. A straight-line cut is a *1-bend* cut if it consists of at most two segments. The 1-bend cuts play the role of the simple cuts in Section II.

Case c1): There is no saturated 1-bend cut through e^*: Remove edge e^* and add net N where N is the path from a to b following the boundary of the trivial face incident to e^*.

Case c2): There is a saturated 1-bend cut through e^*. We use the same cut selection rule as in Section II, i.e., we let X be the leftmost saturated 1-bend cut through e^*.

Let us turn to net selection next. For a net N with $mincross(N, X) > 0$, a decomposition with respect to X and e^* is a triple (N_1, e^*, N_2) such that $N \sim N_1 e^* N_2$ and $mincross(N_1, X) + mincross(N_2, X) = mincross(N, X) - 1$. A decomposition (N_1, e^*, N_2) is *rightmost* if N_2 is right-of M_2 for all decompositions (M_1, e^*, M_2) of nets M with $mincross(M, X) > 0$.

Net selection rule: Choose a net N with a rightmost decomposition (N_1, e^*, N_2) with respect to X and e^*.

Delete edge e^* and replace N by N_1 and N_2.

For the proof of correctness of this algorithm we refer the reader to [KM86]. We turn to the running time of the algorithm next. In order to measure the size of the input we assume that the nets are specified as polygonal paths. Then n denotes the number of vertices of the grid graph plus the total number of bends in the paths. In [KM86] an $O(n^2)$ algorithm to solve even, bounded HRP's was given. This was recently improved.

Theorem 4 [KM88]. *The solvability of even, bounded HRP's can be decided in linear time $O(n)$. Moreover a solution can be determined in time $O(n)$ if there is one.*

What can be done beyond grid graphs? [K87] showed that Theorem 3 is also valid for other types of grids, e.g. the grids shown in Figure 6. In these cases an instance is called bounded if $deg(v) + ter(v) \leq odeg(v)$ for all vertices v, where $odeg(v)$ is the degree of v in the infinite grid of the respective type.

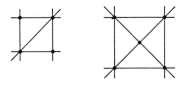

Fig. 6

Theorem 5 [K87]. *Theorem 3 is also valid for subgraphs of the grids shown in Figure 6. The runtime of the algorithm is $O(n^2)$.*

The most general type of graphs for which Theorem 3 is known to hold are *straight-line graphs*. Let G be an embedded planar graph and let $\mathcal{O}$ be the union of some of its faces (including the outer face) considered as open sets. The pair $(G, \mathcal{O})$ is called a straight-line graph if there are line segments $L_1, \ldots, L_t$ such that the endpoints of each L_i lie in the boundary of $\mathcal{O}$, such that the vertices of G are exactly the endpoints and intersections of the line segments, and such that the edges are exactly the induced fragments of the line segments. Boundedness is defined as above; the *odeg*(v) of a vertex v on the boundary of a face in $\mathcal{O}$ is the degree of v plus the number of lines ending in v.

Theorem 6 [S87]. *Theorem 3 holds for straight-line graphs. Moreover, solvability can be decided in polynomial time.*

Open Problem. Design an algorithm for homotopic routing in straight-line graphs with quadratic or even linear running time.

Open Problem. Extend Theorem 3 beyond straight-line graphs. Note that the theorem does not hold for planar graphs as Figure 7 shows. What is the appropriate theorem for planar graphs?

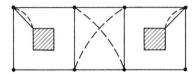

Fig. 7 Nets are indicated by dashed lines and finite non-trivial. This problem is even and satisfies the cut condition. It is not solvable, however

4. The Evenness Assumption

In Theorems 1 and 3 we assumed the routing problem to be even. Figure 8 shows that this assumption is crucial. What can be done without the evenness assumption?

Fig. 8 An unsolvable, non-even problem satisfying the cut condition

For PED this problem was treated by Frank and Becker/Mehlhorn. Let us call an instance of PED half-even if the degree of every vertex v which does not border the outer face is even. Frank extends the cut condition so that it applies to half-even instances; we refer the reader to his paper for the details. Becker/Mehlhorn consider the algorithmic side.

Lemma. *Let P be an instance of PED. $f\,cap(X)$ is even for every cut X iff $deg(v)+$ $ter(v)$ is even for every vertex v.*

We call $deg(v) + ter(v)$ the extended degree of v.

It is easy to see that each solvable half-even instance can be extended by adding some nets (called fictitious) which have vertices with odd extended degree as terminals, such that the new instance is even and solvable. Becker/Mehlhorn give an algorithm to find an appropriate extension of any given solvable half-even instance.

Theorem 7 [BM86]. *The solvability of half-even instances of PED can be decided in time $O(bn)$ where b is the number of vertices bordering the outer face.*

The main idea in their method is captured in the following definition:

Definition. Let P be a solvable half-even instance. Let U be the set of vertices with odd extended degree and let X be a saturated cut. Let $u_1, u_2, \ldots, u_{2k}$ be the vertices in $X \cap U$ in clockwise order along the boundary of the outer face.

a) X is *U-minimal* if $X \cap U \neq \emptyset$ and there is no saturated cut Y with $Y \cap U = \{u_i, \ldots, u_j\}$ where $1 < j - i + 1 < 2k$.

b) *The canonical extension of P with respect to X is given by adding the nets $(u_{2i-1}, u_{2i}), 1 \leq i \leq k$, to the set of nets in P.*

The following lemma is crucial for the correctness of their method.

Lemma. *Let P be a solvable half-even instance. If X is a U-minimal cut, then the canonical extension of P with respect to X is a half-even solvable instance.*

The algorithm determines a U-minimal cut and extends the problem canonically with respect to this cut. This preserves solvability and reduces the size of U. After at most $O(b)$ iterations U is empty. Each iteration takes time $O(n)$.

If the routing region is a grid with no non-trivial inner face we can take advantage of the fact that we only need consider 1-bend cuts to find U-minimal cuts. Hence, each step can be executed in time $O(|U|)$.

Theorem 8 [KM85]. *Half-even instance of PED on grids with n vertices and no non-trivial inner face and $|U|$ vertices of odd extended degree can be extended to even problems in time $O(\log^2 n + |U|^2)$.*

Open Problem. For grid graphs, even instances can be solved in time $O(n^{3/2})$, but extending a half-even instance to an even instance takes time $O(bn)$; b might be as large as $\Omega(n)$. Find a faster algorithm for half-even grid graph problems.

For the homotopic routing problem in grid graphs the situation is more complicated. As before, we call an instance half-even, if the degree of v is even for every vertex not on the boundary of a non-trivial face. We call an instance locally even, if $deg(v) + ter(v)$ is even for every vertex v. If all terminals of nets are on the outer face then a locally even instance is necessarily even, but in general this is not the case.

As before we know that for any solvable half-even instance there exists an extension to a solvable locally even instance by adding some fictitious nets.

Furthermore every solvable locally even instance can be extended to a solvable even instance by adding some non-trivial circular nets. Circular nets are simple cycles of a certain homotopy (Kaufmann/Maley). Unfortunately, the extensions are not easy to find in this case.

Theorem 9 [KMa88]. *The homotopic routing problem in grid graphs is NP-complete for locally even instances.*

Kaufmann/Maley also obtained a positive result. Suppose that we allow to move modules by one unit.

Theorem 10 [KMa88]. *The homotopic routing problem with movable modules is solvable in linear time in the case of locally-even instances and is NP-complete in the case of half-even instances.*

5. Routing Regions Without Holes

In this Section we come back to the problem considered in Section 2, i.e. all terminals lie on the boundary of the outer face. The algorithm discussed in Section 2 solves this problem in time $O(n^2)$ for general planar graphs and $O(n^{3/2})$ for grid graphs. We now turn attention to grid graphs where all bounded faces are trivial. Such graphs are called generalized switchboxes. In the case of generalized switchboxes we only need to consider a special kind of cuts, namely straight cuts (they consist of only one straight path segment) or 1-bend cuts (consist of at most two straight segments).

Lemma [KM85]. *Let P be an even bounded instance of PED on a grid.*
a) If there is an oversaturated cut then there is an oversaturated 1-bend cut.
b) If there is an oversaturated cut then there is an oversaturated 0-bend cut or an oversaturated 1-bend cut connecting two concave corners.

A concave corner is a pair $((v', v), (v, v''))$ of boundary edges of some nontrivial face sharing a vertex v of degree 4. A 1-bend cut connects 2 concave corners $((v', v), (v, v''))$ and $((w', w), (w, w''))$ if the rectangle defined by the two corners v and w is non-empty and does not contain any boundary vertices except v and w. We call two such corners *rectilinear visible*.

Using the algorithm of Kaufmann/Mehlhorn for homotopic edge-disjoint paths we can solve each solvable generalized switchbox problem in time $O(n)$. However simpler and/or faster algorithms are known for some cases of generalized switchboxes.

Kaufmann/Mehlhorn [KM85] give an algorithm to solve PED for generalized switchboxes in time $O(n \log^2 n + |U|^2)$. It works for half-even instances; U is the set of the vertices with odd extended degree. Although the runtime is worse than the runtime of the algorithm for the homotopic case, the implementation is much less complicated. The algorithm is almost the same as in the planar case, but takes advantage of the fact that nets and cuts are given by their endpoints and that the capacities of 1-bend cuts are easily computed by the difference of the coordinates of their endpoints. Nets and possibly critical cuts are represented by intervals and are stored in a range tree. This data stucture supports the necessary operations like "determine the leftmost saturated cut X" and "find the rightmost net crossing X" in time $O(\log 2n)$. This is the most important point to get a runtime $O(n \log^2 n)$.

Simpler and faster algorithms are known for convex grids. Nishizeki/Saito/Suzuki [NSS85] define convex grids as the subclass of generalized switchboxes, where any two vertices can be connected by a path with at most 1 bend. They show that in convex grids only straight cuts have to be considered in this case. Based on this observation they achieve a runtime linear in the size of the routing region by a simple algorithm. Lai and Sprague [LS86] show the correct condition for the solvability of half-even instances in this class.

Kaufmann [K87] extends the notion of convex grids to generalized switchboxes where any horizontal or vertical line crosses the boundary at most twice. He shows that the instances in this class are extremely simple to solve. The algorithm works roughly as follows:

If there is a vertex v with $deg(v) = ter(v) = 2$, route both nets to the adjacent boundary vertices in the obvious way. If not, consider any certain corner v, and an adjacent boundary vertex w. If any net starts at w, then route it on the edge (v, w) and throw (v, w) away. If not, add a net $\{v, w\}$ and throw (v, w) away. Iterate until the routing region is empty.

Note that for this class straight cuts and cuts with 1 bend have to be considered, but only in the correctness proof of the algorithm. In the algorithm itself no cuts are considered. The runtime is $O(n)$.

The same algorithm works also for generalized switchboxes where any horizontal line crosses the boundary at most twice. This problem class is called half-convex grids. A weakness of the algorithm is that it can only be applied to even instances.

Open Problem. Find a condition for the solvability of half-even instances of PED in half-convex grids.

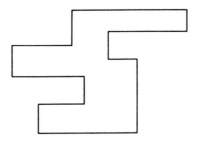

Fig. 9 A half-convex region

All algorithms mentioned so far remove one or two edges per step such that runtime $O(n)$ is optimal in a certain sense. Mehlhorn/Preparata [MP83] show that for rectangular routing regions better results are achievable. They present an algorithm for even and half-even instances with runtime $O(b \log b)$ where b is the perimeter of the rectangle. The main idea is to avoid to produce layouts as shown in Figure 10a (such layouts are typically produced by the algorithm of Section 2) and instead to route as shown in Figure 10b. Note that in the case of n nets the layout of Fig. 10a has $O(n^2)$ bends but the layout of Fig. 10b has only $O(n)$ bends.

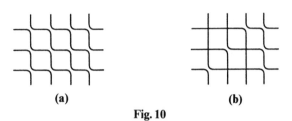

(a)   (b)

Fig. 10

Kaufmann/Klär [KKl88] extend the result to generalized switchboxes without rectilinear visible corners and gets an algorithm for the solution of even and half-even instances of runtime $O(b \log^2 b)$.

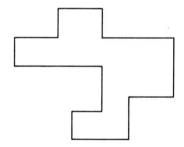

Fig. 11 A generalized switchbox without any rectilinear visible corners

Open Problem. Extend the last algorithms to larger problem classes like the homotopic routing problem.

6. Routing Regions with Exactly One Inner Hole

We consider routing regions with exactly two non-trivial faces one of which is the outer face. Also, all terminals of nets are supposed to lie on the boundary of these faces and no homotopies are given.

Okamura [O83] gives necessary and sufficient conditions for the solvability of such problems on planar graphs in the same style as Okamura/Seymour. In her paper the positions of the terminals are restricted in the following way: For each net both terminals have to lie on the same nontrivial face.

Matsumoto/Nishizeki/Saito give an algorithm with time complexity $O(bn^2 \log n)$ for this class of problems. Here n denotes as usual the size of the graph and b denotes the number of vertices on the boundary of the two non-trivial faces.

From now on we restrict ourselves to a very simple case: The routing region is a grid graph with two non-trivial faces (we call them the inner and the outer face) and the boundaries of both non-trivial faces are rectangles. Different types of problems are obtained by putting different restrictions on the positions of the terminals.

Under the restriction that both terminals of a net lie on the same face Suzuki/Ishiguro/Nishizeki [SIN87] give an algorithm with a running time linear in the size of the graph. It is based on the principles developed by Okamura and Matsumoto et al.

An alternative technique is the following: First determine a fixed homotopy for each net, such that no cut condition is violated. Then use the algorithm for homotopic routing [KM86] or a similar algorithm to find the final edge-disjoint paths.

This techniques can be used in the case where all terminals lie on the inner face. The first algorithm was given by LaPaugh [L82]. We describe a solution due to Suzuki et al. They propose to replace the routing region by a cycle with multiple edges. The length of the cycle is equal to the perimeter of the inner rectangle. The multiplicity of an edge e of the cycle is given by the capacity of the straight-line cut through the corresponding edge of the inner rectangle, cf. Figure 12.

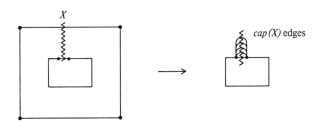

Fig. 12

The reduced problem can be solved by any of the standard algorithms for planar graphs [MNS85,BM86]. This determines the homotopies for the nets. The second step is then a standard homotopic routing problem. In [SNSFT88] a special data structure is used which supports the necessary operation efficiently such that a runtime of $O(k + n)$ for the first step and $O(\min(b_{out}, k \log k))$ for the second step can be achieved. k denotes the number of the nets and b_{out} the number of the vertices on the boundary of the outer face. The correctness proof for this method is based on the observation that only a limited set of cuts have to be considered, namely cuts which consist of two straight-line segments connecting the outer and the inner face. The capacity and density of such cuts is the same for the original and the reduced problem. Furthermore the solutions are 2- resp. 3-layer wireable, cf. Section 7 for a discussion of wireability.

If the terminals may lie anywhere on the two faces but not in the four corners of the outer rectangle (cf. Figure 13), the same technique is applicable, since also in this case only 2-segment cuts have to be considered. In [SIN87] an algorithm with $O(n)$ runtime for this problem class is presented.

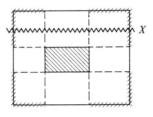

Fig. 13

Kaufmann/Klär ([KK188]) extend their technique and get faster algorithms ($O(k \log k)$) for wider problem classes.

Open Problem. Is there an efficient algorithm for problems without any restrictions on the terminal positions?

7. Layer Assignment

A *layout* is a set of edge-disjoint paths. The problem of wiring a given layout in a grid is as follows: We have k copies of the routing region stacked vertically on top of each other. The k copies of each vertex are connected in the form of a line. We call the obtained graph the wiring graph and each copy of the routing region a layer. A layer assignment or wiring lifts a grid path p into a path P in the wiring graph such that p is the vertical projection of P. We require that the liftings of different paths p and q are *vertex-disjoint*. A layer assignment is a k-layer wiring if only k layers are used. Clearly, if the layout consists of vertex-disjoint paths, it can be wired within one layer. If the paths may cross each other but are not allowed to bend on a common vertex (knock knee), we call the layout Manhattan mode layout. Manhattan layouts can always be wired in two layers by assigning horizontal path segments to layer 1 and vertical segments to layer 2.

For arbitrary layouts in grids the situation is more complex. First of all, it is easy to find examples which are not 2- and 3-layer wirable.

Lipski [L84] showed that it is NP-complete to decide whether an arbitrary layout can be wired using three layers. Brady and Brown [BB84] showed that every layout can be wired using four layers. Both papers use the concept of two-colorable maps which we now briefly discuss. We show first that any layer assignment gives rise to a two-colorable map. We start by shifting the infinite grid by one-half unit in x- and y-direction. The squares of the shifted grid are called wiring tiles, cf. Figure 14. A wiring tile is either used by only one path and is then called trivial or by two paths. In the latter case the two paths either cross in the tile or bend in the tile (knock-knee). Trivial tiles can be removed because arbitrary layer changes can be performed in these tiles. This leaves us with the non-trivial tiles. In a non-trivial tile containing a crossing the horizontal wire segment either runs above the vertical segment (color I) or below the vertical wire segment (color II). In a non-trivial tile containing a knock-knee the tile is divided by a 45° or 135° degree line as shown in Figure 14. In one part of the tile the horizontal wire segment uses a higher-numbered layer (color I) and in the other part the vertical wire segment uses the higher-numbered layer (color II). In this way any layer assignment gives rise to a two-colorable map. The boundary between differently colored regions consists of diagonals which "cut" the knock-knees and horizontal and vertical tile boundaries, cf. Figure 14 c.

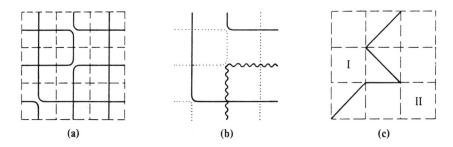

Fig. 14 (a) shows a layout consisting of 16 tiles, (b) shows a 3-layer wiring (layer 1 = ····, layer 2 = ∿∿, layer 3 = ——) and (c) indicates the corresponding 2-colorable map

The problem of layer assignment can now be formulated as follows. Start with the diagonals induced by the knock-knees. Then add horizontal and vertical tile boundaries so as to turn the layout into a two-colorable map. In addition, avoid either the patterns shown in Figure 15 or the patterns shown in Figure 16. In the former case the layout is 3-layer wirable and in the latter case the layout is 4-layer wirable. Brady and Brown have shown that the pattern shown in Figure 16 can always be avoided and hence every layout is 4-layer wirable.

An alternative technique for 4-layer wiring which is not based on two-colorable maps, but on constraint graphs, was recently proposed by Tollis [T88]. His algorithm uses the 4th layer only if it is necessary and thus frequently uses only three layers.

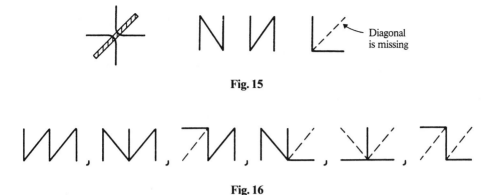

Fig. 15

Fig. 16

A different approach is to produce the layout and the wiring simultaneously and not in separate phases. Preparata/Lipski [PL82] show in the first paper on 3-layer wirability that three layers are sufficient to solve a two-terminal channel routing problem with the minimum number of tracks. Gonzalez/Zheng [GZ88] extend this result to 3-terminal and multiterminal net channel routing.

Open Problems. Prove or disprove that every grid graph routing problem has a three-layer wiring.

Investigate techniques to modify and/or stretch the layout by a small factor to get a 2- resp. 3-layer wirable layout. Some results on the second problem can be found in [GZ88] and [BS87].

8. Multiterminal Nets

In the previous Sections we have considered nets with two terminals. We now allow nets with more than two terminals and call such nets multi-terminal nets. A layout is now a collection of edge-disjoint trees, one for each net. The tree for a net must connect the terminals of the net. The PED-problem with multi-terminal nets is difficult even for the simple case of channel routing. In channel routing the routing region is a rectangle and all terminals lie on the horizontal sides of this rectangle. The height of the rectangle is called the channel width and the goal is to minimize channel width. Sarrafzadeh [Sa87] shows that it is NP-complete to solve multiterminal net problems in knock-knee mode with optimal channel width. Mehlhorn et al. [MPS86] give simple approximation algorithms with unified approach for 2-, 3- and multiterminal net problems. For 2-terminal nets their method achieves optimal channel width, for 3-terminal nets it yields solutions being within a factor 3/2 of the optimum and for multi-terminal nets it achieves a factor of 2. Gao and Kaufmann [GK87] showed recently that the factor $3/2 + o(1)$ can always be achieved.

All these results are based on the same principle: The idea is to split the multiterminal nets into simpler parts such as 2-terminal nets or nets with all terminals on the same side of the channel. These simple parts are then routed independently as indicated in Figure 17.

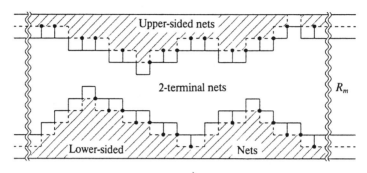

Fig. 17

The one-sided nets are routed close to the shores of the channel and the two-terminals nets are routed in the "middle" of the channel. The crucial step is to find a decomposition which does not increase channel density too much.

Consider a net N with terminals $v, \ldots, v_{l-1}$. Assume that the terminals appear in that order in a clockwise traversal of the boundary of the rectangle. If one replaces each net N by the l two-terminal nets $\{v; v_{(i+1) \bmod l}, 0 \le i < l\}$, then the density of every horizontal or vertical cut is at most doubled. Thus, if the original instance satisfies the cut condition and if one inserts a new grid column and row between every pair of columns and rows, then an instance satisfying the cut criterion is obtained. The obtained instance can be solved by the algorithm of [MP86].

Open Problems. Develop better approximation algorithms for multi-terminal net problems in rectangles.

Give a nontrivial lower bound for the multiterminal net problem in a channel.

References

[BM86]   Becker, M., Mehlhorn, K. (1986): Algorithms for routing in planar graphs. Acta Inf. **23**, 163–176

[BB84]   Brady, M., Brown, D. (1984): VLSI routing: Four layers suffice. M.I.T. VLSI Conference 1984

[BS87]   Brady, M., Sarrafzadeh, M. (1987): Layout stretching to ensure wirability. Technical Report

[F82]   Frank, A. (1982): Disjoint paths in a rectilinear grid. Combinatorica **2**, 361–371

[GK87]   Gao, S., Kaufmann, M. (1987): Channel routing of multiterminal nets. Proc. 28th IEEE FOCS, pp. 316–325

[GZ88]   Gonzalez, T., Zheng, S.-Q. (1988): Simple three-layer channel routing algorithms. In: Reif, J.H. (ed.): VLSI algorithms and architectures. Springer Verlag, Berlin, pp. 237–246 (Lect. Notes. Comput. Sci., Vol. 319)

[K87a]   Kaufmann, M.: Über lokales Verdrahten von 2-Punkt-Netzen. Dissertation

[K87b]   Kaufmann, M. (1987): A linear-time algorithm for routing in a convex grid. Technical Report

[KKl88]   Kaufmann, M., Klär, G.: Routing around a rectangle – towards the general case. Preprint

[KKl88]   Kaufmann, M., Klär, G.: Routing in generalized switchboxes without rectilinear visible corners. Technical Report

[KMa88]   Kaufmann, M., Maley, F.M. (1988): Parity conditions in homotopic knock-knee routing. Preprint

[KM85]   Kaufmann, M., Mehlhorn, K. (1985): Generalized switchbox routing. J. Algorithms **7**, 510–531 (and ICALP 1985)

[KM86]   Kaufmann, M., Mehlhorn, K. (1986): On local routing of two-terminal nets. Technical Report 03/1986, FB 10, Universität des Saarlandes; an extended abstract appeared in STACS 87, LNCS 247, pp. 40-52

[KM88]   Kaufmann, M., Mehlhorn, K. (1988): A linear-time algorithm for the local routing problem. Technical Report, FB 10, Universität des Saarlandes

[KvL82]   Kramer, M.E., van Leeuwen, J. (1982): Wire routing is NP-complete. Technical Report RUU-CS-82-4, Department of Computer Science, Rijksuniversiteit te Utrecht

[LS87]   Lai, T.-H., Sprague, A. (1987): On the routability of a convex grid. J. Algorithms **8**, 372–384

[LaP80]   LaPaugh, A.S. (1980): A polynomial-time algorithm for optimal routing around a rectangle. Proc. 21st IEEE FOCS, pp. 282–293

[L84]   Lipski jr., W. (1984): On the structure of three-layer wirable layouts. In: Preparata, F.P. (ed.): VLSI theory. JAI Press, London, pp. 231–243 (Adv. Comput. Res., Vol. 2)

[MNS86]   Matsumoto, K., Nishizeki, T., Saito, N. (1986): Planar multicommodity flows, maximum matchings and negative cycles. SIAM J. Comput. **15**, 495–510

[MP86]   Mehlhorn, K., Preparata, F.P. (1986): Routing through a rectangle. J. Assoc. Comput. Mach. **33**, 60–85

[MPS86]   Mehlhorn, K., Preparata, F.P., Sarrafzadeh, M. (1986): Channel routing in knock-knee mode: Simplified algorithms and proofs. Algorithmica **1**, 213–221

[NSS85]   Nishizeki, T., Saito, N., Suzuki, K. (1985): A linear-time routing algorithm for convex grids. IEEE Trans. Comput.-Aided Design **4**, 68–76

[O83]   Okamura, H. (1983): Multicommodity flows in graphs. Discrete Appl. Math. **6**, 55–62

[OS81]   Okamura, H., Seymour, P.D. (1981): Multicommodity flows in planar graphs. J. Comb. Theory, Ser B **31**, 75–81

[PL82]   Preparata, F.P., Lipski jr., W. (1982): Three layers are enough. Proc. 23rd IEEE FOCS, pp. 350–357

[Sa87]   Sarrafzadeh, M.: On the complexity of the general channel routing problem in the knock-knee mode. IEEE Trans. Comput.-Aided Design **6**

[S86]   Schrijver, A.: Decompositions of graphs on surfaces and a homotopic circulation theorem. J. Comb. Theory, Ser. B, to appear

[S87]   Schrijver, A. (1987): Edge-disjoint homotopic paths in straight-line planar graphs. Report OS-R8718, Mathematical Centre, Amsterdam (SIAM J. Discrete Math., to appear)

[SIN87]   Suzuki, H., Ishiguro, A., Nishizeki, T. (1987): Edge-disjoint paths in a grid bounded by two nested rectangles. Technical Report, Tohoku University

[SIN88]   Suzuki, H., Ishiguro, A., Nishizeki, T. (1988): Variable-priority queue and doughnut routing. Technical Report, Tohoku University

[SNSFT88] Suzuki, H., Nishizeki, T., Saito, N., Frank, A., Tardos, É. (1988): Algorithms for routing around a rectangle. Technical Report, Tohoku University

[T88]   Tollis, I.G. (1988): A new algorithm for wiring layouts. In: Reif, J.H. (ed.): VLSI algorithms and architectures. Springer Verlag, Berlin, pp. 257–267 (Lect. Notes Comput. Sci., Vol. 319)

Steiner Trees in VLSI-Layout

Bernhard Korte, Hans Jürgen Prömel and Angelika Steger

1. Introduction

Finding Steiner trees in graphs has proven to be one of the most essential tools in attacking the routing problem in VLSI-layout. Roughly speaking, the routing problem can be described as follows: given a graph G (in most practical applications a subgraph of a grid-graph) and a collection $\mathcal{N} = \{\mathcal{N}_0, \ldots, \mathcal{N}_{r-1}\}$ of mutually disjoint subsets of the vertex-set of G. Find pairwise disjoint trees $S_0, \ldots, S_{r-1}$ such that S_i spans $\mathcal{N}_i$ for every i (but possibly contains more vertices than $\mathcal{N}_i$). Such a tree S_i is called a Steiner tree for $\mathcal{N}_i$. The intended interpretation is that the vertices in each of the subsets $\mathcal{N}_i$ form a net which has to be connected by wires. Of course, wirings of different nets have to be disjoint otherwise short circuits would be produced.

Assuming that each of the sets $\mathcal{N}_i$ contains 2 elements the problem reduces to finding a collection of disjoint paths. This problem is well-studied, cf. the survey of Frank (1990) in this volume. Unfortunately, in practical applications typically more than half of the nets contain more than 2 elements. So, generally, one has to find a collection of disjoint Steiner trees to solve the routing problem. This results in some additional difficulties. For example, deciding whether there exist k edge-disjoint paths connecting k pairs of vertices is solvable in polynomial time (provided that k is fixed) but deciding whether there exist two edge-disjoint Steiner trees connecting vertex sets $\mathcal{N}_0, \mathcal{N}_1$ resp. is NP-complete. Moreover, one is not only interested in just finding connections but in finding "short" connections, e.g., to speed up the cycle time of the chip.

It is quite easy, using e.g. Dijkstra's algorithm, to find a shortest path connecting two vertices, but it is NP-complete to decide whether a Steiner tree of a certain length exists spanning a vertex set. Nevertheless, quite a lot of effort has been spent to find results and applicable algorithms for certain aspects of the Steiner tree problem.

The task of this paper is twofold: On the one hand we review results on Steiner trees as well as algorithmic approaches to minimize, resp. to pack Steiner trees in graphs. On the other hand we want to demonstrate how these achievements can be used to find layouts for VLSI-chips.

Our demonstration is led by a real world chip problem of a large sea-of-cells chip. We will not go deeply into technical details here. The only technological facts we need for our purposes are the following: Two layers are available for wiring. The upper layer can only be used for vertical wiring, the second layer, the lower one, only for horizontal wiring. The channels for possible wires are prescribed and are of equal distance on each layer. So the projection of these channels into the plane is a grid-graph. A change of layers is possible using so-called vias which can be viewed as small holes positioned at the grid-points.

There are basically two approaches to model this situation mathematically. Firstly, to consider only the planar projection of the two layers, i.e., considering a two dimensional grid as the appropriate graph model. In this case, the wirings of different nets are allowed to cross because physically these crossing wires run on different layers. But, since each via can only be used by one net, knock-knees (i.e., two different nets bending at the same point) are not allowed. This model, which is widely used (cf., e.g., Burstein (1986), Mehlhorn (1986)), is refered to as to the Manhattan wiring model. The other possibility is, also to model vias by edges, thus to consider a subgraph of the 3-dimensional grid. More precisely, an $n \times m$ two-layer routing graph G is a pair (V, E) where

$$V = \{(x, y, \delta) : x \in n, y \in m, \delta \in 2\} \text{ and}$$
$$\{(x, y, \delta), (v, w, \epsilon)\} \in E \text{ if and only if}$$
$$\text{either} \quad x = v, \ y = w \text{ and } \delta \neq \epsilon \text{ (via-edge)}$$
$$\text{or} \quad x = v, \ |y - w| = 1 \text{ and } \delta = \epsilon = 1 \text{ (vertical edge)}$$
$$\text{or} \quad |x - v| = 1, \ y = w \text{ and } \delta = \epsilon = 0 \text{ (horizontal edge).}$$

Such a two-layer routing graph is shown in Fig. 1.1.

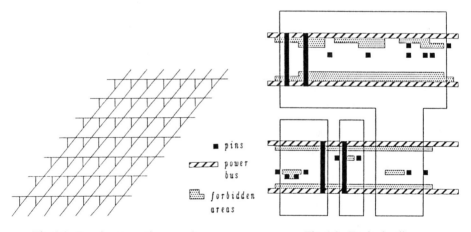

■ pins

▨▨▨ power bus

▨ forbidden areas

Fig. 1.1 Two-layer routing graph   Fig. 1.2 Typical cells

All logic functions are realized by a small library of about 20 basic cells. These cells can be viewed as small rectangles having all the same vertical extension but different width. There is only one exception from this rule, one of these basic cells is a union of two rectangles correspondingly having double "height" of the others. Each cell usually has one output pin and several input pins at fixed positions. Placing a cell means to put it on another layer below the two wiring layers. The effect of such a placement for the routing is twofold: First the placement determines the position of the input and the output pins which have to be connected. Placing the cell results in mapping the pins on some grid points of the projection of the wiring layers. Technically, input pins are located at the first layer, while output pins are located at the second layer. Secondly, after the cells are placed some channels of the second layer are blocked for the wiring, i.e., some forbidden areas are created, cf. Fig. 1.2 .

But, depending on the complexity of the cells, about 75% of the horizontal wiring channels crossing a cell are still useable for logic wiring.

Though a proper placement of the cells is very important for the wireability of the chip, throughout this paper we assume a placement to be given. The problem of placing the cells in a sea-of-cell approach is discussed in detail in Garbers et al. (1990).

To demonstrate the applicability of Steiner trees we use two chips from the new IBM ES/370 chip set. This chip set is a high-performance multi-chip 32-bit processor with S/370 architecture (cf. Spruth (1989) for details). Its realization uses 12.7 mm^2 chips in a 1.0 μ CMOS-technology.

The ES/370 chip set consists of six basic chips. The central processing unit chip (CPU-chip) is the core of this set. It provides the basic CPU-functions and contains logic and a data local store (as a macro) of in total 200,000 transistors.

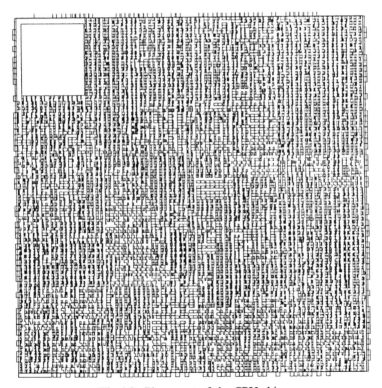

Fig. 1.3 Placement of the CPU-chip

The CPU-chip works in close cooperation with the memory management unit (MMU) chip. This chip performs address translations, contains an 8K byte cache and a 4K byte main storage protection key store, and consists of 800,000 transistors in total.

We will focus our consideration on these two chips. The transistors of the CPU-chip are distributed on roughly 17,000 cells from the standard cell-library plus one macro. These modules contain together 69,709 pins grouped in 19,318 nets (ranging from 2 to 25 pins) which have to be wired. The MMU-chip contains a logic circuit of about 7,700 cells and additionally four macros. The number of pins of this second circuit is 32,442, partitioned into 9,892 nets. Both of these circuits are laid out on 12.7 mm² silicon, or in terms of wiring channels, on 2,340 vertical times 3,564 horizontal channels. A placement of the CPU-chip can be seen in Fig. 1.3. In the left upper corner, the data local store is placed. Fore more details on these chips and the underlying techonology, cf. Koetzle (1987,1987a) and Spruth (1989).

Since we assume the placement to be given we have to solve the following mathematical problem:

Given a subgraph G of a two-layer routing graph (resp. of a grid graph) and a collection $\mathcal{N} = \{\mathcal{N}_0, \ldots, \mathcal{N}_{r-1}\}$ of mutually disjoint subsets of the vertex-set of G. Find r pairwise vertex-disjoint (resp. disjoint in the Manhattan wiring model) Steiner trees $S_0, \ldots, S_{r-1}$ such that S_i spans $\mathcal{N}_i$ in G for every i.

2. The Steiner Problem in Graphs

"A very simple but instructive problem was treated by Jacob Steiner, the famous representative of geometry at the University of Berlin in the early nineteenth century. Three villages A, B, C are to be joined by a system of roads of minimum total length." This remark of Courant and Robbins (1941) led to the name of a problem that originated two hundred years earlier and should have been attributed to the French mathematician Pierre Fermat (1601-1665). He closed his celebrated work on minima and maxima posing the problem: "Étant donnés trois points, en trouver un quatrième tel que la somme de ses distances aux trois points donnés soit minima." (cf. Œuvres de Fermat, 1896). Already around 1640 Evangelista Torricelli and Francesco Cavalieri independently solved the problem geometrically: the fourth point p has to be chosen in such a way that joining the three given points to p creates three angles of exactly 120 degrees - assuming all angles in the triangle induced by the points given are less than 120 degrees.

Over the centuries this problem was rediscovered and generalized by other mathematicians, among them Jacob Steiner (1796-1863). Nowadays the Euclidean Steiner problem is the following problem:

2.1 Euclidean Steiner Problem. Given a (finite) set $T \subset \mathbb{R}^2$. Find a shortest network that interconnects all points of T.

For a more detailed history on the Euclidean Steiner problem see Kuhn (1974). In this paper we will not cover the Euclidean Steiner problem, but refer the interested reader to Gilbert and Pollak (1968), Winter (1985) and for a more popular survey to the recent article by Bern and Graham (1989). Instead we consider a closely related problem first mentioned in Hakimi (1971):

2.2 Steiner Problem in Graphs. Given a graph $G = (V, E)$, a nonnegativ length function $l : E \rightarrow \mathbb{R}_{\geq 0}$ and a set of terminals $T \subseteq V$. Find a connected subgraph of minimal total length containing all terminals, i.e., a *Steiner minimal tree* for T in G.

Notations and Definitions

For a graph $G = (V, E)$ we usually denote the cardinality of the vertex set $V = V(G)$ and the edge set $E = E(G)$ by n and m, respectively. For a vertex $v \in V(G)$ we denote the set of neighbours of v by $\Gamma(v)$ and the degree of v by $d(v)$, i.e., $d(v) = |\Gamma(v)|$. A graph $G = (V, E)$ with an adjoint (nonnegative) length function $l : E \rightarrow \mathbb{R}_{\geq 0}$ is called a *weighted graph*, denoted also as $G = (V, E, l)$. For a subgraph H of G the neighbourhood and degree of v with respect to H are denoted by $\Gamma_H(v)$ and $d_H(v)$, respectively. Given a length function $l : E \rightarrow \mathbb{R}$, the (total) length of a subgraph H is denoted by $l(H)$. For $v, w \in V(G)$ we denote the length of a shortest path from v to w by $p(v, w)$.

Let $T \subseteq V$ be a set of terminals. We denote its cardinality by t. The *complete distance graph* of T with respect to G is given by $G_D(T) = (T, \binom{T}{2}, l_T^D)$, where $l_T^D(\{u, v\}) = p(u, v)$ for $u, v \in T$. A subgraph S of G is called a *Steiner tree* for T in G, if S is a tree containing all terminals, i.e. $T \subseteq V(S)$, and has the property that all leaves of S are terminals. The vertices in $V(S) \setminus T$ are called *Steiner points*.

Special Cases and Complexity

The Steiner Problem in Graphs 2.2 includes as special cases two well known problems from combinatorial optimization. If T contains just two terminals, the Steiner problem reduces to the well studied *shortest path* problem. If on the other hand T consists of the whole vertex set V, the Steiner problem coincides with the *minimal spanning tree* problem. Based on the classical algorithms by Dijkstra (1959) and Prim (1957), Fredman and Tarjan (1987) describe $O(n \log n + m)$ implementations for both problems using Fibonacci heaps. Though Gabow, Galil, Spencer and Tarjan (1986) slightly reduced this bound for the minimal spanning tree problem, this improvement does not help for our purposes.

Observe that these bounds imply that the complete distance graph of a terminal set T and a minimal spanning tree therein can be computed in $O(|T|(n \log n + m))$ and $O(|T|^2)$, respectively.

In contrary to the polynomial solvability of these special cases the general Steiner problem is NP-hard. To see this we consider the following decision problem.

2.3 Steiner Problem in Bipartite Graphs.

Instance: A bipartite graph $G = (V, E)$, a set of terminals $T \subseteq V$ and a number L.

Question: Does there exist a Steiner tree S for T in G such that the number of edges in S does not exceed L, i.e. $|E(S)| \leq L$?

2.4 Theorem (Karp, 1972). *The "Steiner problem in bipartite graphs" (SPBG) is NP-complete.*

Proof. Obviously SPBG $\in$ NP. To see the completeness we transform "Exact cover by 3-sets" (X3C) to SPBG (cf. Garey, Johnson, 1979). Let an arbitrary instance of X3C be given, i.e., a set $X = \{x_0, \ldots, x_{3p-1}\}$ and a set system $\mathscr{C} = \{C_0, \ldots, C_{q-1}\}$ such that $C_i \subseteq X$ and $|C_i| = 3$ for all $i \in q$. Construct an instance of SPBG as follows (cf. Fig. 2.1): let $V = \{v_0\} \cup \mathscr{C} \cup X$, $E = \{\{v_0, C_i\} \mid i \in q\} \cup \{\{C_i, x_j\} \mid x_j \in C_i, i \in q\}$, $T = X$ and $L = 4p$.

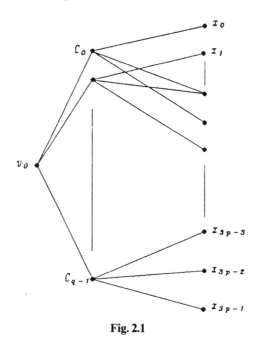

Fig. 2.1

It is easy to see that a Steiner tree S for T in G satisfying $|E(S)| \leq L$ exists if and only if $\mathscr{C}$ contains an exact cover for X. $\square$

Note, that Theorem 2.4 implies that the Steiner Problem in Graphs 2.2 is NP-hard in the strong sense.

Exact Algorithms

We now describe two exponential algorithms which solve the Steiner problem in graphs exactly. They are based on the algorithms for two special cases, i.e. the shortest path problem and the minimal spanning tree problem. For a survey on other exact algorithms, see Winter (1987).

The algorithm of Dreyfus and Wagner (1972) is an application of dynamic programming. Its essence are two recursions, which derive the length of a Steiner minimal tree for a given terminal set from the lengths of the Steiner minimal trees for subsets of the given terminal set. For $K \subseteq V$ and $v \in V \setminus K$ let $s(K \cup \{v\})$

denote the length of a Steiner minimal tree for $K \cup \{v\}$ and let $s_v(K \cup \{v\})$ denote the minimal length of a Steiner tree for $K \cup \{v\}$ satisfying the additional constraint that v has degree at least two. Obviously, $s_v(K \cup \{v\}) \geq s(K \cup \{v\})$.

It is not too difficult to check that the following recursions hold:

(a) $s_v(K \cup \{v\}) = \min_{\emptyset \neq K' \subset K} \{s(K' \cup \{v\}) + s(K \setminus K' \cup \{v\})\}$

and

(b) $s(K \cup \{v\}) = \min\{\min_{w \in K}\{p(v, w) + s(K)\}, \min_{w \notin K}\{p(v, w) + s_w(K \cup \{w\})\}\}$.

These recursions directly imply the correctness of the Dreyfus-Wagner algorithm.

2.5 Dreyfus-Wagner Algorithm.

(1) (initialization)

Compute $p(v, w)$ for all $v, w \in V$. (Observe that for all $w \in T$ we have $s(\{w\} \cup \{v\}) = p(w, v)$.)

(2) (recursion)

Perform the following calculations for all k between 2 and $|T| - 1$:

– For all $K \subseteq T$ such that $|K| = k$ and for all $v \in V \setminus K$ compute

$$s_v(K \cup \{v\}) = \min_{\emptyset \neq K' \subset K} \{s(K' \cup \{v\}) + s(K \setminus K' \cup \{v\})\}.$$

– For all $K \subseteq T$ such that $|K| = k$ and for all $v \in V \setminus K$ compute

$$s(K \cup \{v\}) = \min\{\min_{w \in K}\{p(v, w) + s(K)\}, \min_{w \notin K}\{p(v, w) + s_w(K \cup \{w\})\}\}.$$

The algorithm, as stated above, only computes the length of a Steiner minimal tree S for T. For the explicit construction of S the algorithm has to be supplemented by a (straightforward) backtracking procedure.

For estimating the complexity of the algorithm observe that for the initialisation $O(n^2 \log n + nm)$ steps suffice and that $O(3^t n)$ and $O(2^t n^2)$ are upper bounds for the number of steps executed in recursion (a) and (b), respectively. Summarizing, we have that the Dreyfus-Wagner algorithm computes in $O(3^t n + 2^t n^2 + n^2 \log n + nm)$ steps a Steiner minimal tree S for a given terminal set T.

Observe that the algorithm is polynomial if the number t of terminals is fixed.

Polynomial modifications of the Dreyfus-Wagner algorithm solving the Steiner problem if the graph is planar and all terminals lie on the outer face (or a fixed number of faces) were described independently by Bern (1987), Erickson, Monma, Veinott (1987) and Provan (1988). The basic idea of their algorithms is that the constraint on the position of the terminals allows to order the terminals. Using this ordering one can show that in every step of the recursion it suffices to compute $s_v(K \cup \{v\})$, resp. $s(K \cup \{v\})$, for only a polynomial number of different sets K.

The basic idea of the second algorithm (Lawler (1976)) is to reduce the Steiner problem to a minimal spanning tree problem by specifying all Steiner points with degree at least three: Let S be a Steiner minimal tree for T and let U be the set of Steiner points of degree at least three. Then S corresponds to a minimal spanning tree in the complete distance graph $G(T \cup U)$. The algorithm of Lawler now consists of subsequently testing all possible sets U of $V \setminus T$. Actually, an easy observation shows that not all subsets U have to be considered:

2.6 Observation. *The number of Steiner points of degree at least 3 in a Steiner tree S for T in G is at most $t - 2$.*

Proof. Let s_2 be the number of Steiner points with degree 2 and let s_3 be the number of Steiner points with degree at least 3. Then $|V(S)| = t + s_2 + s_3$ and $2(|V(S)| - 1) = 2|E(S)| = \sum_{v \in V(S)} d_s(v) \geq t + 2s_2 + 3s_3$ implies that $s_3 \leq t - 2$. □

Using this fact, the worst case complexity of the enumerative algorithm can be estimated by $O(2^{n-t}t^2 + n^2 \log n + m)$. Observe that the complexity is polynomial in n, if either t or $n - t$ is constant.

Approximation Algorithms

As we have seen Problem 2.2 is NP-hard. The existence of a polynomial algorithm for its solution is therefore quite unlikely. Because the Steiner problem is also NP-hard in the strong sense and its optimal value is polynomially bounded in the length of the input, even more is true:

2.7 Observation. *Unless $P=NP$ there exists no fully polynomial approximation scheme for the Steiner problem in graphs with integer length function.*

Nevertheless many heuristics and approximation algorithms can be found in the literature. The worst case bounds of the approximation algorithms all rely on the following simple lemma.

2.8 Lemma. *Let $G = (V, E, \ell)$ be a weighted graph and $T \subseteq V$. Let $G_D(T) = (T, E_D, \ell_D)$ be the complete distance graph of T with respect to G, let S_{opt} be a Steiner minimal tree for T in G and let M be a minimal spanning tree in G_D. Then $\ell_D(M) \leq 2(1 - \frac{1}{a}) \cdot \ell(S_{opt})$, where a denotes the number of leaves in S_{opt}.*

Proof. A spanning tree satisfying the above inequality arises from the tour around the outer face of a planar embedding of S_{opt} by omitting the longest path between two neighbouring leaves. □

The graph $G = (V, E, \ell)$ with $V = \{v_0, \ldots, v_k\}$, $E = \{\{v_i, v_{i+1}\} \mid 1 \leq i < k\} \cup \{\{v_k, v_1\}\} \cup \{\{v_0, v_i\} \mid 1 \leq i \leq k\}$, $T = \{v_1, \ldots, v_k\}$ and $\ell(e) = 1$ if $v_0 \in e$ and $\ell(e) = 2 - \epsilon$ for some $\epsilon > 0$ otherwise shows that the bound in Lemma 2.8 is essentially best possible.

We now consider two Steiner tree algorithms based on Lemma 2.8 in greater detail. The first one appeared, independently, at several places in the literature, cf. Winter (1987). Among others it was published by Kou, Markowsky and Berman (1981), who also proved the worst case bound of Lemma 2.8.

2.9 KMB-Algorithm.
(1) Compute the complete distance graph $G_D(T) = (T, E_D, \ell_D)$ for T with respect to G.
(2) Find a minimal spanning tree M_D in G_D.
(3) Transform M_D into a subgraph G_S of G, i.e. replace each edge of M_D by a corresponding shortest path in G.
(4) Find a minimal spanning tree M_S in G_S.
(5) Transform M_S into a Steiner tree S_{KMB} for T by subsequently omitting leaves which are not terminals.

The complexity of the KMB-algorithm is easily observed to be $O(|T| \cdot (n \log n + m))$. Together with Lemma 2.8 this shows that the KMB-algorithm computes in $O(|T| \cdot (n \log n + m))$ a Steiner tree S_{KMB} for Problem 2.2 such that $\ell(S_{KMB})/\ell(S_{opt}) \leq 2(1 - \frac{1}{a})$.

The most time consuming step in the KMB-algorithm is step (1), i.e., the computation of the complete distance graph. Mehlhorn (1988) proposes to substitute this computation by constructing a graph $G'_D = (T, E'_D, \ell'_D)$ such that every minimal spanning tree of G'_D is a minimal spanning tree of G_D. This graph G'_D can be computed by a single (one-to-all) shortest path calculation, yielding an $O(n \log n + m)$ approximation algorithm for the Steiner problem satisfying the same worst case bound. Observe that this time bound is essentially best possible.

A different approach for speeding up the KMB-algorithm was suggested by Wu, Widmayer and Wong (1986). The main idea of their construction is to combine the steps (1) and (2) of the KMB-algorithm. Essentially they proceed similar as Kruskal's minimal spanning tree algorithm (Kruskal (1956)): in the beginning the t terminals are considered as disjoint trees, which are gradually merged into a single tree. Merging of the subtrees occures along shortest paths - similar to the merging in Kruskal's algorithm along shortest edges. In particular this procedure guarantees that the length of the constructed tree does not exceed the length of a minimal spanning tree in the complete distance graph.

The algorithm of Wu, Widmayer and Wong is a fundamental tool in the global routing algorithm which we will present in the next section. We therefore describe it here in greater detail.

We first define three functions $source(v)$, $pred(v)$ and $length(v)$. Thereby $source(v)$ assigns to $v \in V$ a terminal with minimal distance (denoted by $length(v)$) to v, i.e., length $(v) = p(source(v), v) \leq p(s, v)$ for all $s \in T$. The function $pred(v)$ assigns to v the immediate predecessor of v on a shortest path from v to $source(v)$. Similiar as in Kruskal's algorithm a priority queue Q is used containing all edges for possible extensions of the forest already generated. However, in contrary to Kruskal's algorithm two different kinds of edges have to be distinguished: edges connecting two different trees of the forest and edges used for growing a single tree. The entries kept in the priority queue are of the form (v, s, d, p_v, p_s), where

v, p_v, p_s are elements of V, s is a terminal and d the length of a path from v to s. The interpretation of the tupel depends on whether v is a terminal or not.

In case v is a terminal, we have $source(p_v) = v$, $source(p_s) = s$, $\{p_v, p_s\} \in E$ and $d = length(p_v) + length(p_s) + \ell(p_v, p_s)$. That is, different trees of the forest can be merged by inserting the edge $\{p_v, p_s\}$.

In case v is not a terminal, we have $source(p_s) = s$ and p_v is not defined (written as $p_v = nil$), $\{p_s, v\} \in E$ and $d = length(p_s) + \ell(p_s, v)$. That is, the tree containing s can be enlarged by the node v.

The priority queue Q is sorted by increasing d-values of the tupels. For initialization and updating of the queue during the algorithm we need two operations *insert* and *deletemin*. Additionally we implement a data structure supporting the operations *makeset, find* and *link*. These operations are used to decide whether two nodes belong to the same tree and for merging two trees.

2.10 WWW-Algorithm.
(1) (initialization)
 – for all $s \in T$ let $source(s) = s$, $length(s) = 0$, $pred(s) = nil$,
 – for all $v \in V \backslash T$ let $source(v) = nil$, length $(v) = \infty$, $pred(v) = nil$,
 – for all $s \in T$ and all $v \in \Gamma(s)$ insert $(v, s, \ell(v, s), nil, s)$ in the priority queue Q,
 – define t disjoint trees, each consisting of exactly one terminal,
 – define a list ST, containing the edges used in merging two trees.
(2) (determination of the merging edges)
 repeat the following step until all terminals are contained in a single tree: find and delete a tupel (v, s, d, p_v, p_s) from Q having minimal d-value.
 (2.1) $source(v) = nil$:
  – let $source(v) = s$, $lEngth(v) = d$ and $pred(v) = p_s$,
  – for all $w \in \Gamma(v)$ such that $source(w) = nil$ insert $(w, s, d + \ell(v, w), nil, v)$ in the priority queue Q.
 (2.2) $source(v)$ and s belong to the same tree:
  – do nothing.
 (2.3) $source(v)$ and s belong to different trees:
  (2.3.1) $v \in T$:
   – merge the two trees containing the terminals s and t,
   – insert the edge $\{p_v, p_s\}$ in the list ST.
  (2.3.2) $v \in V \backslash T$:
   – insert $(source(v), s, d + length(v), v, p_s)$ in the priority queue.
(3) (construction of the Steiner tree)
 – start with an empty tree S_{WWW},
 – repeat the following step until the list ST is empty:
  Remove the topmost edge from ST and insert it in S_{WWW}. Trace the pred-function of both nodes of this edge backwards until a terminal or a node already contained in S_{WWW} is reached. Insert the encountered edges in S_{WWW}.

Using standard implementations for the two data structures supporting the *insert, deletemin* and *makeset, find* and *link* operations, respectively, (cf. e.g. Tarjan

(1983)), it is easily checked that the complexity of the WWW-algorithm is bounded by $O(m \log n)$.

Grid Graphs

Recall that the Euclidean Steiner problem asks for a shortest network connecting a given set of terminals in the plane. Thereby the distance of two points is determined using the Euclidean or L_2-norm. For applications in VLSI-design the L_1-norm (Manhattan distance) is of particular interest.

2.11 Rectilinear Steiner Problem. Given a finite set $T \subseteq \mathbb{R}^2$ of terminals in the plane. Find a Steiner minimal tree for T with respect to the L_1-norm.

For applicability of the algorithms for Problem 2.2 to this problem a theorem of Hanan (1966) is fundamental. This result shows that the rectilinear Steiner problem may be viewed as a Steiner problem in a grid graph. To state the theorem precisely we need some definitions.

Given two sets $X = \{x_0, \ldots, x_{n-1}\} \subseteq \mathbb{R}$ and $Y = \{y_0, \ldots, y_{m-1}\} \subseteq \mathbb{R}$ we construct an $n \times m$ grid graph with vertex set $X \times Y$ as follows. Two vertices (x_{i_1}, y_{i_1}) and (x_{i_2}, y_{i_2}) are connected by an edge if and only if $|i_1 - i_2| + |j_1 - j_2| = 1$. The length of this edge is then $|x_{i_1} - x_{i_2}| + |y_{j_1} - y_{j_2}|$, i.e. the Manhattan distance of the two vertices. For a terminal set $T \subseteq \mathbb{R}^2$ the grid graph generated by $X = \{x \in \mathbb{R} \mid \text{there exists a } y \in \mathbb{R} \text{ such that } (x, y) \in T\}$ and $Y = \{y \in \mathbb{R} \mid \text{there exists an } x \in \mathbb{R} \text{ such that } (x, y) \in T\}$ is denoted by $G(T)$.

2.12 Theorem (Hanan, 1966). *Let $T \subseteq \mathbb{R}^2$ be the terminal set of a rectilinear Steiner problem with associated grid $G(T)$. Then there exists a Steiner minimal tree S for T which is contained in $G(T)$.*

The proof is not difficult, but a bit technical and will therefore be omitted. As an immediate consequence of Hanan's theorem one observes that given a Steiner problem in a grid graph one can restrict this grid graph to the grid associated to the terminal set. This reduces the number of vertices of the graph and therefore speeds up the algorithms.

Having in mind the applications in VLSI-layout the following generalization (which can be proved along the same lines as Hanan's theorem) is useful.

Recall that a grid graph can be viewed as part of the integer lattice grid $\mathbb{Z} \times \mathbb{Z}$. A rectilinear hole in a grid graph is obtained by deleting all vertices and edges of a rectilinear subgrid with corner points (x_0, y_0), (x_0, y_1), (x_1, y_0) and (x_1, y_1).

2.13 Corollary. *Given a Steiner problem in a grid graph with rectilinear holes. Then a Steiner minimal tree can be found in the grid generated by the rows and columns containing terminals or corners of the holes.*

We mention that a corresponding result holds for the two-layer routing graph as well.

Though at first sight there is hope that the Steiner problem in grid graphs is easier than the general Steiner problem, Garey and Johnson (1976) showed that even this restricted problem is NP-complete. Nevertheless at least for grid graphs without holes the worst case bound of the (presented) approximations algorithms can be improved. This is an immediate consequence of a result due to Hwang (1976). He showed that in a grid graph the length of a minimal spanning tree for a set T is at most $3/2$ times the length of a Steiner minimal tree. For grid graphs with holes on the other hand, one can easily construct Steiner problems so that this ratio is arbitrarily close to 2.

A survey of known results and algorithms for the rectilinear Steiner problem can be found in Richards (1989).

Packing of Steiner Trees

Up to now we only considered the problem of finding a Steiner minimal tree for a single terminal set. A related but more general problem is to consider k terminal sets simultaneously, i.e. to find pairwise vertex- or edge-disjoint Steiner trees for given terminal sets.

2.14 Vertex-(Edge-) Disjoint Packing Problem of Steiner Trees. Given a graph $G = (V, E)$ and k terminal sets $T_0, \ldots, T_{k-1} \subseteq V$. Find k vertex-disjoint (edge-disjoint) Steiner trees $S_0, \ldots, S_{k-1}$ for $T_0, \ldots, T_{k-1}$, respectively.

A special case of the (edge-disjoint) packing problem is the case $T_0 = \ldots = T_{k-1} = V$, i.e. to ask for k edge-disjoint spanning trees. In terms of matroid theory this problem can be stated as finding a base of the matroid obtained by taking the sum of k copies of the circuit matroid of G. Hence, this problem is solved by the greedy algorithm with Edmonds' matroid partition algorithm (Edmonds (1965)) taken as independence oracle. Observe that in a weighted graph the greedy algorithm even finds k edge-disjoint spanning trees with minimal total weight. For a discussion of this problem and its solution see Clausen and Hansen (1980). An efficient implementation is described in Roskind, Tarjan (1985). Their algorithm runs in $O(m \log m + k^2 n^2)$, reducing to $O(k^2 n^2)$ in the cardinality case.

Another special case of the packing problem is to consider only two-element terminal sets, i.e. to pack paths. In contrary to the spanning tree problem packing paths is NP-complete, even if the graph is restricted to be planar.

2.15 Edge-Disjoint Packing of Paths in a Planar Graph.
Instance: A planar graph $G = (V, E)$, a positive integer k, pairs of vertices $(s_0, t_0), \ldots, (s_{k-1}, t_{k-1})$.
Question: Does G contain k mutually edge-disjoint paths $P_0, \ldots, P_{k-1}$ such that P_i connects s_i with t_i?

2.16 Theorem (Kramer, van Leeuwen, 1984). *Edge-disjoint packing of paths in planar graphs (PPP) is NP-complete.*

Proof. We transform "Planar 3-Satisfiability" (P3SAT) to PPP. (For the NP-completeness of P3SAT see Lichtenstein (1982).) Let an arbitrary instance of P3SAT be given by p variables $x_0, \ldots, x_{p-1}$ and q clauses $C_0, \ldots, C_{q-1}$ such that the graph $G = (V, E)$ with vertex set $V = \{x_i \mid i \in p\} \cup \{C_j \mid j \in q\}$ and edge set $E = \{\{x_i, C_j\} \mid i \in p, j \in q, x_i \in C_j \text{ or } \bar{x}_i \in C_j\}$ is planar.

In a first step we replace each vertex x_i by a path such that the graph stays planar and each inner vertex of the path is connected to exactly one "clause" vertex. We now replace each path corresponding to a variable k and each vertex corresponding to a clause by a suitable graph and introduce appropriate vertex pairs (s_i, t_i). The transformation of a path corresponding to a variable is shown in Fig. 2.2.

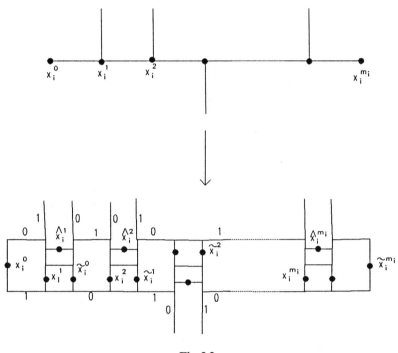

Fig. 2.2

The vertex pairs which have to be connected by paths are $(x_i^0, \tilde{x}_i^0), \ldots, (x_i^{m_i}, \tilde{x}_i^{m_i})$. Observe that choosing the upper or lower path for the pair $(x_i^0, \tilde{x}_i^0)$ uniquely determines all paths for the remaining pairs $(x_i^j, \tilde{x}_i^j)$. Hence, an assignment of "false" or 0 to the upper edge and "true" or 1 to the lower edge carries over to the remaining edges as shown in Fig. 2.2.

Finally, we replace every vertex C_j corresponding to a clause containing the variables x, y and z, say, by a graph as shown in Fig. 2.3 and introduce the vertex pairs $(\hat{x}, \tilde{x}), (\hat{y}, \tilde{y}), (\hat{z}, \tilde{z}), (s_j, \tilde{s}_j)$ and $(t_j, \tilde{t}_j)$. (The vertices $\hat{x}, \hat{y}$ and $\hat{z}$ are contained in the subgraph for the paths corresponding to the variables x, y, and z, cf. Fig. 2.2.)

The assignment of the edges in Fig. 2.3 with 0 and 1 is the one induced by the assignment in the paths corresponding to variables. The vertices $\tilde{x}, \tilde{y}$ and $\tilde{z}$ are placed on the 0-edge if the corresponding variable is negated in the clause C_j and on the 1-edge otherwise, i.e. Fig. 2.3 shows the transformation of the clause $x \vee \bar{y} \vee z$.

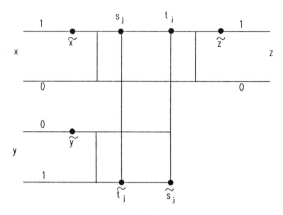

Fig. 2.3

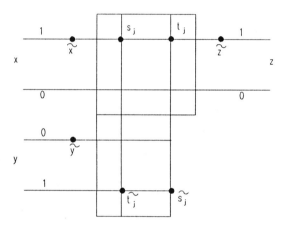

Fig. 2.4

It is easily checked that the paths for $(s_j, \tilde{s}_j)$ and $(t_j, \tilde{t}_j)$ exist if and only if at least one of the paths for x, y and z enters the clause-subgraph on the "correct" edge. Hence, edge-disjoint paths exist if and only if there is a truth-assignment for the satisfiability problem.   □

In fact, a slight modification of this proof can be used to show that also vertex-disjoint packing of paths in planar graphs is NP-complete. Namely, replace every vertex corresponding to a clause by a subgraph as shown in Fig. 2.4.

Polynomially solvable versions of Problem 2.15 are obtained by further restricting the position of the terminals and the class of graphs. Frank (1985) and Becker, Mehlhorn (1986), for example, describe a polynomial algorithm for planar graphs in which all inner vertices have even degree and all terminals lie on the boundary of the outer face.

For the NP-completeness of the packing problem of paths it is essential that the number k of paths is part of the input. For fixed k, Robertson and Seymour devised a polynomial algorithm, cf. their paper "An outline of a disjoint paths algorithm" in this volume. In fact they even solved a more general problem (see also Robertson, Seymour (1986)).

2.17 Theorem (Robertson, Seymour, 1986). *Let K be fixed. Given a graph $G = (V, E)$, k terminal sets $T_0, \ldots, T_{k-1} \subseteq V$ such that $\sum_{i \in k} |T_i| \leq K$. Then one can find k vertex-disjoint Steiner trees $S_0, \ldots, S_{k-1}$ for $T_0, \ldots, T_{k-1}$, respectively, in $O(n^2 m)$ steps.*

Note, that this theorem implies the existence of an $O(nm^3)$ algorithm for the edge-disjoint problem. To see this, replace each vertex v by a clique of size $d_G(x)$ and connect each vertex of the clique with exactly one vertex in $\Gamma(v)$. For each terminal add an additional vertex connected with all vertices of the corresponding clique.

If, on the other hand, one restricts the number of terminal sets but not the number of terminals, we will show that the problem remains NP-complete:

2.18 Edge-Disjoint Packing of Two Steiner Trees in Planar Graphs.
Instance: A planar graph $G = (V, E)$, two terminal sets $T_0, T_1 \subseteq V$.
Question: Does G contain two edge-disjoint Steiner trees S_0 and S_1 for T_0 and T_1, respectively?

2.19 Theorem. *Edge-disjoint packing of two Steiner trees in planar graphs (P2P) is NP-complete.*

Proof. We transforms P3SAT to P2P. Let an arbitray instance of P3SAT be given by p variables $x_0, \ldots, x_{p-1}$ and q clauses $C_0, \ldots, C_{q-1}$ with $C_i \subseteq \{x_0, \ldots, x_{p-1}, \overline{x}_0, \ldots, \overline{x}_{p-1}\}$ and $|C_i| = 3$ such that the graph $G = (V, E)$ with

$$V = \{x_0, \ldots, x_{p-1}, \overline{x}_0, \ldots, \overline{x}_{p-1}, y_0, \ldots, y_{p-1}, z_0, \ldots, z_{p-1}, C_0, \ldots, C_{q-1}\} \text{ and}$$

$$E = \{\{y_i, x_i\} \mid i \in p\} \cup \{\{y_i, \overline{x}_i\} \mid i \in p\} \cup$$
$$\{\{z_i, x_i\} \mid i \in p\} \cup \{\{z_i, \overline{x}_i\} \mid i \in p\} \cup$$
$$\{\{z_i, y_{i+1}\} \mid i \in p-1\} \cup \{\{z_{p-1}, y_0\}\} \cup$$
$$\{\{x_i, C_j\} \mid i \in p, j \in q, x_i \in C_j\} \cup \{\{\overline{x}_i, C_j\} \mid i \in p, j \in q, \overline{x}_i \in C_j\}$$

is planar.

We replace every vertex x_i and $\overline{x}_i$ by a path of length q and connect the adjacent clause vertices to a vertex of the path in such a way that every vertex of the path has degree at most 3 and that the graph stays planar. Furthermore, we double every edge $\{z_i, y_{i+1}\}$ and replace the edge $\{z_{p-1}, y_0\}$ by two vertices u and v and two pairs of parallel edges $\{u, y_0\}$ and $\{v, z_{p-1}\}$. Finally, we insert an additional vertex C_j^k, $j \in 3$, on each edge connecting a clause vertex C_j and a vertex of a path corresponding to a variable, cf. Fig. 2.5.

The new graph is denoted by $\overline{G}$. Observe that $\overline{G}$ is still planar. We define a packing problem in $\overline{G}$ by choosing $T_0 = \{u, v\}$ and $T_1 = \{u, v\} \cup \{C_j^i \mid j \in q, i \in 3\}$.

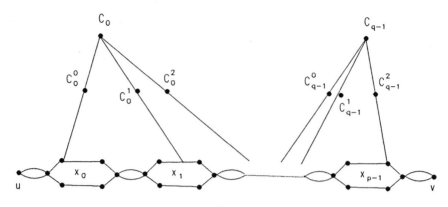

Fig. 2.5

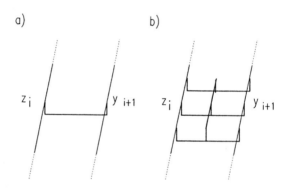

Fig. 2.6

It is not difficult to see that $\overline{G}$ contains edge-disjoint Steiner trees S_0 and S_1 for T_0 and T_1 if and only if there exists a truth-assignment for the satisfiability problem.   □

We close this section by showing how Theorem 2.19 implies that packing Steiner trees in two-layer routing graphs is difficult. (Note that in two-layer routing graphs the edge-disjoint and vertex-disjoint version of Problem 2.14 coincide.)

2.20 Packing Two Steiner Trees in Two-Layer Routing Graphs.
Instance: A two-layer routing graph $G = (V, E)$ and two terminal sets $T_0, T_1 \subseteq$
V.
Question: Does G contain two disjoint Steiner trees S_0 and S_1 for T_0 and T_1, respectively?

2.21 Corollary. *Packing of 2 Steiner trees in two-layer routing graphs is NP-complete.*

Proof. Proceed as in the proof of Theorem 2.19. If one replaces in $\overline{G}$ all parallel edges by single edges, the graph obtained is planar and has degree at most 3. It is not difficult to see that this graph can be embedded in a two-layer routing graph in such a way that z_i and y_{i+1} (and similiarly u, y_0 and v, z_{p-1}) are connected as shown in Fig. 2.6 a). This connection is then replaced by the subgraph shown in Fig. 2.6 b).

One easily checks that this subgraph correctly models the use of the two parallel edges in $\overline{G}$. □

3. A Routing Algorithm for VLSI-Chips

In this final part of the paper we show how ideas and results presented in the last chapter can be incorporated in a procedure to solve the routing problem in VLSI-layout. As mentioned before, we will discuss the features of this algorithm at layouts for the central processing unit (CPU) chip and the memory managing unit (MMU) chip of the new IBM ES/370 chip set.

Recall the mathematical formulation of the routing problem: given a subgraph G of a two-layer routing graph and a collection $\mathcal{N} = \{\mathcal{N}_0, \ldots, \mathcal{N}_{r-1}\}$ of mutually disjoint subsets of the vertex-set of G, i.e. the nets in the layout-problem. Find pairwise vertex disjoint Steiner trees $S_0, \ldots, S_{r-1}$ in G such that S_i is a wiring for net $\mathcal{N}_i$. Note that in both layout examples the corresponding graphs have $2340 \times 3564 \times 2 = 18,679,520$ vertices.

The order of magnitude of these kind of problems suggests an hierarchical approach. We have chosen a top-down scheme proceeding in two phases. In a first step we perform a global routing were the wiring of the nets is only determined roughly. At this stage, the original graph is partitioned into subgraphs called pages. For each net it will be specified through which of these pages the wiring goes. Then, in a second step, the local (or detailed) routing step, this global information is used to find, page by page, the actual layout of the wires. For a general discussion and other approaches of the global routing problem we refer to Kuh and Marek-Sadowska (1986) and Sangiovanni-Vincentelli (1987).

In more detail our global routing approach can be described as follows: starting from the original two-layer routing graph G we construct a new two-layer routing graph by contracting pages. Fix positive integers k and l. For $\delta \in 2$ and positive integers i, j let $V_{i,j,\delta} = \{(x, y, \delta) \mid ki \le x < k(i+1),\ lj \le y < l(j+1)\}$ be a partition of the vertex-set of G. We condense each of these $V_{i,j,\delta}$ to a single vertex, omit loops and replace parallel edges by a single edge. Of course, the resulting graph is again a two-layer routing graph. On this new graph we impose additionally a capacity function associating to every edge a number not exceeding the number of parallel edges condensed to this edge. In fact, determining the "right" capacity function for this global routing graph is quite important for a successful local routing.

The induced global routing problem then is given by the collection $\overline{\mathcal{N}} = \{\overline{\mathcal{N}}_0, \ldots, \overline{\mathcal{N}}_{r-1}\}$ of global nets, where $(i, j, \delta) \in \overline{\mathcal{N}}_k$ if there exists $p \in V_{i,j,\delta}$ with $p \in \mathcal{N}_k$. Again, projecting the problem to the planar grid graph yields the corresponding Manhattan wiring problem. The task of the global routing step is to find a wiring for each net $\overline{\mathcal{N}}_k$ such that the number of nets using a particular edge of the global routing graph does not exceed the capacity of this edge. Additionally, the total net length should be as small as possible. Ideally, the wiring for each net is a Steiner minimal tree. There are some arguments for this: It is intuitively clear that a shorter net length of the global routing makes it easier to get a feasible local routing. Even more, a short global routing usually implies a short local routing and hence a short wire length at the end. This is important for two reasons: Firstly, decreasing the total wire length decreases the fault-susceptibility of the chip. Secondly and even more important, the cycle-time of a chip, which basically determines the speed of the processor, is strongly influenced, besides by the delay of each cell, by the length of the interconnections. Therefore, a shorter wiring may considerably reduce the cycle-time and consequently speeds up the processor. However, not all nets are equally important for the cycle-time. Observe that minimizing the length of one particular net and obeying the capacity restrictions may cause an increase of the length of other nets. Thus, we will introduce weights associated with the nets indicating how important a short realization of this net is with respect to the cycle-time.

In our examples we have chosen $k = 60$ and $l = 108$, i.e., pages have 108 horizontal channels and 60 vertical channels. This size was motivated by physical side constraints, viz. the distance of the power supply and the standard "height" of the cells which cover 27 horizontal channels. In both cases this leaves us with a 39×33 two-layer routing graph for the global routing with edge lengths (scaled to) 1.0 for horizontal edges and 1.2 for vertical edges. The length of the via edge was put to 0.1 by empirical reasons. Actually, this length is a parameter which can be used to penalize vias. If this value is zero we obtain the Manhattan model.

The number of (nontrivial) nets in the global routing problem was 10196 for the CPU-chip and 5204 for the MMU-chip. Thus, in both cases nearly half of the nets are completely contained in one single page. Precise net list statistics for these two global-routing problems are given in Table 3.1.

The capacities of the global routing graph are determined by a function considering the number of edges connecting the two corresponding pages as well as the number and the distribution of pins in these pages. This function is quite involved and we will not go into details. Instead we show in Fig. 3.1 the capacities as chosen for the global routing of the CPU-chip. Here, the via edges are skipped since their capacities are large and not restrictive.

Table 3.1

	2	3	4	5	6	7	8	9	10	11	12	13
CPU-chip	6010	1822	808	503	353	248	129	91	66	61	28	21
MMU-chip	3245	866	395	225	165	113	64	29	45	11	17	13
	14	15	16	17	18	19	20	21	22	23	24	25
CPU-chip	12	7	1	1	4	11	7	4	7	1	0	1
MMU-chip	6	10	1	1	1	0	1	1	0	0	0	0

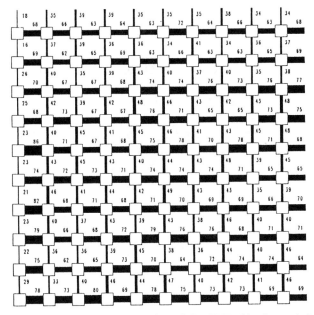

Fig. 3.1 Capacities for the global routing of the CPU-chip (lower left corner)

Observe that some of the edges may have capacity zero, for example those edges which are incident to a vertex corresponding to a page which is covered by a macro. Note that these holes particularly rule out the validity of Hwang's bound of $\frac{3}{2}$ for approximating the Steiner-minimal tree by a spanning tree for the corresponding distance graph, cf. Section 2.

Before focusing on how to solve these global routing problems we will briefly go one step further indicating how the information obtained in the global routing will be used to obtain the final wiring. The global routing determines the boundaries of the pages which are intersected by a certain net. Now starting in the left upper corner of the global routing graph we blow up the vertices and wire the corresponding pages row by row. For a single page the terminal set of a net is specified firstly by the intersection of the pin-set with the page under consideration. Secondly by the transfer-pins at the upper and the left border arising from the local routing in previous pages. Finally, by sets of vertices corresponding to boundary lines on the right and lower side. Of course, concerning these last sets of terminals a Steiner tree has to be found for each net connecting the internal pins with an arbitrary vertex of the boundary set.

Note that one may view the local routing problem for a single page as a special case of the global routing problem in which all edge capacities are 0 or 1. In the remainder of this paper we will focus on the global routing discussing special aspects of the local routing only briefly at the end.

Lower Bounds

As we have seen, cf. Lemma 2.8, determining the length of a minimal spanning tree in the distance graph for a certain set T of terminals yields a bound for the length of a Steiner minimal tree for T which is at most $\left(2 - \frac{1}{|T|}\right)$ times this desired value. But in practice the bound given ty the minimal spanning trees is much better, cf. Table 3.2. We have computed a Steiner minimal tree for each net in the global routing problem of the MMU-chip not taking into consideration the capacities (after omitting all edges with capacity zero). These computations are done by a version of the Dreyfus-Wagner algorithm 2.5. Observe that though we have implemented several reductions, e.g., Corollary 2.13, the actual computation took a few days of computing time. By a more expensive implementation this time could easily be reduced but will definitely stay in a range which is beyond any practical use. In the second column of Table 3.2 the total length of the corresponding spanning trees is indicated yielding for example a factor of 1.03 instead of 1.21 which is the average bound coming from Lemma 2.8. Finally, in column four, the total length of the "embedded" spanning trees for all nets is given. These lengths plus the corresponding embeddings were obtained with an implementation of the Kou-Markowsky-Berman algorithm 2.9.

Table 3.2

	Steiner minimal trees (lower bound) total length	Minimal spanning trees		KMB-algorithm	
		total length	percentage of lower bound	total length	percentage of lower bound
CPU-chip	79437.2	81476.5	102.57	80684.0	101.57
MMU-chip	42455.8	43326.0	102.05	42958.7	101.18

The KMB-algorithm and the WWW-algorithm

All these lengths are computed by considering the nets independently, i.e., not regarding the capacity function. These numbers do not tell too much about the expected wire length for the corresponding packing problem and in particular not whether this problem is solvable at all. It can be obtained from the results that, for example in case of the KMB-algorithm, by adding up all trees the edges are used up to 340% of the allowed capacity. Obeying the capacity constraints in the most straightforward way, viz. the nets are routed subsequently and an edge gets "blocked" as soon as the number of nets using this edges reaches the capacity of this edge, yields e.g. an increase of the total net length from 42958.7 to 44174.3, thus by 2.8%, still leaving 34 nets unrouted, cf. Table 3.3. Alternatively, the Wu-Widmayer-Wong (WWW-) algorithm 2.10 was implemented and used in the same way.

Cost Functions

A common tool to minimize congestions in such a sequential a-net-at-a-time approach is to penalize edges which are heavily used, cf. e.g. Sangiovanni-Vincentelli (1987). This means that the current length of an edge e is no longer just the original edge-length $l(e)$ but the original edge-length plus a penalty term

Table 3.3

	KMB-algorithm		WWW-algorithm	
	total length	number of nets not routed	total length	number of nets not routed
CPU-chip	80961.7	14	81519.2	16
MMU-chip	44174.3	34	44980.9	21

depending on the capacity $c(e)$ of this edge and the number of nets which use already the edge e at the present time t. This last quantity is denoted by $f(e,t)$. There are several cost functions that can be used to achieve the goal described above. We start discussing two kinds of cost functions. Later on we will introduce a third one. These first two types of cost functions are the following:

(A) $k_\alpha^A(e,f) = \alpha \cdot l(e) + (1-\alpha)\frac{c(e)}{c(e)-f(e,t)}$ and

(B) $k_\beta^B(e,f) = \beta \cdot l(e) + (1-\beta)\frac{c(e)}{2c(e)-f(e,t)}$, if $f(e,t) < c(e)$

　　 arbitrarily large otherwise.

The first cost function heavily penalizes a high load of an edge while in the second cost function the original edge-length dominates and the penalty term has only the effect of choosing the "less used" path in a class of paths having the same original length.

Queueing the Nets

Solving a packing problem for Steiner trees by simply solving a sequence of Steiner tree problems is certainly a very straightforward heuristic. However, different sequences may drastically influence the solvability in general and the total net length in particular. An empirical observation is, cf. e.g. Froleyks, Korte and Prömel (1987), that "shortest first" is a good strategy, which means that the nets are sorted according to lengths of the minimal spanning trees in their corresponding distance graphs and then they are routed in increasing order.

Here we renounced sorting the nets with respect to these lengths but according to their priority for the timing of the chip. This leads to shortening particularly critical paths and so to improve the performance. To overcome the drawbacks of sequential strategies we emphasize on a simultaneous routing procedure.

Modifying the WWW-Algorithm

A first step in building up trees simultaneously is to grow successively a forest for each net only containing edges up to a given length. More precisely: given a sorted net list and a strictly increasing sequence of numbers $\alpha_0, \ldots, \alpha_{v-1}$. Use the WWW-algorithm 2.10 to grow a Steiner tree for the first net. Stop this procedure as soon as the length of the generated edge in the complete distance graph exceeds α_0. Thus, at this stage in step (3) of the algorithm only paths, resp. forests, corresponding to previously generated edges of the complete distance graph are constructed. This yields a partially routed first net. Next we proceed in the same way with the second net, then with the third one and so on, finishing the first stage. At the i-th stage each net is successively extended using a (modified) WWW-algorithm with α_{i+1} as the limit. Finally, at the $(v+1)$st stage we complete all nets.

The following straightforward modification of the WWW-algorithm allows to handle trees (the sets of terminals which are already connected) similiar as single terminals. We just treat the terminals of each such tree as one root.

3.1 Modified WWW-Algorithm.
Input: A Steiner problem with t sets of vertices $T_0, \ldots, T_{t-1}$. Let $T = \bigcup_{i \in t} T_i$.
Output: A Steiner tree connecting these sets of terminals.

Replace step (1) of the WWW-algorithm 2.10 by the following step (1′).

(1′) (initialization)
- for all $s \in T$ let $source(s) = s$, $length(s) = 0$, $pred(s) = nil$,
- for all $v \in V \setminus T$ let $source(v) = nil$, $length(v) = \infty$, $pred(v) = nil$,
- for all $s \in T$ and $v \in \Gamma(v)$ insert $(v, s, l(v, s), nil, s)$ in the priority queue Q,
- define t disjoint (imaginary) trees, each tree consisting of the terminals of one set T_i,
- initialize the list ST.

Note that modifying the WWW-algorithm in this way does not increase its worst-case complexity.

Connecting Three Parts
The proposed KMB-, resp. WWW-algorithm guarantees a fast and good approximation of a Steiner minimal tree for a given terminal set. Observe that in case of two terminal nets both of these algorithms degenerate to finding a shortest path between these two terminals. But already in case of three terminal nets each of these heuristics may produce a solution which is far from being optimal. On the other hand, Steiner minimal trees for 3 terminals are still quite fast to determine.

Recall that a solution of a Steiner problem with only three terminals needs at most one Steiner point with degree at least 3, cf. Observation 2.6, and therefore a Steiner minimal tree can be computed in $O(n^2 \log n)$ steps choosing successively each vertex as such a Steiner point. In fact usually the number of candidates is much smaller. This idea can easily be extended to connect forests containing exactly three trees. Here we reduce the edge cost on the edges in these three trees to zero, pick one terminal out of each tree and perform the same procedure as before. This procedure is called 3EXACT and was very effectively inserted between the different stages of the modified WWW-algorithm.

Before the WWW-algorithm starts with its first round the sorted list of nets is sequentially checked for three terminal nets. If a net consists of 3 terminals only, it is routed by 3EXACT and removed from the list, provided its length is smaller than α_0. Otherwise it is marked as a three terminal net, i.e., these nets will no longer be considered by further iterations of the WWW-algorithm, but they will only be routed by 3EXACT at a later stage. This is also motivated by the idea of routing short nets first.

Generally, before the modified WWW-algorithm starts its i-th round, the net list is checked for partially routed nets having exactly three components. These components are connected by 3EXACT, provided the connection is shorter than α_i. Otherwise the net is marked and will be treated as indicated before.

Additionally, a 4EXACT algorithm was designed and inserted analogously into the WWW-algorithm. But here the savings did no longer pay for the additional computing time.

Post-Optimizing

After having routed all nets we try to shorten the net-lengths in a post-optimizing procedure. This is done by going step by step through the net list and removing successively in each net all pathsegments between terminals and/or Steiner points with degree at least 3. Deleting such a path leaves a forests with two components. Now these two components are reconnected with aid of a standard shortest path algorithm yielding a new tree which is at most as long as the old one. This procedure can be repeated several times.

Global Routing Algorithm – Version A

Collecting all the ingredients described above we obtain the following global routing algorithm.

3.2 Global Routing Algorithm (A).

Input: A subgraph G of a two-layer routing graph and a collection $\mathcal{N} = \{\mathcal{N}_1, \ldots, \mathcal{N}_{r-1}\}$ of subsets of the vertex set of G. A length function l which associates to every edge of G its length and a capacity function c associating to each edge of G a positive integer, viz. the capacity $c(e)$ of this edge.

Output: Pairwise Steiner trees $S_0, \ldots, S_{r-1}$ in G such that S_i gives a wiring for the net $\mathcal{N}_i$ and that the number of trees using edge e is at most $c(e)$ for each edge. Moreover $\sum l(S_i)$, were the summation is taken over all nets is "small".

(1) (initialization)
   - sort the nets in decreasing priority for the timing of the chip,
   - choose integers μ and v, a cost-function k, a sequence of numbers $\alpha = (\alpha_0, \ldots, \alpha_{v-1})$,
(2) (iterative wiring)
   - execute the modified WWW-algorithm and the 3EXACT algorithm alternatingly $(v + 1)$-times with the parameters k and $(\alpha_0, \ldots, \alpha_{v-1})$.
(3) (post-optimization)
   - repeat the post-optimization routine μ-times.

For analysing the complexity we assume that μ, v are bounded by constants and that the cost-function k can be evaluated within a fixed number of steps. The initialization step is dominated by sorting the net list which can be done in $O(r \log r)$ steps if the weights of the nets are known. The modified WWW-algorithm has complexity $O(m \log n)$ whereas the 3EXACT algorithm can be executed in $O(n^2 \log n)$ steps. Applying each of these routines $O(r)$-times yields an $O(rn^2 \log n)$ bound for the number of steps in the iterative wiring. The final post-optimization phase can be performed in $O(rn^2 \log n)$ steps as well. So the global routing algorithm (A) has a worst-case complexity of $O(rn^2 \log n + r \log r)$ provided that the assumptions above are satisfied. We have observed that the cost-function k^B often yields slightly better results compared to k^A. This lead us to a small modification of the global routing algorithm (A) which turned out to be quite successful. Consider the cost function

$$(C) \qquad k_\gamma^C(e, f) = \begin{cases} \gamma \cdot l(e) + (1 - \gamma)\frac{c(e)}{2c(e) - f(e,t)}, & \text{if } f(e, t) < 2c(e) \\ \gamma \cdot l(e) + (1 - \gamma)(c(e) + f(e, t)) & \text{otherwise.} \end{cases}$$

Applying global routing algorithm (A) with cost function k^C, it may happen that after termination some edges are overloaded, i.e., the number of nets using this edge exceeds the capacity of the edge. To overcome this drawback an additional pull-down routine was used similar to the post-optimization procedure described before.

A Pull-Down Routine

Sort the nets by increasing priority for the timing (again it can easily be seen that the order of the net list heavily influences the final result). Now we consider step by step the edges. From each edge which is overloaded we remove successively (according to the sorted net list) the net segments which use this edge. Each such deletion leaves a forest with two components. These two components are reconnected using a standard shortest path algorithm with respect to a cost-function k blocking overloaded edges. For a similar approach, see Shragowitz and Keel (1987). As soon as the capacity function is respected on all edges, another execution of the post-optimization procedure is performed with still surprisingly great success.

Global Routing Algorithm – Version (B)

Summarizing the ideas mentioned above we get the following slight variation of the global routing algorithm (A).

3.3 Global Routing Algorithm (B).
Add in global routing algorithm (A) the following steps

(4) (second initialization)
  – choose an integer μ' and a cost-function k' which satisfies that $k'(e)$ is infinity whenever $f(e, t) \geq c(e)$ for some edge e.
(pulling-down)
  – execute the pull-down routine with respect to the cost-function k'.
(6) (second post-optimization)
  – repeat the post-optimization routine μ'-times.

In the worst case the global routing algorithm (B) has the same running time as the global routing algorithm (A), provided that μ' is assumed to be a constant and also the cost-function chosen in (4) is easy to compute: Each net uses at

Table 3.4

	lower bound	global routing algorithm (A) with cost function k^A		global routing algorithm (A) with cost function k^B	
		total length	percentage of lower bound	total length	percentage of lower bound
CPU-chip	79437.2	79912.9	100.60	79877.8	100.55
MMU-chip	42455.8	43638.5	102.79	43727.0	102.99

	lower bound	global routing algorithm (B) with cost function k^C	
		total length	percentage of lower bound
CPU-chip	79437.2	79509.7	100.09
MMU-chip	42455.8	43220.6	101.80

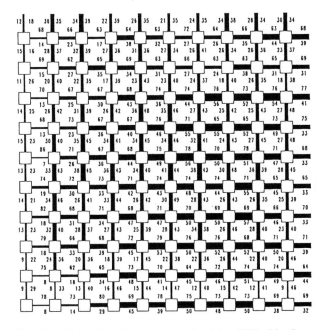

Fig. 3.2 Utilization of the global routing graph of the CPU-chip (lower left corner)

most $O(n)$ many edges and, hence, at most $O(r \cdot n)$ applications of a shortest path finder have to be performed in the pulling-down routine. This gives the desired bound.

Practical Results

The implementation of the global routing algorithms described above was done in PASCAL on an IBM 3081-KX under VM/CMS. In Table 3.4 we show the results

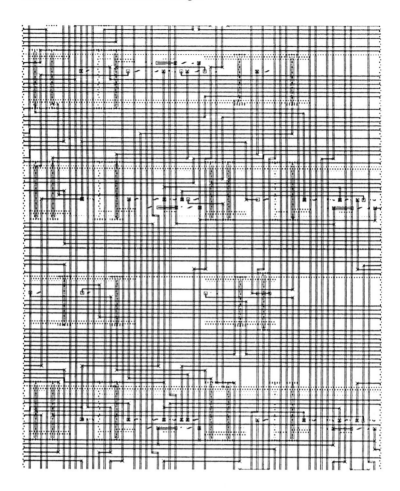

Fig. 3.3   Wiring of a typical page of the MMU-chip

obtained for the two global routing problems by using the three cost functions k^A, k^B and k^C. Table 3.4 also compares these results with the lower bound from Table 3.2. In particular the following parameters were used in computing these results: For the cost functions k^A, k^B and k^C we chose $\alpha = \beta = \gamma = 0.9$. The number of iterations of the post-optimization procedure was set to $\mu = \mu' = 10$. Finally, $\nu = 4$ number of iterations of the WWW-algorithm and the 3EXACT-algorithm were executed, the bounds were $\alpha_0 = 1.5$, $\alpha_1 = 3.5$, $\alpha_2 = 5.5$ and $\alpha_3 = 7.5$. These values were chosen by empirical testing. The running time of the algorithm varied between one hour for global routing algorithm (A) with cost function k^A or k^B and three hours for global routing algorithm (B) with cost function k^C.

Fig. 3.2 shows the utilization of the global routing graph of the MMU-chip after application of the global routing algorithm (B).

COORDINATES: X: 1 ... 2343.Y: 1 ... 3516.Z: 0 ...   1

Fig 3.4 Final wiring of the MMU-chip

Local Routing

After having such a global routing at hand, a local routing is performed in several steps as described earlier. For this purpose a few special features were invented into our algorithm which we will not discuss in detail. The most important change is a new criterion for queueing the nets: no longer wire length or timing aspects are the leading parameters but wireability. The wire length and so the timing are nearly determined after the global-routing and, in fact, the final wire length can be predicted with a precision of $\pm 3\%$ at this stage.

On the other hand it quite easily happens that if we follow for example the shortest first strategy that a certain pin get "blocked" by the wiring of earlier, shorter, nets, i.e., this pin is no longer reachable by wires from other pins of the same net. In this case the whole routing would be incomplete. To avoid this we assign to every pin p a critical value $cr(p)$ which is, roughly, a function of the edges in the neighborhood of p which are already used by other nets. Then we sort the nets according to their most critical pins. We proceed wiring the most critical net until its most critical pin is connected to some other part of this net. Then the critical values of the pins are updated.

The total running time of the local routing part of the routing algorithm was about 8 hours of CPU-time for each of the two chips under consideration. The total wire length for the CPU-chip was 28.2 m (vias not counted) plus 140,676 vias, and for the MMU-chip 15.3 m plus 64,330 vias. Fig. 3.3 shows the wiring of one typical page of the MMU-chip and Fig. 3.4, then, exhibits the final wiring of the MMU-chip.

Acknowledgment. The authors would like to thank students of the Research Institute of Discrete Mathematics for their programming work. Particularly Ulf Dietmar Radicke contributed a lot of effort, skill and ideas while implementing and testing the global routing algorithm.

References

Becker, M., Mehlhorn, K. (1986): Algorithms for routing in planar graphs. Acta Inf. **23**, 163–176

Bern, M.W. (1987): Network design problems: Steiner trees and spanning k-trees. Dissertation, University of California, Berkeley

Bern, M.W., Graham, R.L. (1989): The shortest-network problem. Sci. American, No. 1/89, 66–71

Burstein, M. (1986): Channel routing. In: Ohtsuki, T. (ed.): Layout design and verification. North-Holland, Amsterdam, pp. 133–167

Clausen, J., Hansen, L.A. (1980): Finding k edge-disjoint trees of minimum total weight in a network: An application of matroid theory. Math. Program. Study **13**, 88–101

Courant, R., Robbins, H. (1941): What is Mathematics? Oxford University Press, London

Dijkstra, E.W. (1959): A note on two problems in connexion with graphs. Numer. Math. **1**, 269–271

Dreyfus, S.E., Wagner, R.A. (1972): The Steiner problem in graphs. Networks **1**, 195–207

Edmonds, J. (1965): Minimum partition of a matroid into independent subsets. J. Res. Natl. Bur. Stand. **69B**, 67–72

Erickson, R.E., Monma, C.L., Veinott, A.F. (1987): Send-and-split method for minimum-cost network flows. Math. Oper. Res. **12**, 634–664

Fermat, P. (1896): Maxima et minima IV: Méthode du maximum et minimum. In: Tannery, P., Henry, C. (eds): Œuvres de Fermat. Gauthier-Villars et Fils, Paris, pp. 131–136

Frank, A. (1985): Edge-disjoint paths in planar graphs. J. Comb. Theory, Ser. B **39**, 164–178

Frank, A. (1989): This volume

Fredman, M.L., Tarjan, R.E. (1987): Fibonacci heaps and their uses in improved network optimization algorithms. J. Assoc. Comput. Mach. **34**, 596–615

Froleyks, B., Korte, B., Prömel, H.J. (1987): Routing in VLSI-layout. Report No. 87494-OR, Institut für Operations Research, Universität Bonn

Gabow, H.N., Galil, Z., Spencer, T., Tarjan, R.E. (1986): Efficient algorithms for finding minimum spanning trees in undirected and directed graphs. Combinatorica **6**, 109–122

Garbers, J., Korte, B., Prömel, H.J., Schwietzke, E., Steger, A. (1990): VLSI-Placement based on routing and timing information. Proc. European Design Automation Conference 1990, 317–321

Garey, M.R., Johnson, D.S. (1977): The rectilinear Steiner tree problem is NP-complete. SIAM J. Appl. Math. **32**, 826–834

Garey, M.R., Johnson, D.S. (1979): Computers and intractability: A guide to the theory of NP-completeness. W.H. Freeman, San Francisco, CA

Gilbert, E.N., Pollak, H.O. (1968): Steiner minimal trees. SIAM J. Appl. Math. **16**, 1–29

Hakimi, S.L. (1971): Steiner's problem in graphs and its implications. Networks **1**, 113–133

Hanan, M. (1966): On Steiner's problem with rectilinear distance. SIAM J. Appl. Math. **14**, 255–265

Hwang, F.K. (1976): On Steiner minimal trees with rectilinear distance. SIAM J. Appl. Math. **30**, 104–114

Karp, R.M. (1972): Reducibility among combinatorial problems. In: Miller, R.E., Thatcher, J.W. (eds.): Complexity of computer computations. Plenum Press, New York, NY, pp. 85-103 (IBM Res. Symp. Ser., Vol. 4)

Koetzle, G. (1987): Hierarchisches Designkonzept für anwendungsspezifische VLSI Chips mit über 1000000 Transistoren. In: Informationstechnische Gesellschaft im vde (ed.): Großintegration. vde-verlag, Berlin, pp. 55–58 (ITG-Fachber., Vol. 98)

Koetzle, G. (1987a): System implementation on a highly structured VLSI master image. In: Proebster, W.E., Reiner, H. (eds.): Proc. VLSI and Computers. IEEE Computer Society Press, Washington, DC, pp. 604–609

Kou, L., Markowsky, G., Berman, L. (1981): A fast algorithm for Steiner trees. Acta Inf. **15**, 141–145

Kramer, M.E., van Leeuwen, J. (1984): The complexity of wire-routing and finding the minimum area layouts for arbitrary VLSI circuits. In: Preparata, F.P. (ed.): VLSI theory. JAI Press, London, pp. 129–146 (Adv. Comput. Res., Vol. 2)

Kruskal, J.B. (1956): On the shortest spanning subtree of a graph and the traveling salesman problem. Proc. Am. Math. Soc. **7**, 48–50

Kuh, E.S., Marek-Sadowska, M. (1986): Global routing. In: Ohtsuki, T. (ed.): Layout design and verification. North-Holland, Amsterdam, pp. 169–198

Kuhn, H.W. (1974): Steiner's problem revisited. In: Dantzig, G.B., Eaves, B.C. (eds.): Studies in optimization. MAA, Washington, DC (Stud. Math., Vol. 10)

Lawler, E.L. (1976): Combinatorial optimization: Networks and matroids. Holt, Rinehart and Winston, New York, NY

Lichtenstein, D. (1982): Planar formulae and their uses. SIAM J. Comput. **11**, 329–343

Mehlhorn, K. (1986): Über Verdrahtungsalgorithmen. Inf.-Spektrum **9**, 227–234

Mehlhorn, K. (1988): A faster approximation algorithm for the Steiner problem in graphs. Inf. Process. Lett. **27**, 125–128

Prim, R.C. (1957): Shortest connection networks and some generalizations. Bell Syst. Tech. J. **36**, 1389–1401

Provan, J.S. (1988): Convexity and the Steiner tree problem. Networks **18**, 55–72

Richards, D. (1989): Fast heuristic algorithms for rectilinear Steiner trees. Algorithmica **4**, 191–207

Robertson, N., Seymour, P.D. (1986): Graph minors XIII: The disjoint paths problem. Preprint (J. Comb. Theory, Ser. B, to appear)

Robertson, N., Seymour, P.D. (1989): This volume

Roskind, J., Tarjan, R.E. (1985): A note on finding minimum-cost edge-disjoint spanning trees. Math. Oper. Res. **10**, 701–708

Sangiovanni-Vincentelli, A. (1987): Automatic layout of integrated circuits. In: de Micheli, G., Sangiovanni-Vincentelli, A., Antognetti, P. (eds.): Design systems for VLSI circuits. Martinus Nijhoff Publishers, Dordrecht, pp. 113–195 (NATO ASI Ser., Ser. E, Vol. 136)

Shragowitz, E., Keel, S. (1987): A global router based on a multicommodity flow model. Integration **5**, 3–16

Spruth, W. (1989): The design of a microprocessor. (Springer Verlag, Berlin, Heidelberg)

Tarjan, R.E. (1983): Data structures and network algorithms. SIAM, Philadelphia, PA

Winter, P. (1985): An algorithm for the Steiner problem in the Euclidean plane. Networks **15**, 323–345

Winter, P. (1987): Steiner problem in networks: A survey. Networks **17**, 129–167

Wu, Y.F., Widmayer, P., Wong, C.K. (1986): A faster approximation algorithm for the Steiner problem in graphs. Acta Inf. **23**, 223–229

Cycles Through Prescribed Elements in a Graph

Michael V. Lomonosov

1. Introduction

In these notes we discuss a possible approach to the following *A*-Cycle Problem: Given a graph G with a distinguished subset A of $V(G) \cup E(G)$ find a cycle of G containing A (shortly, an *A-cycle*) or a proof that it does not exist.

'Cycle' here is always simple, i.e. a connected 2-regular graph. A graph with terminal (1-valent) vertices is always regarded as "open": we assume that these vertices - called *ends* - do not belong to the graph. In this way we speak about chains, trees, flows, etc. In particular, disjoint graphs may have common ends.

For a subgraph H of G define $\mathrm{bd}_G(H)$ (or, shortly, $\mathrm{bd}(H)$) as the set of vertices of H incident to edges from $E(G) - E(H)$, and put $\mathrm{in}_G(H) = E(H) \cup \{v \in V(H) :$ all neighbours of v belong to $H\}$ or in (H), for short. No inconvenience arises when a vertex lies both in $\mathrm{bd}(H)$ and in (H).

Except this, we use the terminology and notations of [2]. A pair (G, A) is called *cyclic*, if G has an *A*-cycle and *acyclic* otherwise.

In its general form the A-Cycle Problem includes (as a particular case $A = V(G)$) the Hamilton Cycle Problem and therefore is NP-complete [12]. So we confine ourselves to partial solutions of the problem in some special cases when $|A|$ is not too large compared with connectivity $\kappa(G)$. One of them is solved by the following well-known theorem:

1.1 *Suppose that* $|A| = 2$. *Then* (G, A) *is acyclic iff G has a cut-vertex disconnecting A* (Whitney, 1932).

The following theorems (1.2)-(1.6) can be regarded as generalizations of this fundamental result.

1.2 *Every r vertices of an r-connected graph lie on a common cycle* (Dirac [3]).

The example of $K_{r,r+1}$ shows that $|A|$ in this theorem cannot be increased.

1.3 *If G has a set X of vertices such that* $\omega(G-X) > |X|$ *then G is non-hamiltonian* (see [2]; ω denotes the number of components).

1.4 *Let G be r-connected and A be a set of r + 1 vertices of G; (G, A) is acyclic iff there exists a set X of r vertices such that G − X has r + 1 components each containing a member of A* (Watkins & Mesner [17], see also [8]). One may notice that this result includes the Whitney theorem as a particular case r = 1.

1.5 *Suppose that G is r-connected and A is an independent set of r − 1 edges (an (r − 1)-matching) of G; then (G, A) is cyclic* (Haggkvist & Thomassen [5]).

1.6 *If A contains an odd cocycle of G then (G, A) is acyclic.*

The following situation seems to be of different nature:

1.7 *Every 9 (or less) vertices in a cubic 3-connected graph lie on a common cycle* (Holton, McKay, Plummer and Thomassen [6], see also [9]).

The Petersen graph provides a "limiting example" with 10 vertices: in Section 5 this graph is shown to give a complete solution for 10 vertices in a cubic 3-connected graph, in terms of "similarity".

It can be noticed, further, that in most of the above cases, namely (1.1), (1.3), (1.4) and (1.6), the acyclicity has the same simple reason. Indeed, an A-cycle has the property that for every subset B of A it contains a B-*system* of size $|B|$, i.e. a collection of as many as $|B|$ disjoint chains each connecting two members of B. Define $\mu(G, B)$ as the maximal size of a B-system in G, for arbitrary $B \subset V(G) \cup E(G)$. Then obviously,

1.8 *If there exists $B \subseteq A$ such that $\mu(G, B) < |B|$ then (G, A) is acyclic*

and this is, really, the reason why an A-cycle does not exist in all the above cases, except (1.7). The Watkins-Mesner theorem (1.4) asserts that no other reason for (G, A) to be acyclic exists when $|A| − \kappa(G) \leq 1$.

Applying the Mader theorem [14] (see also [13]), the condition (1.8) can be formulated in a dual form. Suppose, for simplicity, that the vertices from A are pairwise non-adjacent, and instead of each edge of A we consider a 2-valent vertex put into the middle of this edge. We say that a pair (X, Y), where $X \subseteq V(G − A)$ and $Y \subseteq E(G − A − X)$, *disconnects* A, if $G − X − Y$ has no chain connecting two members of A (assuming that A contains no incident pair {edge, vertex}, one may be sure that a disconnecting pair (X, Y) always exists). For any such pair (X, Y) define

1.9 $c(X, Y) = |X| + \sum_{i=1}^{n} \lfloor |\mathrm{bd}_{G−X}(Q_i)| / 2 \rfloor$

where $\lfloor x \rfloor$ denotes the maximal integer $\leq x$, and $Q_1, \ldots, Q_n$ are the components of the edge-induced subgraph $G(Y)$. Clearly,

1.10 $\mu(G, A) \leq c(X, Y)$

for any (X, Y) disconnecting A. The Mader theorem [14] asserts that

1.11 $\mu(G, A) = \min c(X, Y)$ *over all* (X, Y) *disconnecting* A.

Now we see that (1.8) is equivalent to

1.12   *There exist* $B \subseteq A$ *and a pair* (X, Y) *disconnecting* B, *such that*

$$|B| > c(X, Y).$$

A pair (X, Y) satisfying (1.12) is called an A-*separator* (for every A containing B). Given a class $\mathscr{P}$ of pairs (G, A) we say that the *cycle-separator alternative* (*CSA*) holds in $\mathscr{P}$ if G has either an A-cycle or an A-separator whenever $(G, A) \in \mathscr{P}$. Figure 1 presents various types of an A-separator with non-empty Y for 5 vertices in a 3-connected graph.

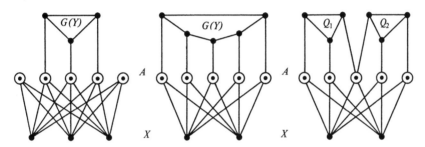

Fig. 1 Various forms of A-separator with non-empty Y; $|A| = 5$, $\mu(G, A) = 4$

An important problem is to determine limits of the *CSA*-region. In Section 2 it is done in terms of $|A|$ and connectivity of G, for the case when A consists of vertices only. Really interesting solutions are expected just outside the *CSA*-region; two such results are presented in Sections 5 and 6.

In general, the following sequential approach to the A-cycle Problem seems natural. Consider a class $\mathscr{P}$ of (G, A)'s assuming for simplicity that $A \subseteq V(G)$. The first question should be:

Q.1: *what is the value of*

$$N_0(\mathscr{P}) = \min\{|A| ; (G, A) \in \mathscr{P} \text{ is acyclic }\}?$$

and, if $N_0(\mathscr{P}) < \infty$, the next question naturally is:

Q.2: *what is the reason for* $(G, A) \in \mathscr{P}$ *to be acyclic when* $|A| \geq N_0(\mathscr{P})$?

The notion of "reason" is undefined; the question 'why, say, are $K_{r,r+1}$ and the Petersen graph non-hamiltonian?' may be answered: 'because one is $K_{r,r+1}$ and the other is Petersen'. In Sections 5 and 6 we shall see that even answers like this are not always senseless. Here we only mention that theorem (1.4) -

the prototype of CSA - can be stated in exactly this form: for an r-connected G and A consisting of $r + 1$ vertices, (G, A) is acyclic iff it is "similar" to $K_{r,r+1}$ in the sense which can be easily defined. The similarity approach is based on a phenomenon which can be informally described as follows.

1.13 The set of *minimal* acyclic representatives of a given class $\mathscr{P}$ of pairs (G, A) is finite (they may appear as "limiting examples" in the theorem characterizing $N_0(\mathscr{P})$); "minimal" means that every acyclic member of $\mathscr{P}$ contains a subdivision of one of such representatives.

1.14 Each minimal acyclic (G_0, A_0) is *unstable* in the following sense. Let H be a subdivision of G_0, and F denote the set of new edges $e = (u, v)$, $u, v \in V(H)$ which can be added to H. Put $F_0 = \{e \in F; (H + e, A_0)$ is acyclic$\}$. Thus, F_0 always contains the edges with u, v lying on the same thread (subdivided edge) of H; further, if H has an A-separator (X, Y), then the edges incident to X, or having both ends in the same component of $H(y)$, or in the same component of $H - X - Y$, also belong to F_0. The *less* is F_0 compared with F the *more* unstable is (G_0, A_0).

1.15 The set F_0 severely restricts the structure of every acyclic (G, A) from $\mathscr{P}$, containing a subdivision of (G_0, A_0); we say then that (G, A) is *similar* to (G_0, A_0), which notion needs to be properly defined in each case.

2. A-Cycles in r-Connected Graphs

Let $\mathscr{P}_r$ denote the class of pairs (G, A) where G is r-connected and A consists of vertices. Our two questions concerning $\mathscr{P}_r$ are answered by the Dirac theorem (1.2) which establishes the value $N_0(\mathscr{P}_r) = r + 1$, and the Watkins-Mesner theorem (1.4) establishing the CSA for $|A| = r + 1$. We now proceed to answer these questions for a more interesting class $\mathscr{P}_r^0 = \{(G, A) \in \mathscr{P}_r; G$ has no A-separator$\}$, and, thus to describe the CSA-region in terms of graph connectivity and $|A|$ (see [8]).

For arbitrary subsets $X \subseteq V(G) - A$ and $Y \subseteq E(G - A - X)$ let $d(A; X, Y)$ denote the number of components of $G - X - Y$ meeting A, and put $h(G, A) = \min(c(X, Y) - d(A; X, Y))$ over all (X, Y) satisfying $c(X, Y) \geq \kappa(G)$. It can be verified that $h(G, A) \geq \min(\mu(G, B) - |B|), B \subseteq A$. Further, we call (G, A) *deficient*, *balanced* or *excessive* when $h(G, A)$ is, respectively, negative, zero or positive. Suppose that G has an A'-cycle for every proper $A' \subset A$. then, clearly, G has an A-separator iff (G, A) is deficient.

For the balanced (G, A) the following obvious criterion is sometimes useful. Let (X, Y) be as above, and $H_1, \ldots, H_m$ be the components of $G - X - Y$ with non-empty $A_i = A \cap H_i$, $m = d(A; X, Y)$. For $i = 1, \ldots, m$ put $H_i' = G(X \cup V(H_i))$, so that $H_i \subset H_i'$ and $\text{bd}_G(H') = X \cup \text{bd}_{G-X}(H_i)$, and define F_i as the set of edges $e = (u, v)$ with $u, v \in \text{bd}_G(H')$ such that there exists a chain $L \subseteq \text{in}(H_i')$ connecting u, v and passing through A_i.

2.1 Theorem. *A balanced pair (G, A) is cyclic iff for some (X, Y) satisfying $d(A; X, Y) = c(X, Y)$ $[m \geq \kappa(G)]$ the family $\{F_i; i = 1, \ldots, m\}$ has:*

(2.1.1) *the Hall property* $|\cup\{F_i; i \in J\}| > |J|$ *for all* $J \subseteq \{1,\dots,m\}$;
(2.1.2) *a system B of distinct representatives such that* $G(Y \cup B)$ *has a B-cycle.*

In this Section the value $N_0(\mathscr{P}_r^0) = r+3$ is obtained; this includes the following two results.

2.2 Theorem. *If G is r-connected and* $|A| \le r + 2$ *then G has either an A-cycle or an A-separator.*

2.3 Example. For every $r \ge 4$ Figure 2(a) presents (G, A) with an *r-connected G and* $A \subseteq V(G)$ *shown by circles,* $|A| = r + 3$, *such that G has neither an A-cycle nor an A-separator.*

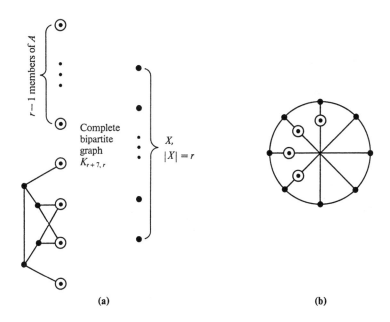

(a)   (b)

Fig. 2 Acyclic non-deficient pairs (G, A);
(a) $r \ge 4$, $|A| = \mu(G, A) = r + 3$, (b) $r = 2$, $|A| = \mu(G, A) = 4$

Only the following incomplete answer is given here to the question Q.2 concerning $\mathscr{P}_r^0$:

2.4 Theorem. *For* $r \ge 4$ *every acyclic* $(G, A) \in \mathscr{P}_r^0$ *with* $|A| = r + 3$ *is balanced (more precisely, there exists an r-subset* $X \subseteq V(G) - A$ *such that* $d(A; X, \emptyset) = r$), *and* (G, A) *contains a subdivision of the example (2.3).*

2.5 Example. Figure 3 (in the particular case $r = 3$) presents *an excessive acyclic* (G, A) *with a 3-connected G and a set A of six vertices (shown by circles).*

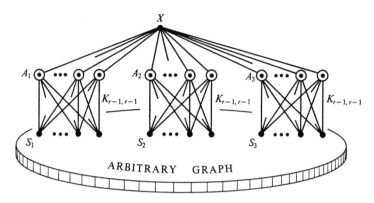

Fig. 3 An excessive acyclic (G, A) with $\kappa(G) = r \geq 3$; $A = A_1 \cup A_2 \cup A_3$, $|A_i| = |S_i| = r - 1$, $i = 1, 2, 3$; $\mu(G, A) = 3r - 2$

The proofs are only outlined. First, in the spirit of (1.13), the following construction is introduced:

2.6 Proposition. *Let (G, A) be an acyclic pair, with an r-connected G and $A \subseteq V(G)$, such that $|A| \leq r + 3$ for $r \geq 4$, and $|A| \leq r + 2 = 5$ for $r = 3$. Suppose that (G, A') is cyclic for every proper $A' \subset A$. Then G has a subgraph W containing A, with the following properties:*

(2.6.1) $\deg_W(a) \geq r$, $a \in A$;

(2.6.2) *W is a union of A and disjoint trees $T_1, \ldots, T_n$ with the ends in A, each T_i being a subdivision of a "star" or a "multipod";*

(2.6.3) *let X_i denote the (unique) set of vertices of T_i disconnecting the ends of T_i; then $\sum_{i=1}^{n} |X_i| = |A| - 1$;*

(2.6.4) *(W, A') is cyclic for every proper $A' \subset A$.*

Here a star is always of degree ≥ 4 and a multipod is a ternary tree with an odd number of ends (a star of degree 3 is called here "tripod"); hence, $|X_i| = 1$ for a star and $= m$ for a multipod with $2m + 1$ ends; edges incident to A are considered as one-end multipods, with $m = 0$; put $S = X \cup_1 \ldots \cup_n X$; it follows from (2.6.1-4) that $(S, \emptyset)$ is an A-separator of W. An easy calculation shows that, under the assumptions $r \geq 4$ and $|A| < r + 3$, there can be at most two multipods at all, no multipod with more than 5 ends, and at most one with exactly 5.

Proof (an outline). Choose $a_0 \in A$, put $A_0 = A - \{a_0\}$, and consider an A_0-cycle C of G. Surely $a_0 \notin C$. By the Menger theorem, there exist at least r disjoint chains ("rays") connecting a_0 with distinct vertices of C; let F_0 denote a union of $\{a_0\}$ and a maximal number of such rays (a "maximal flow" from a_0 to C), and S denote the set of ends of F_0 in C. Put $W_0 = C + F_0$. Now, S divides C into $|S| \geq r$ open segments; since (W_0, A) is acyclic, each segment contains a member of A. We say that $a \in A_0$ is excised by S if a is the single member of A_0 in some segment. In W_0 excised are $\geq r - 2$ members of A_0. Launch the following procedure:

(P.1) Start with $W = W_0$.

(P.2) For $k \geq 0$ let $W = C + F_0 + \ldots + F_K$; choose an excised $a_{K+1} \in A$ of degree 2 in W; if there is no then stop.

(P.3) Let Q denote the segment of C containing a_{K+1} and q', q'' - the ends of Q. Construct a maximal flow F_{K+1} from a a_{K+1} to $C - Q$, with the following properties:

    (P.3.1) q', q'' are among the ends of F_{K+1};

    (P.3.2) F_{K+1} has as few as possible ends in $S - \{q', q''\}$. (These requirements can be satisfied by applying the well-known maxflow techniques [4].) Now, let S_{K+1} denote the set of ends of F_{K+1} and Q' denote the part of F_{K+1} terminating in $\{q', q''\}$. Put $C := C - Q + Q'$, $W := W - Q + F_{K+1'}$ and $S : S \cup S_{K+1}$. Go to (P.2).

The assumed acyclicity of (G, A) implies that

(*)   the flows $F_0, F_1, \ldots$ are pairwise disjoint, and

(**) S divides C into segments each of which meets A (in particular, $|S| < |A| - 1$).

Assume that a_0 was chosen so as to maximize $|S|$. If $|S| = |A| - 1$ the required subgraph W is ready. Consider the case $|S| < |A| - 1$ which means that either $|S| = r$ and $|A| = r + 2$ or $r + 3$, or $|S| = r + 1$ and $|A| = r + 3$. The subgraph W obtained in (P.2) then is a union of A, trees $T_1, \ldots, T_m$ ($m = |S| = r$ or $r + 1$) and "lines" $L_1, \ldots, L_p$ ($p = 1$ or 2) where each T_i is a (subdivision of a) star or tripod with the center in S, and each L_j is a chain of C connecting two members of A_0 unseparated by S.

Consider external chains with respect to W, i.e chains P of G with the ends $u, v \in V(W)$, satisfying $P \cap W = \emptyset$. Each end of P may lie in any of the following sets: $A, S, T_i - X_i, L_j$. We say that P is a *bridge* (for a given W) when it connects distinct components of $W - S$. If P is not a bridge we have $\mu(W + P, A) = \mu(W, A) = |S| \leq |A| - 2$. For a bridge there are two possibilities:

(i) if P connects distinct rays of a star or distinct trees T_i, T_k, then $\mu(W + P, A) = \mu(W, A) + 1$;

(ii) in all other cases $\mu(W + P, A) = \mu(W, A)$.

For W obtained in (P.2) it is easy to show that (i) leads to an A-cycle. On the other hand, the assumption that (G, A') is cyclic for proper $A' \subset A$, implies that $(S, \emptyset)$ with $|S| < |A| - 2$ cannot be an A-separator. Therefore there inevitably exists a bridge P connecting some tripod T with a line, L (T cannot be a star, by the choice of a_0; this means, in particular, that we have now either $r = 3$, $|S| = 3$ and $|A| = 5$, or $r \geq 4$, $|S| = r + 1$ and $|A| = r + 3$). Now, $T + P + L$ is a 5-end multipod, and the end of P lying on L should be added to S. Put $W := W + P$ and return to (P.2). The assertion (2.6) follows.   □

We proceed to prove theorems (2.2) and (2.4). Again, we consider bridges with respect to W, but now W has the form (2.6.1-4).

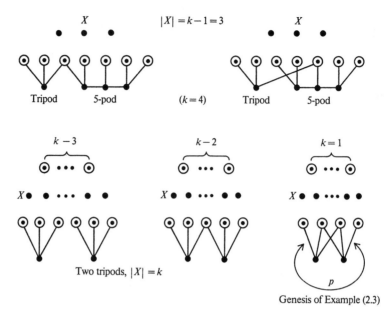

Genesis of Example (2.3)

Fig. 4

2.7 Proposition. *Let P be a bridge satisfying* (i); *then W has an A-cycle, unless P connects two tripods with two common ends, as is shown in Figure 4.*

To prove this, all possible combinations of T_i, T_k connected by P should be examined. Each such configuration is reducible to a balanced one, and Theorem (2.1) can be applied. $\qquad\square$

The exclusive case in (2.7) (see Figure 4) occurs only when $|A| = r + 3$. Suppose now that the bridges are of type (ii) only (which is, for instance, the case when $|A| = r + 2$). Then Theorems (2.2) and (2.4) can be proved by induction, G being reconstructed from W by adjoining external chains one after another: $G = \lim G_k$, where G_k has an A-separator (X, Y), and $G_{k+1} = G_k + P$, P being an external chain of G with respect to W, $P \not\subseteq G_k$, and $G_0 = W$, with an A-separator $(X, Y) = (S, \emptyset)$. To see this in more detail, note that we can, assuming that W is well subdivided in G, find in W one more A-separator, (X', Y') : let $T_1, \ldots, T_m$ be stars and $T_{m+1}, \ldots, T_n$-multipods, and let x_i denote the center of T_i, $i = 1, \ldots, m$, and Y_j denote the set of edges of T_j not incident to A, $j = m + 1, \ldots, n$. Put $X' = \{x_1, \ldots, x_m\}$ and $Y' = Y_{m+1} \cup \ldots \cup Y_n$; clearly (X', Y') is an A-separator of W. Since the exclusive case of (2.7) is ruled out, it is true, at least for $k = 0$, that the external chains with respect to W are of the following three types only:

(a)   with both ends in the same component of $G_k - X - Y$;
(b)   with an end in X;
(c)   with both ends in the same component of $G(Y')$.

If all external chains are as (a) or (b) then (X, Y) is an A-separator in G. Otherwise let P be of type (c). This means that the ends of P lie in some T_j, $j \in \{m+1, \ldots, n\}$. Let C denote the circuit of $T_j + P$. Put $G_{k+1} := G_k + P$, $Y := Y \cup E(C)$, $X := X - V(C)$. Get convinced that (a), (b), (c) are preserved, and repeat the step. Both (2.2), (2.4) follow. □

3. Weakly Separable Graphs

In the light of the previous results it is natural to ask what solution the A-cycle Problem may have for a graph in which no (X, Y) dissociates A into too many parts compared with $c(X, Y)$ (see [10]).

3.1 Definition. G is weakly separable if for $X \subset V(G)$, $Y \subseteq E(G - X)$ there holds

(3.1.1) $\omega(G - X - Y) \leq \begin{cases} \kappa(G) - 1, & \text{when } c(X, Y) = \kappa(G) \\ \kappa(G) + 1, & \text{when } c(X, Y) = \kappa(G) + 1. \end{cases}$

Let $\mathcal{W}_r$ denote the class of pairs (G, A) where G is r-connected and weakly separable, and $A \subseteq V(G)$. From (2.2) and (2.4) we have

3.1.2 Corollary. *If $r \geq 3$ then $N_0(\mathcal{W}_r) = r + 3$, and CSA holds for $(G, A) \in \mathcal{W}_r$ when $|A| = r + 3$.*

3.2 Example. A-cycle is a polytopal graph. A graph is called k-*polytopal* if it is isomorphic to the *skeleton* (vertices +edges) of a k-dimensional convex finite polyhedron. Polytopal graphs are weakly separable as it is seen from the following

3.2.1 Proposition. *A k-polytopal graph G with $k \geq 3$ is k-connected, and*

$$\omega(G - X - Y) \leq \begin{cases} k - 1 & \text{when } c(X, Y) = k \\ k + 1 & \text{when } c(X, Y) = k + 1 \end{cases}$$

for $X \subset V(G)$, $Y \subseteq E(G - X)$.

Together with (3.1.2) this implies the following property of convex polyhedra:

3.2.2 Theorem. *Every $k + 2$ vertices of a k-polytopal graph lie on a common cycle. For every $(k + 3)$-set A of its vertices G has either an A-cycle or an A-separator.*

A-separators in k-polytopal graphs really exist for $|A| \geq k + 3$, as we shall now see.

3.2.3 Example. For $k \geq 3$ define a graph H_k as follows. Let S, A be disjoint sets, $|S| = k + 2$ and $k + 3 < |A| < 2k$, and let S', S'' be $(k + 1)$-subsets of S with $|S' \cap S''| = k$. Choose a family $\{X_a; a \in A\}$ of distinct k-subsets X_a of S, where $X_a \neq S' \cap S''$, $a \in A$, and define H_k as the following union of complete graphs (here $K(Z)$ denotes a complete graph with the vertex-set Z):

$$H_k = K(S') \cup K(S'') \cup (\cup\{K(X_a \cup \{a\}); a \in A\}$$

A polytope with this skeleton has the same structure, with k-dimensional simplices instead of complete graphs. Finally, $(S, \emptyset)$ is easily seen to be an A-separator.

It seems interesting to investigate the possible forms of an A-separator in polytopes, at least for small values of $|A|$, and the limits of the CSA-region for polytopal graphs, including examples of polytopes with neither A-cycle nor A-separator for some A. For $k = 3$ and $|A| = 6$ the example (3.2.3) was shown in [10] to exhaust all acyclic cases, up to a certain similarity relation. Recall first that '3-polytopal' means 'planar 3-connected'. Further, it can be noticed that the polytope of Example (3.2.3) defines H_k only when each $a \in A$ is close enough to the hyperplane of the corresponding X_a. Certain edges of H_k disappear when some $a \in A$ takes a marginal position where the polytope is on the brink of losing convexity. There exists *the* minimal subgraph H_k^0 of H_k defined by this polytope; H_3^0 is exactly the Herschel graph (see Figure 5).

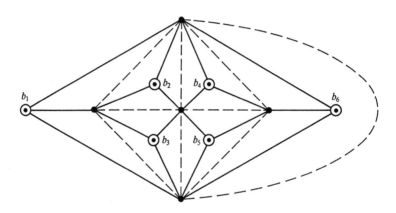

Fig. 5 The Herschel graph (bold edges) in the graph H_3 from Example (3.2.3)

3.2.4 Theorem. *Let A be a set of 6 vertices in a planar 3-connected graph G. G has no A-cycle iff (G, A) is similar to (H, B) where $H_3^0 \subseteq H \subseteq H_3$ and $B = \{b_1, \ldots, b_6\}$.*

(see Figure 5 where H_3^0 is shown by bold edges, the dotted edges completing it to the graph H_3 of example (3.2.3)).

The notion of similarity used here should be explained. A subgraph T of G is called an A-tripole, if $|\text{bd}(T)| = 3$, $\text{bd}(T) \cap A = \emptyset$ and $|T \cap A| < 1$. *Reducing an A-tripole T consists of :* replacing the interior of T with a new vertex, t, and three edges connecting t with $\text{bd}(T)$, and, in the case $A \cap T \neq \emptyset$, replacing the element of $A \cap T$ with t. We say that (G, A) is similar to (H, B) if the second can be obtained from the first by reducing tripoles and erasing 2-valent vertices when they appear.

4. Regular Graphs

4.1 Proposition. *An r-regular r-connected graph G has no A-separator for whatever* $A \subseteq V(G)$.

Proof. The case $r = 2$ is trivial. Suppose that $r \geq 3$. For arbitrary $X \subset V(G)$ and $Y \subseteq Y(G - X)$ let $Q_1, \ldots, Q_t$ denote the components of $G(Y)$, and put $|X| = p$, $|\mathrm{bd}(Q_j)| = q_j$, $j = 1, \ldots, t$, so that

4.1.1
$$c(X, Y) = p + \sum_{j=1}^{t} \lfloor q_j/2 \rfloor$$

Now, let $H_1, \ldots, H_n$ denote the components of $G - X - Y$. Each H_i is adjacent to at least r vertices from $X \cup \mathrm{bd}(Y)$ [$= \mathrm{bd}(G(Y))$], since G is r-connected. On the other hand, each $x \in X$ is adjacent to at most r components H_i, since G is r-regular, and each $y \in \mathrm{bd}(Y)$ is adjacent to at most one H_i. This leads to

(4.1.2) $nr \leq pr + \sum_{j=1}^{t} q_j$

which, together with (4.1.1), provides

(4.1.3) $c(X, Y) - n \geq \sum_{j=1}^{t} \left[\lfloor q_j/2 \rfloor - q_j/r \right]$

The right-hand side of (4.1.3) is always non-negative, whence $c(X, Y) \geq n$.   □

From (4.1.3) one can also derive a structure of balanced pairs (G, A) with r-regular r-connected G: they are in the obvious sense "similar" to r-regular bipartite graphs, with one of the two parts serving as A. Indeed, the right-hand side of (4.1.3) is zero iff the following three conditions are satisfied:

(4.1.4) each H_k is adjacent to *exactly* r vertices from $X \cup \mathrm{bd}(Y)$;
(4.1.5) each $x \in X$ is adjacent to *exactly* r subgraphs H_i;
(4.1.6) Y is empty, if $r \geq 4$; for $r = 3$, either $Y = \emptyset$ or each Q_j has $|\mathrm{bd}(Q_j)| = 3$.

The minimal $|A|$ for which there exists an acyclic (G, A) with an r-regular r-connected G is expected to be large. For $r = 3$ this number is known to be 10 (see (1.7)). In general it cannot exceed $10(r - 1)$, as the example of the Meredith graph shows (Figure 6).

5. *A*-Cycles in Cubic 3-Connected Graphs

It was mentioned (see (1.7)), that any 9 vertices of a cubic 3-connected graph lie on a common cycle, but not any 10 (the Petersen graph). In this Section we find the reason for a cubic 3-connected graph G to have no A-cycle when $|A| = 10$ and 11.

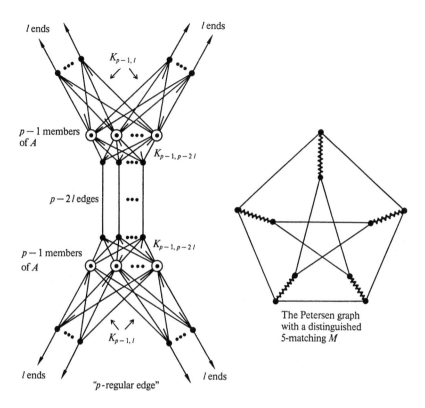

Fig. 6 The Meredith graph. The "p-regular edge" is inserted instead of each edge from M; the obtained graph is p-connected provided $\lceil p/4 \rceil < l < \lfloor p/2 \rfloor$ ($|A| = 10(p-1)$)

5.1 Theorem. *Let G be a cubic 3-connected graph, and A be a set of 10 vertices of G. (G, A) is acyclic iff $V(G)$ is a union of pairwise disjoint sets $Z_1, \ldots, Z_{10}$, each containing a member of A, such that:*

(5.1.1) *any two of them are connected by at most one edge, and*
(5.1.2) *contracting all the Z_i's transforms G into the Petersen graph.*

(Shortly, (G, A) is *similar* to the Petersen graph.)

Proof (an outline). 'If' is obvious. To establish the 'only if' part of the theorem suppose that (G, A) is acyclic. The following assertion is true, on the line of (1.13).

5.2 Proposition. *G contains a subgraph, W, homeomorphic to the Petersen graph, such that A is the set of 3-valent vertices of W.*

To show this, consider reducing (G, A) in the following two ways:

(5.2.1) Define an *A-tripole* as a 2-connected subgraph T of G such that $|\mathrm{bd}(T)| = 3$ and $|A \cap T| \le 1$. *Reducing* an *A*-tripole T (in the case of cubic graphs) consists of contracting T into a vertex, say t, and, in the case $A \cap T \ne \emptyset$, in replacing the member of $A \cap T$ with t.

(5.2.2) *Erasing* an edge $e = (u, v)$ of G not incident to A consists of removing e and erasing the 2-valent vertices u, v of $G - e$.

Now, starting from $(W', A') = (G, A)$, do as follows:

(5.2.3) reduce all W'-tripoles in the current (W', A'), if any;
(5.2.4) if A' covers all edges of W' then stop; otherwise erase arbitrary edge not incident to A' and go to (5.2.3)

At any moment (W', A') is homeomorphically embedded into (G, A), and, therefore, acyclic. The fact that W' is always 3-connected follows from

5.2.5 Proposition. *Let G be cubic 3-connected and let $|A| \le 10$. If G has a 3-cocycle F such that each component of $G - F$ contains at least two members of A then (G, A) is cyclic.*

To see this, consider the vertex-sets X, Y of the components of $G - F$, and put $F = \{(x_i, y_i); \; i = 1, 2, 3\}$ where $x_i \in X$, $y_i \in Y$. Let G_X and G_Y denote the graphs obtained from G by contracting X and Y respectively, so that $G_X = G(Y) \cup \{x\} \cup \{(x, y_i); \; i = 1, 2, 3\}$ where x represents the contracted X, and similarly for G. Finally, put $A_X = (A \cap Y) \cup \{x\}$ and $A_Y = (A \cap X) \cup \{y\}$; without loss of generality, $|A_X| \le |A_Y|$. The following statement is obvious.

(5.2.6) (G, A) is cyclic iff there exists $i \in \{1, 2, 3\}$ such that both $(G_X - (x, y_i), A_X)$, $(G_Y - (y, x_i), A_Y)$ are cyclic.

Sometimes the following sufficient condition is more preferable:

(5.2.7) If for each $i \in \{1, 2, 3\}$ G_X has an A-cycle passing through (x, y_i) and G_Y has an A-cycle passing through (y, x_i) then (G, A) is cyclic.

Since $|A| \le 10$ and $|A \cap X|$, $|A \cap Y| \ge 2$, the conditions of (5.2.6) or (5.2.7) can be verified using the following

5.2.8 Lemma. *Consider G, B and e, where G is a cubic 3-connected graph, $B \subseteq V(G)$ and $e \in E(G)$. If $|B| \le 5$ then G has a B-cycle passing through e, and a B-cycle avoiding e.*

This property was established [11] using a similar method based on tripole reductions and erasing edges.

Returning to proof of (5.2.5), suppose first that $|A \cap Y| \le 4$. Since $|A_Y| \le 9$, G_Y has an A_Y-cycle, say C, by (1.7). Without loss of generality, $(x, y_1) \notin C$. By the Lemma, there exists an A-cycle of G avoiding (y, x_1), so (G, A) is cyclic, by

(5.2.6). If $|A \cap X| = |A \cap Y| = 5$, Lemma (5.2.8) guarantees that the requirement of (5.2.7) is fulfilled (when applied to $A \cap X$ and $A \cap Y$), so that (G, A) is cyclic in this case too.

Return now to proof of (5.2). Let (W', A') denote the result of the procedure (5.2.3-4). Then W' is cubic 3-connected, with the property that each edge of W' is incident to A'. This means, in particular, that $|V(W')| \leq 2|A'| = 20$. All such graphs were shown [11] (without using a computer) to have an A'-cycle, except the Petersen graph with the vertex-set A'. This completes proof of (5.2). $\square$

Now, in the spirit of (1.14), the following instability property of the Petersen graph can be easily established.

5.3 Proposition. *Let W be a subdivision of the Petersen graph, A be the set of 3-valent vertices of W, and u, v be distinct 2-valent vertices of W. Introduce a new edge $e = (u, v)$. $(W + e, A)$ is acyclic iff u and v lie either on the same thread (subdivided edge) of W or on adjacent threads.*

Finally, the acyclicity assumption, together with (5.3), restricts the structure of (G, A) as follows, Since G is 2-connected it is a union of W and a number of external chains L connecting distinct 2-valent vertices of W ('external' means that $L \cap W = \emptyset$). It follows now from (5.3) that G is a union of W and ten subgraphs $G(a)$, $a \in A$, defined as follows. Let $W(a, b)$ denote the thread of W connecting the vertices a, b from A, and $T(a)$ denote the union of $\{a\}$ and the three threads incident to a. For an external chain L with the ends in $T(a)$ let $C(L, a)$ denote the circuit of $T(a) \cup L$ (so that if the ends of L belong to $W(a, b)$ we have $C(L, a) = C(L, b)$). For every $a \in A$ define $C(a)$ as the union of $\{a\}$ and the circuits $C(L, a)$ (so that $C(a) = \{a\}$ if no $C(L, a)$ exist). Now, $G(a)$ is defined as the block of $C(a)$ containing a, or as $\{a\}$ when a is an isolated vertex of $C(a)$. In the latter case $G(a)$ is called trivial. For a non-trivial G put $T'(a) = T(a) \cup G(a)$.

5.4 Proposition. $G(a) \cap G(b) = \emptyset$ *for* $a \neq b$.

Proof. Suppose $G(a) \cap G(b) \neq \emptyset$. Then both $G(a)$ and $G(b)$ are non-trivial. Let x, y, z denote the 3-valent vertices of W neighbouring to a (i.e. the ends of $T(a)$), and u, v, w denote those of b. Since the Petersen graph has no 3- and 4-circuits, we may assume that $x, y \notin \{b, u, v, w\}$ and $u, v \notin \{a, x, y, z\}$. Now, there are two possibilities.

If a and b are adjacent in the Petersen graph, consider the subgraph $H = G(a) \cup G(b) \cup W(a, b)$ of G. Let x', y' denote the ends of $T'(a)$ corresponding to x, y and $Q(x', y')$ denote the chain of $T'(a)$ connecting x', y'; similarly define u', v' and $Q(u', v')$ in $T'(b)$. H has a chain, namely $W(a, b)$, connecting $Q(x', y')$ and $Q(u', v')$ and incident to a, b. Since $G(a)$ meets $G(b)$, H is 2-connected; therefore H has two disjoint chains, J and J', connecting two vertices of $Q(x', y')$ with two vertices of $Q(u', v')$ and incident to a, b. But $(W - W(a, b)) \cup J \cup J'$ is easily seen to have an A-cycle, by (5.2) - contradiction.

Suppose that a, b are non-adjacent in the Petersen graph, and put $H = G(a) \cup G(b)$. Since, by the assumption, H is connected, it has a chain, J, connecting

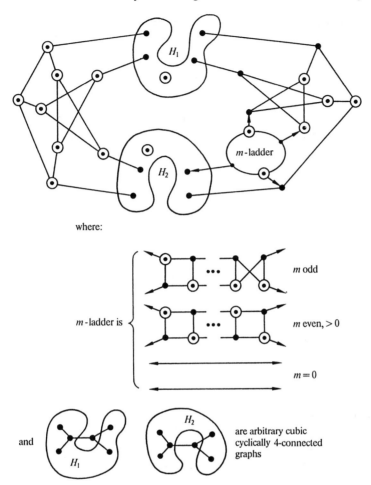

Fig. 7 An acyclic (G, A) with a cubic cyclically 4-connected G and $|A| = 14 + m$, $m \geq 0$ (A. Kelmans)

$T'(a)$ and $T'(b)$. Suppose, without loss of generality, that the ends of J lie on $W(a, z)$ and $W(b, w)$. Then $z = w$, by (5.2). This means that $J \subset G(z)$, whence $G(a)$ meets $G(z)$ contradicting what has been proved above.   □

To complete the proof of Theorem, note that $G(a)$ is either $\{a\}$ or an A-tripole whose boundary coincides with the ends of $T'(a)$. Finally, $G(a)$ and $G(b)$ are connected by at most one edge, which cannot be but an edge of $W(a, b)$.   □

The following slightly stronger result can be obtained using similar method [11].

5.5 Theorem. *Let G be a cubic 3-connected graph, and A be a set of 11 vertices of G. (G, A) is cyclic iff (G, B) is cyclic for every 10-subset B of A.*

What is the minimal value of $|A|$ for which no finite characterization of acyclic pairs (G, A) in terms of similarity can exist? The example in Figure 7, due to A. Kelmans, shows that it cannot exceed 14: for every $k \geq 14$ there exist infinitely many acyclic pairs (G, A) with $|A| = k$ and G cubic and cyclically 4-connected.

We have seen that the solution of A-cycle Problem given in the above two theorems is based on listing all non-hamiltonian cubic 3-connected graphs with a bounded number of vertices. In this connection the following results were obtained [12], concerning cubic bipartite 3-connected graphs (abbreviated as CB-graphs):

5.6 Theorem. *All CB-graphs with 30 or less vertices are Hamiltonian.*

(Proved without using a computer)

5.7 Theorem. *There exists a non-hamiltonian CB-graph with 50 vertices.*

The only known example of such a graph, due to A. Kelmans [7], is shown in Figure 7.

6. A Cycle Through Four Given Edges

The A-cycle Problem seems to be much more difficult when A is a set of edges. Here we consider the case when A consists of at most four edges, and G is an arbitrary graph. The culmination is Theorem (6.6) discovered by N. Robertson [16].

Consider an acyclic non-deficient pair (G, A) where G is a 2-connected graph and A is a set of at most 4 edges of G. Suppose that (G, A') is cyclic for every proper $A' \subset A$. The following property is immediate:

6.1 *G has no 2-cut partitioning A into two 2-subsets.* [Suppose that such a 2-cut Z exists, i.e. $G = G' \cup G''$ where $G' \cap G'' = Z$ and $|A'| = |A''| = 2$, where $A' = A \cap G'$ and $A'' = A \cap G''$. Choose $a' \in A'$ and $a'' \in A''$, and let C' be an $(A - \{a''\})$-cycle, and C'' be an $(A - \{a'\})$-cycle of G, by (6.2). Then $C' \cup C''$ contains an A-cycle of G - contradiction.]

Further, let a subgraph H of G be called an A-*dipole*, if $|\mathrm{bd}(H)| = 2$ and $|A \cap \mathrm{in}(H)| \leq 1$. An A-dipole can be replaced with an edge connecting its boundary vertices, with the correspondent correction of A, if needed. Together with (6.1) this justifies

6.2 Assumption. *G is 3-connected.*

Finally, the following two assumptions are aimed to eliminate situations reducible to a smaller (G, A) with a 2-connected G and $|A| \leq 3$.

6.3 Assumption. *A is a matching* (i.e. the members of A are pairwise non-adjacent).

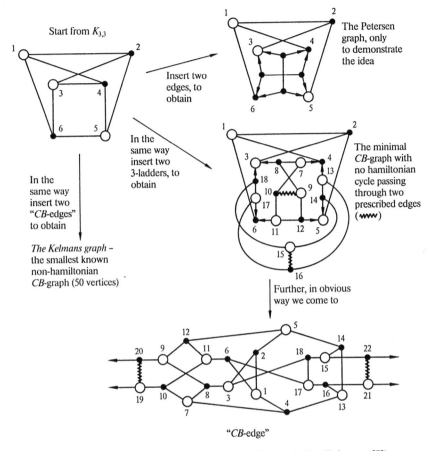

Fig. 8 The 50-vertex non-hamiltonian CB-graph (A. Kelmans [7])

6.4 Assumption. *G has no 3-cut which partitions A into two 2-subsets.*

[Suppose that $Z = \{z_1, z_2, z_3\}$ is such a 3-cut, i.e. $G = G' \cup G''$ with $G' \cap G'' = Z$ and $|A'| = |A''| = 2$. Introduce new vertices, v' and v'', and put $H' = G' \cup \{v'\} \cup \{(v', z_i); \ i = 1, 2, 3\}$, $B' = A' \cup \{v'\}$, and similarly for H'', B''. Then (G, A) is cyclic iff there exists $i \in \{1, 2, 3\}$ such that (up to the notation) $(H' - z_i, B')$ and $(H'' - (v'', z_i), B'')$ are both cyclic]

6.5 Example. (H, D), where $H = C \cup D$, $C = (z_1, \ldots, z_8, z_1)$ is a 8-circuit and $D = \{(z_i, z_{i+4}); \ i = 1, 2, 3, 4\}$, is acyclic and satisfies all the above conditions.

6.6 Theorem (N. Robertson [16]). *Let G be a 3-connected graph and A be a set of four pairwise non-adjacent vertices of G. Suppose that G has no 3-cut separating A into two 2-subsets. If G has no A-cycle then (G, A) is similar to (H, D) of Example (6.5) or to $(H + R, D)$ where R is an arbitrary graph satisfying $V(R) \cap V(H) \subseteq \{z_i; \ i = 1, 3, 5, 7\}$.*

Here similarity is defined as follows: (G, A) is similar to (G', A') if either $(G, A) = (G', A')$ or G has subgraphs $T_1, \ldots, T_m$, $m \geq 1$, with the properties: $|\text{bd}(T_i)| = 3$, $A \cap \text{in}(T_i) = \emptyset$, and $\text{in}(T_i) \cap \text{in}(T_j) = \emptyset$ for $i, j = 1, \ldots, m$, $i \neq j$, so that replacing each $\text{in}(T_i)$ with a 3-star and subsequent erasing 2-valent vertices, if needed, transfers (G, A) into (G', A').

Proof. According to the scheme (1.13-15), we start with the following

6.7 Proposition. *G contains a subgraph W such that $A \subset E(W)$ and (W, A) is a subdivision of (H, D).*

Proof. Divide A into two pairs, A' and A'', and let Z' (Z'') denote the set of ends of A' (A''). Since $|Z'| = |Z''| = 4$ and no 3-cut separates Z' from Z'', there exist four disjoint chains of G, with distinct ends, connecting Z' with Z'', by the Menger theorem. Let C denote the union of these chains and $Z' \cup Z'' \cup A$; C is a 2-regular graph each of whose components contains at least two members of A. Since (C, A) is acyclic, C cannot be but a union of two disjoint circuits, C' and C'', with $|A \cap C'| = |A \cap C''| = 2$. Again, by the Menger theorem, G has four disjoint chains, with distinct ends, connecting C' and C''. Let W denote the union of C and these chains. Since (W, A) is acyclic, it cannot be but a subdivision of (H, D) of Example (6.5).   □

Introduce the following notations: $Z = \{z_i; \; i = 1, \ldots, 8\}$ denotes the set of 3-valent vertices of W ordered along some orientation of the circuit of $W - A$, and $W(i, j)$ denotes the thread of W connecting z_i and z_j. A lies on the diametral threads $W(i, \; i + 4)$, $i = 1, 2, 3, 4$. For $i \in \{1, \ldots, 8\}$ $T(i)$ denotes the union of z_i and the three threads incident to z_i.

6.8 Proposition (Instability of (H, D)). *Let u, v be two vertices of W, and P an external chain connecting u and v. $(W + P, A)$ is acyclic iff u, v are incident to the same $T(i)$, and moreover, the circuit of $T(i) + P + Z$ does not meet A unless both u, v are incident to the diametral thread.*

Now, G is a union of W and a number of external chains. For every external chain P let $C(P, i)$ denote the circuit of $T(i) + P + Z$, when the ends of P are incident to $T(i)$. (So that $C(P, i) = C(P, j)$ when u, v are incident to $W(i, j)$.) Let $C(i)$ denote the union of $\{z_i\}$ and all circuits $C(P, i)$, and define $G(i)$ as $\{z_i\}$ if z_i is a cut-vertex of $C(i)$ and as the block of $C(i)$ containing z_i otherwise.

6.9 Proposition. *G is the union of W and the subgraphs $G(i)$, $i \in \{1, \ldots, 8\}$.*

[Follows from the assumption that G is 3-connected.]

6.10 Proposition. *$A \cap G(i) = \emptyset$, for all i.*

Proof. Suppose that $G(1)$ contains an edge a from A. Put $A' = A - \{a\}$ and consider the A'-cycle C' of W passing through z_1. (Then $z_1 \notin C'$.) Let e be an edge from $C' \cap G(1)$ and C'' be an $\{a, e\}$-cycle of $G(1)$. Then $C' \cup C''$ contains an A-cycle - contradiction. $\qquad\square$

6.11 Proposition. *Suppose that $i - j$ is odd. Then $G(i) \cap G(j) \neq \emptyset$ iff $|i - j| = 1$ (or 7), and, moreover, this meet consists of a single vertex from $W(i, j)$.*

Proof. Put $Q(i, j) = G(i) \cap G(j)$, and suppose that $Q(1, i) \neq \emptyset$ fo some $i \in \{2, 4, 6, 8\}$. Consider two cases.
(6.11.1) $Q(1, i) \nsubseteq W$. Then there exists an external chain in $G(1) \cup G(i)$ violating the instability condition (see (6.8)).
(6.11.2) $Q(1, i) \subseteq W$. This is possible only for $i = 2$ or 8. Assume $i = 2$. If $Q(1, 2)$ contains an edge of $W(1, 2)$ then $G(1) \cup G(2)$ is 2-connected.
Now, $W - \cup\{W(i, i + 1);\ i = 1, 3, 5, 7\}$ is a union of two circuits, say C' and C'', each containing two members of A. Then $G(1) \cup G(2) \cup C' \cup C''$ contains an A-cycle - a contradiction. $\qquad\square$

6.12 Proposition. *Suppose that $Q(i, j) \nsubseteq Z$, where, say, $i, j \in \{1, 3, 5, 7\}$, $i \neq j$. Then $G = R \cup G'$ where $R \cap G' \subseteq \{z;\ i = 2, 4, 6, 8\}$ and $G' = W \cup (\cup\{G(i);\ z_i \notin R \cap G'\})$.*

Proof. Suppose there exists a vertex $x \in Q(1, i)$, $i \in \{3, 5, 7\}$, such that $x \notin W$. Let K, L be two disjoint chains of $G(1)$ connecting x with two distinct vertices of W, and K', L' be the same for $G(i)$. Then it is possible to chose one of K, L and one of K', L' so that their union, together with x, contains an external chain P with ends in $G(1)$ and $G(i)$. By (6.11), P belongs to no $G(j)$ with even j. On the other hand, P satisfies the instability condition (6.8). This means that there exists a subgraph R of G such that $G = R \cup G'$ with $R \cap G' \subseteq \{z_i;\ i = 2, 4, 6, 8\}$. Finally, (G', A) has the same properties as (G, A) plus an additional one, that $G'(i) = \{z_i\}$ for each i such that $z_i \in R \cap G'$. $\qquad\square$

Theorem (6.6) follows from (6.7-12).

References

[1] Bondy, J.A., Lovász, L. (1981): Cycles through specified vertices of a graph. Combinatorica **1**, 117–140
[2] Bondy, J.A., Murty, U.S.R. (1978): Graph theory with applications. Macmillan, London
[3] Dirac, G.A. (1960): 4-chrome Graphen und vollständige 4-Graphen. Math. Nachr. **22**, 51–60
[4] Ford Jr., L.R., Fulkerson, D.R. (1962): Flows in networks. Princeton University Press, Princeton, NJ
[5] Haggkvist, R., Thomassen, C. (1982): Circuits through specified edges. Discrete Math. **41**, 29–34
[6] Holton, A., McKay, B.D., Plummer, M.D., Thomassen, C. (1982): A nine-point theorem for 3-connected graphs. Combinatorica **2**, 53–62
[7] Kelmans, A.K. (1986): Constructions of non-Hamiltonian cubic bipartite 3-connected graphs. The Institute for Systems Studies, Moscow (Collect. Pap. Issue 10, in Russian)

[8]   Kelmans, A.K., Lomonosov, M.V. (1982): When m vertices in a k-connected graph cannot be walked around along a simple cycle. Discrete Math. **38**, 317–322

[9]   Kelmans, A.K., Lomonosov, M.V. (1981): Problems on cycles through prescribed vertices in a graph. In: Hajnal, A., Lovász, L., Sós, V. (eds.): Finite and infinite sets I. North Holland, Amsterdam (Colloq. Math. Soc. Janos Bolyai, Vol. 37)

[10] Kelmans, A.K., Lomonosov, M.V. (1983): On cycles through prescribed vertices in weakly separable graphs. Discrete Math. **46**, 183–189

[11] Kelmans, A.K., Lomonosov, M.V.: A cubic bipartite 3-connected graph with at most 30 vertices is Hamiltonian. (Unpublished)

[12] Kelmans, A.K., Lomonosov, M.V.: A cubic 3-connected graph with no cycle through some 10 vertices is similar to the Petersen graph. (Unpublished)

[13] Lovász, L. (1980): Matroid matching and some applications. J. Comb. Theory, Ser. B **28**, 208–236

[14] Mader, W. (1978): Über die Maximalzahl kreuzungsfreier H-Wege. Arch. Math. **31**, 387–402

[15] Reingold, E.M., Nievergelt, J., Deo, N.: Combinatorial algorithms: Theory and practice. Prentice-Hall, Englewood Cliffs, NJ

[16] Robertson, N. (Private communication)

[17] Watkins, M.E., Mesner, D.M. (1967): Cycles and connectivity in graphs. Can. J. Math. **19**, 1319–1328

[18] Woodall, D.R. (1977): Circuits containing specified edges. J. Comb. Theory, Ser. B **22**, 274–278

Communication Complexity: A Survey

László Lovász

0. Introduction

Complexity is one of the crucial scientific phenomena of our times. We are far from understanding all its aspects; but the first elements of a comprehensive theory do begin to emerge. To define a measure of complexity, we consider a certain (computing) task, and ask: what is the minimum amount of certain resources that is needed to carry out this task? The resources one considers depend on the device we are using and on other circumstances. The most common complexity measures are time and space, but many others can be considered.

The increasing importance of distributed computing, networking, VLSI and the use of computers in telecommunication have pointed out the significance of **communication** as a resource. In many devices, communication is significantly slower and costlier than local computation, and it is the real bottleneck in solving certain problems. One may mention the obvious example of a rocket approaching Jupiter or Halley's comet: local computation on the Earth can use the most powerful devices, and even the rocket can be equipped with very reasonable computers; but the communication is extremely slow and unreliable (which means that to make it more reliable we have to introduce more redundancy and thereby sacrifice even more time). If we go to more everyday examples it becomes less and less justified to concentrate solely on the problems of communication, but (as we shall see) even within a single chip, communication between various parts may be an important task which has to be solved efficiently in non-trivial ways.

Communication complexity also plays an important role in theoretical studies: many of the known "lower-bound" results in complexity theory are obtained by analyzing the communication between various parts of the input and output. This is perhaps only a temporary phenomenon, since such techniques do not seem to be powerful enough to lead to much-hoped-for results like P≠NP; but it does indicate that separating just this one factor contributing to the complexity of various tasks can lead to important information about the whole task.

In the following sections we give some examples from the theory of VLSI that lead to problems concerning communication; then we give a survey of the theory of communication complexity. I hope also to "communicate" my feeling that this lovely theory brings together a surprising number of ideas from classical and modern mathematics, and it also illustrates in a rather clear way basic notions from complexity theory like non-determinism and randomization.

1. Time-Area Tradeoff in VLSI: A Simple Example

Suppose that we want to design a "chip" that checks whether two strings of n bits, arriving simultaneously on $2n$ wires, are the same. More exactly, by a "chip" we mean a rectangular grid graph, with processors sitting in some of the nodes, and edge-disjoint paths (called "wires") connecting some of these processors. On the upper edge of the array, we have $2n$ specified nodes called "input ports", and somewhere else on the boundary we have one more specified node called "output port". We feed $2n$ bits $x_1, \ldots, x_n, y_1, \ldots, y_n$ into the input ports, and we want the gadget to produce a single bit at the output port, which is 1 iff $x_i = y_i$ for all i. We want the "chip" to be fast and small.

We make the following assumptions. The "chip" is *systolic*, i.e., it works in distinct steps. At each step, each processor reads the bits sent to it, computes a bit for some of the wires starting from it, and sends it to the processor at the other end of the wire. It will be convenient to assume that it has a memory of a constant number of bits, but it would not make any essential difference to exclude this.

Our simple task can be solved by a chip with the simplest possible topology: a single path of length $2n$ (Fig. 1). Let $p_1, \ldots, p_{2n}$ be the nodes of this path, each containing a processor. These receive, in order, bits $x_1, \ldots, x_n, y_1, \ldots, y_n$. Let each node except p_{n+1} "sleep" at the beginning. In the first step, p_{n+1} sends bit y_1 to the left and an "alarm" bit to the right. This wakes up p_{n+2}, who sends y_2 to the left and an "alarm" bit to the right. Each processor receiving a bit from the right will transmit it to the left in the next step. So the bits $y_1, \ldots, y_n$ will march to the left with one space left between any two.

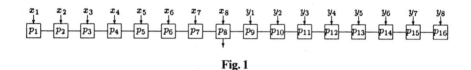

Fig. 1

Now when p_1 receives y_1, it compares it with x_1. He sends 1 to the right if they are equal and a 0 otherwise. Processor p_2 receives this bit simultaneously with y_2; he compares y_2 with x_2 and sends a 1 to the right if he received a 1 and also $x_2 = y_2$; else, he sends a 0. This goes on like this until p_n receives a bit from the right (which is 1 if and only if $x_1 = y_1, \ldots, x_{n-1} = y_{n-1}$) and simultaneously the bit y_n from the left. He compares x_n and y_n, and produces the output bit.

It is clear that for the simplicity of the topology we had to pay with time: this shifting back and forth took $2n$ steps. It is perhaps more natural to allow more space and solve the problem in about $O(\log n)$ steps, using the chip in Fig. 2. In the first step, processors $q_1, \ldots, q_n$ make the comparisons $x_1 = y_1, \ldots, x_n = y_n$; their findings are collected by a binary tree in $\log n$ further steps (all log's are base 2).

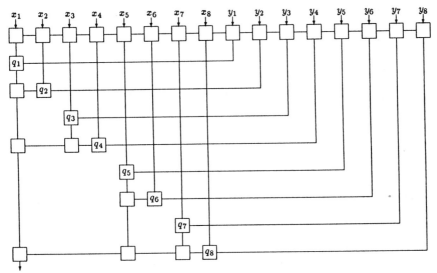

Fig. 2

But we had to make the chip much bigger: it has $2n^2 - 2n$ nodes; or, if we define the "area" of the chip as the total length of wires, it has area larger than n^2. We could do a little better: combining the ideas from the two designs shown above one obtains a chip working in $O(\log n)$ time, with area $O(n^2/\log n)$. This construction is left to the reader as an exercise.

Now this factor of $\log n$ is not too much gain; could we do better? Let us prove that we cannot. More exactly, we prove the following "area-time tradeoff" theorem, which says that to reduce the area, we have to sacrifice time. This theorem is a special case of the results of Thompson (1979):

1.1 Theorem. *If a chip decides the equality of two 0-1 sequences of length n (in the sense described above), has area A, and works in time T, then*

$$AT \geq n^2.$$

(Note that equality can hold, up to a constant, with $A = O(n)$ and $T = O(n)$, and also with $A = O(n^2/\log n)$ and $T = O(\log n)$, and in fact for every pair of values inbetween satisfying the inequality.)

Proof. Let us draw a vertical line cutting the chip into two, so that one half contains the ports $x_1, \ldots, x_n$ and the other, the ports $y_1, \ldots, y_n$. Let us forget about what happens inside the two parts, and concentrate on the communication across this line. "Obviously" (we'll come back to this!) at least n bits of information must cross this line. Since we only have T steps, there must be a step when simultaneously at least n/T bits cross the line. But one wire can transmit in one step only one bit of information, so there must be at least n/T wires that cross the line.

Now if we shift the line to the right by one position, x_1 and y_1 may be on the same side, so the amount of information sent across the line may be less. But we can argue that the computation still has to decide whether $x_1 \ldots x_{n-1} = y_1 \ldots y_{n-1}$ and so by the same argument, the shifted line must cross at least $(n-1)/T$ wires. Similarly, the line shifted by 2 positions must cross at least $(n-2)/T$ wires etc.

If we add up these numbers, we get exactly the total length of horizontal pieces of the wires, which, in turn, is a trivial lower bound on the area. Hence

$$A \geq \frac{n}{T} + 2\frac{n-1}{T} + 2\frac{n-2}{T} + \ldots = \frac{n^2}{T}. \qquad \square$$

We have used in the proof that at least n bits have to cross a line separating the x-ports from the y-ports. Note that this assertion is now independent of VLSI: the two parts of the chip may be viewed as two processors, each getting a 0-1 sequence of length n, and they have to decide whether these are equal. We want to show that they have to communicate at least n bits.

This seems natural, and it is even true, but an exact proof is not quite obvious. To warn the reader, let us mention that if they are allowed to use a random number generator, and they want the answer only with a certainty of 0.9999999999, then it suffices to communicate $O(\log n)$ bits (see Section 6).

In the following sections we develop the theory of communication complexity, which will allow us to prove these kinds of lower bounds. This example has served to expose the problem and we are not going to treat the interesting and important topic of area-time tradeoffs in VLSI in detail. However, we shall return briefly to VLSI problems in Section 8, to see which related communication problems are raised by them.

2. The Notion of Communication Complexity, Deterministic and Non-Deterministic

We consider a rather simple model of communication. We have two processors and a direct link between them. Each processor knows some information, called the *input* of that processor; and they want to compute a value which is a function of both inputs. For simplicity, we shall assume (except for one case in Section 8) that this value to be computed is a single bit; we call this the *output bit* of the communication problem.

We assume that both processors have unlimited computing power and that local computation is free. However, we are charged for every single bit transmitted from one processor to the other. Our aim is to find methods to solve such problems with a minimum number of bits transmitted.

When talking about this model, it is nicer to imagine that the two processors are two teenagers, Alice and Bob. Alice lives in Budapest, Bob lives in New Zealand. To appreciate the importance of minimizing the length of communication between them, imagine that we (the parents) have to pay the telephone bill for their communication.

The communication analogue to the fundamental notion of an *algorithm* in "ordinary" complexity theory is the notion of a *protocol*. Informally, a *communi-*

cation protocol is a set of rules specifying the order and "meaning" of messages sent. So the protocol prescibes who is to send the first bit; depending on the input of that processor, what this bit should be; depending on this bit, who is to send the second bit, and depending on the input of that processor and on the first bit sent, what this second bit should be etc. The protocol terminates when one processor knows the output bit and the other one knows this about the first one. This definition seems strange, but it turns out more convenient technically than the obvious rule that the protocol ends when both know the result. There is no real difference between the two: if Alice knows the result, and Bob knows this about her, then she only needs to send one more bit. Conversely, in every protocol which ends with both of them knowing the answer, the situation before the last bit (which is sent, say, by Alice) is that Alice must know the answer and that Bob must know that she knows it.

The *complexity* of a protocol is the number of bits communicated in the worst case.

There is always a *trivial protocol*: Alice can send her input to Bob. Then Bob has all the information he needs to compute the answer, and Alice knows this about him. Of course, they can switch roles if this is better. (We shall see that sometimes there is no better protocol than this trivial one).

Let us formalize this notion, following Yao (1979). Let $a_1, \ldots, a_n$ be the possible inputs of Alice and $b_1, \ldots, b_m$, the possible inputs of Bob. (Note that since local computation is free, we don't have to worry about how these are encoded. Two inputs for Alice are clearly equivalent if they give the same results with any input for Bob. So two such inputs can be identified from our point of view.) Let c_{ij} be the value they want to determine for inputs a_i and b_j. The 0-1 matrix $C = (c_{ij})_{i=1}^{n}{}_{j=1}^{m}$ determines the communication problem; we call this the *communication matrix* associated with the problem. Both players know the matrix C; Alice knows the index i of a row, Bob know the index j of a column, and they want to determine the entry c_{ij}. The trivial protocol (in which, say, Alice sends the index of her row to Bob) takes $\lceil \log_2 n \rceil$ bits (of course if $m < n$ then the reverse trivial protocol is better).

A protocol has a very simple interpretation in terms of this matrix. First, it determines who sends the first bit; say, Alice does. This bit is determined by the input of Alice; in other words, the protocol partitions the rows of the matrix C into two classes, and the first bit of Alice tells Bob which of the two classes contains her row. From now on, the game is limited to the submatrix C_1 formed by the rows in this class. Next, the protocol describes a partition of the rows or columns of C_1 into two classes (depending on who is to send the second bit), and the second bit itself specifies which of these two classes contains the line (row or column) of the sender. This limits the game to a submatrix C_2 etc.

If the game ends after k bits then the remaining submatrix C_k is the union of an all-1 submatrix and an all-0 submatrix. (We shall call this an almost homogeneous matrix.) In fact, if (say) Alice knows the answer then her row in C_k must be all-0 or all-1, and since Bob knows for sure that Alice knows the answer, this must be true for every row of C_k.

So the determination of the communication complexity is reduced to the following combinatorial problem: given a 0-1 matrix C, in how many rounds can we partition it into almost-homogeneous submatrices, if in each round we can split each of the current submatrices into two (either horizontally or vertically). We shall denote this number by $\kappa(C)$.

Note that in each step, the maximum rank of the submatrices is decreased by a factor of 2 or less, and hence we obtain the following inequality due to Mehlhorn and Schmidt (1982):

2.1 Lemma. *If C has rank r then $\kappa(C) \geq \log r$.*

In particular, we obtain

2.2 Corollary. *If C has full row rank then the trivial protocol is optimal.*

Here is a simple communication problem to which the previous corollary applies directly. Suppose that Alice and Bob want to solve the following important problem: is the recent edition of the Encyclopædia Britannica that Alice has identical – letter for letter – with that of Bob? An obvious solution to this problem is that one of them reads the Britannica to the other over the phone. While this may be a pleasure under the circumstances, clearly the parents object. But clearly they cannot stop an important project like this; can they come up with a protocol that is more efficient, i.e., that reduces the telephone bill[1]?

To formalize this problem, let us assume that we know the size of both editions – say, n bits. Then there are 2^n possible inputs for Alice and the same number of possible inputs for Bob, and so C is a $2^n \times 2^n$ identity matrix. This matrix has rank 2^n, so it follows from Corollary 2.2 that for this problem there is no protocol better than the trivial one: transmitting the whole input of one of the players. This also completes the proof of Theorem 1.1. (We shall return to this question in the next sections; among others, we shall see that randomization is extremely helpful.)

Let us also point out that this example shows that the matrix C can be of enormous size, and typically only implicitly given – the problem specification tells us how to compute any given element. In particular, the rank of C is not always an easily computable parameter (even though it was in this last example).

A slightly more complicated communication problem is *Disjointness*: each player has a subset of a specified set S, and they have to decide if their inputs are disjoint. In other words, they both know a 0-1 vector of length n, and they want to know whether the inner product is 0 or not. Various versions of this problem will play an important role later in this paper, but now we can settle the communication complexity of the basic version easily. In fact, if we arrange

---

[1] This problem is not quite as artificial as it sounds. In the example of communication with a spacecraft, it may be important to check (say, before a new operation is started) whether the program of the computer on board is still correct. This may require the comparison of strings longer than the Britannica – and with much more costly communication.

the subsets appropriately, the matrix associated with this problem will be lower triangular with 1's in its main diagonal, and so non-singular. It follows that *the trivial protocol is optimal for the Disjointness problem.*

Let us mention one more useful example, where the computation of the rank is not so immediate. Let S be a set with n elements, and assume that Alice has a subset $X \subseteq S$ and Bob has a subset $Y \subseteq S$. They want to know the parity of $|X \cap Y|$. In other words, they both know a 0-1 vector of length n, and they want to compute the inner product modulo 2.

If we write up the corresponding matrix C, it will be symmetric and will have a row (and column) of 0-s: the row corresponding to $\emptyset$. Let us drop this row and column; we claim that the remaining matrix C' is non-singular. To see this, compute $(C')^2$: it turns out to have 2^{n-1} in its main diagonal and 2^{n-2} elsewhere. It is elementary linear algebra that this matrix is non-singular.

Hence $\mathrm{rk}(C) = \mathrm{rk}(C') = 2^n - 1$, and so (at least if $n \geq 2$)

$$\kappa(C) \geq \lceil \log_2 \mathrm{rk}(C) \rceil = n.$$

So we obtain that *for computing the inner product modulo 2, the trivial protocol is optimal.*

It is interesting to remark that the rank over GF(2), which would perhaps seem more natural to use in this problem, would give a very poor bound here: the GF(2) rank of C is only n. (In these examples, the trivial protocols were optimal. In Section 4, we shall see that allowing alternation in the direction of communication may reduce the complexity to the logarithm of the length of the trivial protocol. Duris, Galil and Schnitger (1984) showed that for every k, the best $(k + 1)$-round protocol may be just the logarithm of the best k-round protocol.)

The rank of C (not its logarithm!) bounds $\kappa(C)$ from above. It is easy to see that a 0-1 matrix with rank r has at most 2^r distinct rows. Since repeated rows of C can clearly be suppressed without changing the communication complexity of the problem, we obtain

2.3 Proposition. *If C has rank r then $\kappa(C) \leq r$.*

In this proposition, one could take the GF(2) rank as well, and our previous remark shows that then the bound is sometimes sharp. For the real rank, however, I don't know of any communication problem for which $\kappa(C)$ would be anywhere near the bound r. In particular it appears to be unsettled whether $\kappa(C)$ can be bounded by any polynomial of $\log r$.

In the complexity theory of algorithms, non-deterministic algorithms play a crucial role; they specify important complexity classes like NP. Analogously, non-deterministic protocols play an important role in the complexity theory of communication. They were introduced by Lipton and Sedgewick (1981).

Assume that there is an extra-terrestrial super-being, called E.T., who can monitor the inputs and communication of Alice and Bob, and of course knows the answer to the problem right away. He can also make any announcement on the phoneline. Can he speed up communication by giving the right announcements (hints)? For example, if he gets bored with Bob's voice reading the

Encyclopædia to Alice, can he shortcut the communication with an appropriate short announcement from which both Alice and Bob can be certain what the answer is?

In this case, the answer depends on whether the outcome is that the two editions are equal or that they are different. If they are different, E.T. can simply announce 'the 37th character in the 14th line of page 568 of volume 24 is "," in Alice's edition but ";" in Bob's.' At the cost of these few bits, the problem is settled. On the other hand, if the two editions are identical, then even the E.T. cannot save a single bit of communication! This will become clear as soon as we formalize the notion of a non-deterministic protocol.

Given a communication problem (in the form of a matrix C), there will be two non-deterministic protocols, one for "1" and one for "0". A non-deterministic protocol for (say) "1" has to specify a set P of possible *proofs*. Further, it has to include a rule for Alice that tells her which proofs she should accept depending on her input, and a similar rule for Bob. For this non-deterministic protocol to be correct, we require

(∗) for a given pair of inputs the answer to the communication problem is "1" if and only if there exists a proof $p \in P$ which is accepted by both Alice and Bob.

Without loss of generality we may assume that P consists on 0-1 strings, and then the maximum length of these strings is the *complexity* of the protocol. Of course, with appropriate encoding, this will be $\lceil \log |P| \rceil$. We are interested in finding the protocol for which this complexity is minimum; this value is denoted by $\kappa_1(C)$. A non-deterministic protocol for "0" and $\kappa_0(C)$ are defined analogously.

Consider a non-deterministic protocol for "1", a particular proof $p \in P$, and the entries of C (i.e. inputs for Alice and Bob) for which this proof "works". By (∗), these must be all 1's; also, from the fact that Alice and Bob have to verify the proof independently, we see that these 1's must form a submatrix. So a non-deterministic protocol corresponds to a covering of C by all-1 submatrices. Conversely, every such covering gives a non-deterministic protocol: E.T. can simply announce the name of the submatrix containing the given entry. So we obtain:

2.4 Lemma. *The non-deterministic communication complexity $\kappa_1(C)$ of a matrix C is the least natural number t such that the 1's in C can be covered by at most 2^t all-1 submatrices.*

Note that the all-1 submatrices in the covering do not have to be disjoint. Therefore, there is no immediate relation between the non-deterministic communication and the rank of C.

Yannakakis (1988) introduces two further concepts that further illuminate the connection between rank and communication complexity (deterministic and non-deterministic). He defines the *unambigous communication complexity* $\bar{\kappa}_1(C)$ of C as the least natural number a for which the 1's in C can be covered by 2^a disjoint all-1 submatrices; equivalently, it is the minimum complexity of a non-deterministic protocol in which the "proof" in (∗) is unique. Obviously,

$$\kappa_1(C) \le \bar{\kappa}_1(C) \le \kappa(C),$$

and

$$\log \mathrm{rk}(C) \le \bar{\kappa}_1(C).$$

It is easy to derive analogous results for the complexity of the answer "0".

There is also a way to formulate, in some sense, an upper bound on the complexity in terms of a certain "rank". Let C be an $n \times m$ 0-1 matrix. Yannakakis (1988) defines the *positive rank* $\mathrm{rk}_+(C)$ of C as the least p for which there exists a non-negative $n \times p$ matrix A and a non-negative $p \times m$ matrix B such that $C = AB$. (Note that without the non-negativity restriction, this would define just the ordinary rank.) Trivially

$$\mathrm{rk}(C) \le \mathrm{rk}_+(C).$$

It is also easy to show that

$$\kappa_1(C) \le \log \mathrm{rk}_+(C).$$

In fact, If $C = AB$ where A and B are non-negative matrices with p columns and rows, respectively, then let a_i denote the ith column of A and b_i^T, the ith row of B. Then we have

$$C = \sum_{i=1}^{p} a_i b_i^T.$$

Let Z_i be the support of the matrix $a_i b_i^T$, then clearly Z_i is an all-1 submatrix of C and the Z_i's must cover all the 1's in C. Hence $\kappa_1(C) \le \log p$ as claimed. Finally, we also have

$$\log \mathrm{rk}_+(C) \le \kappa(C),$$

which follows by a similar argument as for the rank. Unfortunately, the positive rank of the matrix does not seem to be easy to handle (in particular, no polynomial-time algorithm is known to compute it).

Let us mention one further, rather trivial bound on the non-deterministic complexity. Assume that C has a 1's but every all-1 submatrix of C has at most b entries. The trivially $\kappa_1(C) \ge \log a - \log b$. While I do not know any good applications of this observation, a version of it plays an important role in obtaining lower bounds for randomized protocols (see Section 6).

Returning to the Encyclopædia Britannica example, it is obvious that the 1's in the the $N \times N$ identity matrix I_N cannot be covered by fewer that N all-1 submatrices. Hence

$$\kappa_1(I_N) = \lceil \log N \rceil.$$

For our Encyclopædia Britannica example this says that even the E.T. cannot prove that the two editions are equal, any cheaper than reading the whole text!

3. A General Upper Bound on Communication Complexity

We have seen three lower bounds on the communication complexity of a matrix:

(3.1) $$\log \mathrm{rk}(C) \le \kappa(C),$$

(3.2) $$\kappa_0(C) \le \kappa(C),$$
        and
(3.3) $$\kappa_1(C) \le \kappa(C).$$

The identity matrix shows that it can happen that the (3.2) is very week: the right hand side can be exponentially large compared with the left. Interchanging the roles of 1 and 0, we obtain that the bound in (3.3) can also be very far from $\kappa(C)$. We do not know such a bad example for (3.1), but it is likely that the situation is similar. Note that (3.1) is also essentially symmetric in 0's and 1's: if we interchange the 0's and 1's in a C, the rank of the resulting matrix $\overline{C}$ differs from the rank of C by at most 1.

However, it is a surprising fact that the product of any two of these three lower bounds is an upper bound on the communication complexity. The first part of the following theorem is due to Aho, Ullman and Yannakakis (1983), the second (and third) to Lovász and Saks (1988b):

3.4 Theorem. *For every matrix C,*
$(a)\kappa(C) \le (\kappa_0(C) + 1)(\kappa_1(C) + 1);$
$(b)\kappa(C) \le \log \mathrm{rk}(C)(\kappa_0(C) + 1);$
$(c)\kappa(C) \le \log \mathrm{rk}(\overline{C})(\kappa_1(C) + 1).$

It is clear that part (b) of this theorem implies part (c). We shall give the proof of a result that is stronger than either one of (a) or (b). Let $\rho_1(C)$ denote the size of the largest square submatrix of C such that (ordering the rows and columns appropriately) every diagonal entry is 1 but every entry above the diagonal is 0. We can define $\rho_0(C)$ analogously.

It is clear that

$$\rho_1(C) \le \mathrm{rk}(C)$$

and

$$\rho_0(C) \le \mathrm{rk}(\overline{C}) \le \mathrm{rk}(C) + 1.$$

Furthermore,

$$\rho_1(C) \le 2^{\kappa_1(C)},$$

since in every covering of C with all-1 blocks, the diagonal entries in the submatrix in the definition of $\rho_1(C)$ must belong to different blocks. Hence the following theorem implies all three parts of Theorem 3.4:

3.6 Theorem. *For every matrix C,*

$$\kappa(C) \le \big(\log \rho_1(C) \big)\big(\kappa_0(C) + 1 \big).$$

Proof. We use induction on $\rho_1(C)$. Let $k = \kappa_0(C)$; then the 0-s of C can be covered by all-0 submatrices $C_1, \ldots, C_N$ where $N \leq 2^k$. Let A_i [B_i] denote the submatrix formed by those rows [columns] of C that meet C_i. Then observe that

$$\rho_1(A_i) + \rho_1(B_i) \leq \rho_1(C).$$

We may assume that for $i = 1, \ldots, M$, we have $\rho_1(A_i) \leq \rho_1(C)/2$ but for $i = M + 1, \ldots, N$, we have $\rho_1(B_i) \leq \rho_1(C)/2$. We may also assume that $M \geq N/2$.

Now we can describe the following protocol.

First, Alice looks at her row to see if it intersects any of the submatrices $C_1, \ldots, C_M$. If so, she sends a "1" and the name of such a submatrix. If not, she sends a "0".

If Bob recieves a "0", he looks at his column to see if it intersects any of the submatrices $C_{m+1}, \ldots, C_N$. If so, he send a "1" and the name of such a submatrix. If not, he send a "0".

This describes one round of the protocol. This round may end in three ways:

Case 1. Alice found an appropriate submatrix. Then they both know that Alice's row belongs to A_i ($1 \leq i \leq M$). Since $\rho_1(A_i) \leq \rho_1(C)/2$, by recurrence they can find the answer by communicating at most

$$\log(\rho_1(A_i))(\kappa_0(A_i) + 1) \leq (\log(\rho_1(C)) - 1)(\kappa_0(C) + 1)$$

bits. Since Alice's message took $1 + \lceil \log M \rceil \leq 1 + \kappa_0(C)$ bits, this is altogether at most $(\log \rho_1(C))(\kappa_0(C) + 1)$ bits.

Case 2. Bob found an appropriate submatrix. This is similar to Case 1. Then they both know that Bob's column belongs to B_i ($M = 1 \leq i \leq N$), and so by recurrence they can find the answer by communicating at most $(\log(\rho_1(C)) - 1)(\kappa_0(C) + 1)$ bits. The count of bits for the round itself is slightly different: 1 for Alice's message and $1 + \lceil \log(M - N) \rceil \leq \kappa_0(C)$ for Bob's message. The final count is the same.

Case 3. Both Alice and Bob failed to find an appropriate submatrix. Then the intersection of their lines cannot belong to any C_i, and so it must be a "1". So they have found the answer.   □

As a further corollary of this theorem we mention the following result of Yannakakis (1988), bounding the deterministic complexity in terms of the unambiguous (but non-deterministic) complexity:

3.5 Corollary. $\kappa(C) \leq \bar{\kappa}_1(C)^2$.

Similarly, we obtain a relation between the rank, positive rank and communication complexity of the matrix:

3.7 Corollary. $\kappa(C) \leq (\log \text{rk}(C))(\log \text{rk}_+(C))$.

Theorem 3.4(a) has an interesting interpretation. Define a *communication problem* as a class $\mathcal{H}$ of 0-1 matrices; for simplicity, assume that they are square matrices. The communication complexity of any $N \times N$ matrix is at most $\log N$. We say that $\mathcal{H}$ is in P_{comm} if it can be solved substantially better: if there exists a constant $c > 0$ such that $\kappa(C) \leq (\log \log N)^c$ for each matrix $C \in \mathcal{H}$, where N is the dimension of C. Similarly, we say that $\mathcal{H}$ is in NP_{comm}, if there exists a constant $c > 0$ such that for each $C \in \mathcal{H}$, $\kappa_1(C) \leq (\log \log N)^c$. We can define co-$NP_{comm}$ analogously. Just as for the analogous computational complexity classes, we have the trivial containment

$$P_{comm} \subseteq NP_{comm} \cap \text{co-}NP_{comm}.$$

However, for the communication complexity classes we also have the following, rather interesting facts:

$$P_{comm} \neq NP_{comm},$$

$$NP_{comm} \neq \text{co-}NP_{comm}$$

(both follow from the example of checking identity), but

$$P_{comm} = NP_{comm} \cap \text{co-}NP_{comm}$$

(by the theorem of Aho, Ullman and Yannakakis 3.4(a)). This idea was developed by Babai, Frankl and Simon (1986), who defined and studied communication analogues of many other well-known complexity classes like #P, PSPACE, BPP etc.

4. Some Protocols

We have seen that the Disjointness problem, i.e., the problem of deciding whether two given sets are disjoint, is trivial from the communication point of view: the trivial protocol (Alice sends her input to Bob) is optimal. Often, however, the sets the players have are restricted in one way or the other. We shall discuss some general types of restrictions in the next section; here we analyse two special examples. In both examples the rank lower bound is very far from the complexity of the trivial protocol; and in both cases it will turn out that this rank bound is quite close to the truth. In particular, we shall describe some examples of non-trivial communication protocols.

4.1 Problem. Let T be a tree and let Alice be given a subtree T_A and Bob a subtree T_B of T. Their task is to decide whether T_A and T_B have a node in common.

Suppose that T has n nodes. The number of subtrees of T can be between $\binom{n}{2} + n + 1$ (if T is a path) and $2^{n-1} + n$ (if T is a star). Typically, however, it is exponential in n; for example, if T has no nodes of degree 2 then it has more than $2^{n/2}$ subtrees.

Let us write up the matrix C_T corresponding to this communication problem (rows and columns indexed by subtrees, with a 1 in a position if and only if the corresponding subtrees are disjoint). The size of C_T can vary, but it is typically exponential. It is perhaps more surprising that the rank of C_T is determined by n, and it is in fact very low:

$$\mathrm{rk}(C_T) = 2n.$$

In the next section, we shall see the general reason behind this lemma; at this moment, we leave it to the reader as an exercise.

The non-deterministic complexity of our problem is easy to find. To exhibit that the two subtrees intersect, E.T. can announce a common node. This takes $\lceil \log n \rceil$ bits. To exhibit that they do not intersect, E.T. can either announce that one or the other is empty, or he can announce an edge uv such that if we delete uv from T, then in the remaining forest, T_A is in the component containing u but T_B is in the component containing v. So there are $2n$ possible announcements of E.T., and – in appropriate encoding – this takes $\lceil \log n \rceil + 1$ bits. It is not difficult to see that these are optimal non-deterministic protocols, i.e.

$$\kappa_1(C_T) = \lceil \log n \rceil + 1, \qquad \kappa_0(C_T) = \lceil \log n \rceil.$$

From this, we obtain by the general results in the previous section the following bounds on the communication complexity of Problem 4.1:

$$\lceil \log n \rceil + 1 \le \kappa(C_T) \le (\lceil \log n \rceil + 1)^2.$$

We show that the lower bound is near the truth (Lovász and Saks (1988a)).

4.2 Theorem. $\kappa(C_T) \le \log n + \log \log n + 1.$

Proof. First, we describe a protocol that gives a somewhat worse result: about $2 \log n$ as an upper bound. The protocol consists of two parts:

$1°$: Alice Alice selects any node $x \in V(T_A)$, and sends its name to Bob (we have to reserve one message to indicate if her subtree is empty; but in this case they are done).

$2°$: Bob determines the (unique) node $y \in V(T_B)$ nearest to x, and sends this to Alice (we have to reserve one message to indicate if his tree is empty).

Now it is easy to argue that the two subtrees T_A and T_B are node-disjoint if and only if $y \notin V(T_A)$. So at this point Alice knows the answer (and Bob knows that she knows).

As it stands, this protocol involves the communication of $2 \log(n+1)$ bits (the names of two nodes). To improve it, we use the following lemma:

4.3 Lemma. *The nodes of T can be labelled by $1, 2, \ldots, n$ so that if we delete the nodes labelled $1, \ldots, k - 1$, every connected component of the remaining forest has no more than $2n/k$ nodes.*

The players agree on such a labelling in advance (it is part of the protocol). Now we replace the first step by

$1^{\circ\circ}$: Alice selects the node $x \in V(T_A)$ with the least label, and sends its label to Bob (she sends 0 is her subtree is empty; in this case, they are done).

Suppose the label of x is k. Receiving this, Bob will know that Alice's tree does not contain the nodes labelled $1, \ldots, k-1$, so he can delete these from the tree. Each of the components of the remaining forest has at most $2n/k$ nodes, and Bob also knows which of these contains T_A, since he knows a node of this tree. So from now on they are restricted to a tree T_0 with at most $2n/k$ nodes, and he can carry out the second step more economically as follows:

$2^{\circ\circ}$: If $V(T_B) \cap V(T_0) \neq \emptyset$, then Bob determines the (unique) node $y \in V(T_B) \cap V(T_0)$ nearest to x, he counts the number l of nodes in T_0 with label not larger than the label of y. He sends to Alice the number l. If $V(T_B) \cap V(T_0) = \emptyset$, he sends 0.

Since this protocol differs from the previous one only in the encoding, its correctness is clear. The number k has $\log k$ bits; the number l is at most $2n/k$, so it has at most $\log(2n/k) = \log n + 1 - \log k$ bits. This is altogether $\log n + 1$ bits.

There is one catch: while sending the bits of k, Alice has to indicate that she is finished. (In other word, she has to use a prefix-free encoding of the integers up to n.) One solution is to start with $\log \log n$ further bits, announcing the length of k. This gives the bound in the theorem. (It can be shown that no encoding trick can get rid of these extra bits.)   □

Our second example is again a special disjointness problem. It was formulated by Yannakakis (1988) in connection with a complexity problem concerning the vertex packing polytope (see Section 7).

4.4 Problem. Let G be a graph and let Alice be given a set A of independent nodes, and let Bob be given a clique B. Their task is to decide whether A and B have a node in common.

Let G have n nodes. The number of independent sets/cliques in G can be exponentially large. The matrix C_G associated with the problem has a row for each independent set, a column for each clique, and (say) a "1" at a position where the corresponding clique and independent set intersect (this is now conversely to the convention in the previous problem, but here this will be the more convenient). The rank of this matrix is very low:

$$\mathrm{rk}(C_G) = n.$$

(This is now quite easy to see: this matrix is the product of the independent set-node incidence matrix and the node-clique incidence matrix.)

The non-deterministic communication complexity for non-disjointness is again $\lceil \log n \rceil$. So we have

$$\kappa_1(C_G) = \log n$$

and hence Theorem 3.4 implies the following bound, which was first proved by more direct means by Yannakakis (1988):

$$\lceil \log n \rceil \leq \kappa(C_G) \leq (\lceil \log n \rceil + 1)^2.$$

We show a protocol due to A. Hajnal that reduces the upper bound by a factor of 2:

4.5 Theorem. $\kappa(C_G) \leq \frac{1}{2}(\log n + 1)^2.$

Proof. We describe the protocol. First, Alice checks if her independent set A contains a node that has degree at least $n/2$. If it does, then she sends the name of such a node v to Bob. Now they both know that Alice's set is contained among the nodes non-adjacent to v (including v), and so the problem is reduced to one on a graph with at most $n/2$ nodes.

If Alice does not find such a node, then Bob looks for a node in his clique B to see if B contains a node with degree less than $n/2$. If it does, he sends the name of such a node u to Alice. Then they both know that Bob's set is contained among the nodes u and those adjacent to u, and so the problem is again reduced to one on a graph with at most $n/2$ nodes.

If neither one is succesful, they know that every node of Alice's set has degree less than $n/2$ while every node of Bob's set has degree larger than $n/2$, and hence they know that the two sets are disjoint.

It is easy to estimate the complexity of this protocol and get the bound as given.   □

Note that in this case the nondeterministic communication complexity for disjointness is not easy to find. A non-deterministic protocol for disjointness corresponds to a family $\mathcal{H}$ of subsets of $V(G)$ such that for each pair (A, B) where A is an independent set, B is a clique, and $A \cap B = \emptyset$, there exists an $H \in \mathcal{H}$ such that $A \subseteq H \subseteq V(G) - B$. We call such a system $\mathcal{H}$ a *separating family*. The size of the smallest family is not known; in particular, we don't know if a separating family of polynomial size exists for each graph. As a corollary to Theorem 4.5, the non-deterministic communication complexity of disjointness is at most $\frac{1}{2}(\log n)^2$. Hence we obtain the following, purely graph-theoretic result:

4.6 Corollary. *Every graph on n vertices has a separating family of size* $n^{(\log n)/2}$.

5. Möbius Functions and the Rank of the Communication Matrix

Our previous discussions have shown that the rank of the communication matrix is very intimately related to the communication complexity. They also show, however, that it is in general not an easy task to compute this rank. In this section we are going to study a rather general situation in which this rank can be computed, or at least, reduced to the study of a well-known function in combinatorics.

To motivate our abstractions, let us discuss one further communication problem in graph theory.

5.1 Problem. Let V be a finite set and assume that both Alice and Bob are given a graph with node set V. Their task is to decide whether the union of these graphs is connected.

Hajnal, Maass and Turán (1988) proved that for this problem, the trivial protocol is optimal. One should notice that the trivial protocol is not that Alice transmits the whole graph; this is redundant information for Bob. All he needs to know is which pairs of nodes are connected in Alice's graph. In other words, he needs the partition of V defined by the connected components of Alice's graph. Since the total number of partitions is the Bell number B_n, this trivial protocol takes $\lceil \log_2 B_n \rceil \approx n \log_2 n$ bits.

To prove that this trivial protocol is optimal, Hajnal, Maass and Turán use the rank bound, but to compute the rank of the corresponding communication matrix is by no means obvious. Their method involves the use of the "Möbius function" of the partition lattice. Lovász and Saks (1988a) showed that this method of computing the rank extends to a large class on problems, which can be formulated as follows. Let S be a finite set. A *filter* on S is a non-empty family of subsets of S such that $U \in \mathscr{F}$, $U \subseteq V \subseteq S$ implies $V \in \mathscr{F}$.

5.2 Problem. Let S be a finite set and $\mathscr{F}$ any filter in 2^S. Let Alice be given a set $X \subseteq S$ and Bob, a set $Y \subseteq S$. Their task is to decide whether $X \cup Y \in \mathscr{F}$.

Of course, one can formulate the "dual" problem, in which $\mathscr{F}$ is an ideal and Alice and Bob have to determine whether $X \cap Y \in \mathscr{F}$. This is trivially equivalent to 5.2 under complementation. To get Problem 5.1, we take all the edges of the complete graph on V as elements of S, and let $\mathscr{F}$ consist of all connected graphs with this set of vertices.

Let us recall a problem from the previous section.

5.3 Problem. Let T be a tree and let both Alice and Bob be given a subtree of T. Their task is to decide whether these subtrees have a node in common.

We have seen that in this case the trivial protocol is by far not optimal. Let us formulate a generalization of this problem. Let S be a finite set. An *alignment* on S is a collection $\mathscr{A}$ of subsets of S closed under intersection.

5.4 Problem. Let S be a finite set and $\mathscr{A}$ an alignment on S. Let Alice be given a set $X \in \mathscr{A}$ and Bob a set $Y \in \mathscr{A}$. Their task is to decide whether $X \cap Y = \emptyset$.

Both of these general problems, and several other natural problems of this kind, are equivalent to the following, which sounds perhaps more special but will be easier to handle.

5.5 The Meet Problem. Let $\mathscr{L}$ be a finite lattice. Let Alice be given an element $x \in \mathscr{L}$ and Bob an element $y \in \mathscr{L}$. Their task is to decide if $x \wedge y = 0$ (here 0 is the zero element of the lattice).

Before showing how the other two problems can be reduced to this, we have to define what we mean by reduction. Fortunately, this is much easier here than in "machine-based" complexity theory. Given an instance of any communication problem, i.e., a 0-1 matrix C, check if there are two equal rows or columns in C. Clearly, one of such a pair of rows can be deleted without changing the complexity (deterministic or non-deterministic) of the problem. Carry out this until every pair of rows and every pair of columns will be different. The remaining matrix will be called the *core* of C.

Now we say that a class $\mathscr{A}$ of 0-1 matrices *can be reduced* to a class $\mathscr{B}$ of 0-1 matrices, if for every $A \in \mathscr{A}$ there exists a $B \in \mathscr{B}$ such that the core of A is the same (up to permutation of rows and columns) as the core of B. Two classes of matrices are *equivalent* if they can be reduced to each other. Lovász and Saks (1988a) proved:

5.6 Lemma. *Problems 5.2, 5.4 and 5.5 are equivalent.*

Proof. Most of the reductions needed to show this are straightforward. We sketch the (perhaps) least trivial fact that 5.2 can be reduced to 5.5. Let S be a finite set and $\mathscr{F}$, a filter on S. Define a relation $\sim$ on 2^S by saying that $X \sim Y$ iff for every $Z \subseteq S$, we have $X \cup Z \in \mathscr{F}$ if and only if $Y \cup Z \in \mathscr{F}$ (i.e., if the rows of the communication matrix corresponding to X and Y are identical). Trivially, this is an equivalence relation. It is equally trivial, but somewhat unexpected, that the following holds:

Claim 1. If $X \sim Y$ then $X \sim X \cup Y$.

This claim implies that the union $\overline{X} = \cup\{Y : Y \sim X\}$ also satisfies $\overline{X} \sim X$. Now it is not difficult to check that

Claim 2. The operation $X \to \overline{X}$ is a closure operator on 2^S. Furthermore, $X \in \mathscr{F}$ if and only if $\overline{X} = S$.

As usual, call a set X *closed* if $\overline{X} = X$. Then it follows easily that

Claim 3. The closed sets form a co-atomic lattice $\mathscr{L}$.

(A lattice is co-atomic if every element of it is the meet of co-atoms, i.e., elements covered by the top element. This property will, however, play no role in the sequel.)

Now the core of the communication matrix is just the submatrix formed by rows and columns corresponding to closed sets. The problem is equivalent to the modified problem where both players are given a *closed* set and they have to decide if the closure of the union (i.e., the join in the lattice $\mathscr{L}$) is the "top" element S. This is just the same problem as 5.5, but "upside down". $\square$

So from now on we shall restrict our attention to the Meet Problem. The corresponding matrix $C = (c_{ij})$ has its rows and columns indexed by the elements of the lattice $\mathscr{L}$, and for $x, y \in \mathscr{L}$ we have

$$c_{xy} = \begin{cases} 1, & \text{if } x \wedge y = 0, \\ 0, & \text{otherwise.} \end{cases}$$

Now this matrix is well-known in algebraic combinatorics! To formulate its main property important for us, we have to introduce a few further matrices associated with the lattice. Let

$$\zeta_{xy} = \begin{cases} 1, & \text{if } x \le y, \\ 0, & \text{otherwise.} \end{cases}$$

and $Z = (\zeta_{xy})$. This matrix Z is sometimes called the *zeta-matrix* of the lattice. If we order the rows and columns of Z compatibly with the partial ordering defined by the lattice, it will be upper triangular with 1's in its main diagonal. Hence it is invertible, and its inverse $M = Z^{-1}$ is an integral matrix of the same shape. This inverse is a very important matrix, called the *Möbius matrix* of the lattice. Let

$$M = \big(\mu(x, y)\big)_{x, y \in \mathscr{L}}.$$

The function μ is called the *Möbius function* of the lattice. From the discussion above we see that $\mu(x, x) = 1$ for all $x \in \mathscr{L}$, and $\mu(x, y) = 0$ for all $x, y \in \mathscr{L}$ such that $x \not\le y$. Moreover, the definition of M implies that for every pair of elements $a \le b$ of the lattice,

$$\sum_{a \le x \le b} \mu(a, x) = \begin{cases} 1, & \text{if } a = b, \\ 0, & \text{otherwise;} \end{cases}$$

and

$$\sum_{a \le x \le b} \mu(x, b) = \begin{cases} 1, & \text{if } a = b, \\ 0, & \text{otherwise.} \end{cases}$$

These identities provide a recursive procedure to compute the Möbius function. It is easy to see from this procedure that the value of the Möbius function $\mu(x, y)$, where $x \le y$, depends only on the internal structure of the interval $[x, y]$. Also note the symmetry in these two identities. This implies that if μ^* denotes the Möbius function of the lattice turned upside down, then

$$\mu^*(x, y) = \mu(y, x).$$

We cannot survey here the many properties and applications of the Möbius function. We shall restrict ourselves to those issues relevant for determining the rank of C. For more see Rota (1964), Lovász (1979), Chapter 2, or Stanley (1986), Chapter 3.

Let us introduce one further matrix: D is a diagonal matrix defined by $(D)_{xx} = \mu(0, x)$. Now it is not difficult to verify the following identity (Wilf (1968)):

5.7 Lemma. $C = Z^T D Z$.

Since Z is invertible, and the rank of the diagonal matrix D is the number of non-zeros in its diagonal, we obtain

5.8 Corollary. *The rank of C is the number of elements $x \in \mathscr{L}$ such that $\mu(0, x) \neq 0$. In particular, if $\mu(0, x) \neq 0$ for all $x \in \mathscr{L}$ then the trivial protocol is optimal for the Meet Problem.*

We shall call an element x of the lattice $\mathscr{L}$ satisfying $\mu(0, x) \neq 0$ a *Möbius element*. To be able to use this corollary, we have to find the Möbius elements. Unfortunately, the Möbius function can be quite complicated, and there is no easy characterization of the Möbius elements. Let us collect a few facts.

It is easy to see that if x is an atom in $\mathscr{L}$ then $\mu(0, x) = 1$, and so every atom is a Möbius element. On the other hand, we have (Ph. Hall (1936)):

5.9 Theorem. *Every Möbius element is a join of atoms.*

Unfortunately, the converse is not true, an element that is a join of atoms is not necessarily Möbius. One important condition on the non-vanishing of the Möbius function is the following (Rota (1964)). Recall that a lattice is *geometric* if it is semimodular and atomic. Geometric lattices are just lattices of flats of matroids.

5.10 Theorem. *If $\mathscr{L}$ is geometric and $x, y \in \mathscr{L}$, $x \leq y$, then $\mu(x, y) \neq 0$. In particular, every element is Möbius.*

From the theorem we obtain the result of Hajnal, Maass and Turán immediately.

5.11 Corollary. *The communication complexity of problem 5.1 is at least $n \log n$.*

Proof. The lattice associated with Problem 5.1 by the general construction is just the partition lattice turned upside down. Now the partition lattice is geometric, and so the Möbius function does not vanish on any interval of it. But then it does not vanish on any interval of the "upside down" lattice either, and hence the trivial protocol for the communication problem is optimal.   □

There are other classes of lattices on which the Möbius function does not vanish. For such a lattice, the trivial protocol is optimal for the disjointness problem. One interesting class to mention here are face lattices of convex polytopes.

In our other starting example, we consider the lattice $\mathscr{L}_T$ of subtrees of a tree T. Using Theorem 5.8 (upside down!), it is easy to compute the Möbius function of this lattice:

5.12 Lemma. *For the Möbius function of $\mathscr{L}_T$,*

$$\mu(0, x) = \begin{cases} 1, & \text{if } x = \emptyset \text{ or } x \text{ is a subtree with 2 nodes,} \\ -1, & \text{if } x \text{ is a subtree with a single node,} \\ 0, & \text{otherwise.} \end{cases}$$

(This way of determining the Möbius function extends to a large class of alignments called *antimatroids*. See Lovász and Saks (1988a).)

From this lemma we see that the rank of C is $2n$. We have seen in the previous section that the bound for this problem is very close to being optimal.

Now let us return to the general Meet Problem, and discuss the non-deterministic complexity of the problem. If E.T. wants to exhibit that $x \wedge y \neq 0$, it suffices to exhibit an atom of $\mathscr{L}$ that lies below both x and y. Hence

$$\kappa_1(C) \leq \log_2(\text{number of atoms}).$$

(It is easy to see that in fact equality holds here.) Hence by Lemma 2.1 and Theorem 3.5, we have the following bounds on the communication complexity of the Meet Problem:

5.13 Theorem. *Let C be the communication matrix associated with the Meet Problem for a lattice $\mathscr{L}$. Let $\mathscr{L}$ have a atoms and b Möbius elements. Then*

$$\log_2 b \leq \kappa(C) \leq (\log_2 a)(\log_2 b).$$

Since every atom is a Möbius element, and $b = \text{rk}(C)$, we obtain:

5.14 Corollary. *For the Meet Problem,*

$$\log_2 \text{rk}(C) \leq \kappa(C) \leq \left(\log_2 \text{rk}(C)\right)^2.$$

6. Randomized Protocols

So far, our protocols have not used randomization. If we do, many of the previously discussed problems become substantially easier. In a *randomized protocol* we allow the players to throw dice, and we only want to result be correct with probability at least $2/3$. The smallest k for which such a protocol exists involving the transmission of at most k bits will be denoted by $\kappa^{\text{rand}}(C)$.

We could require in place of $2/3$ any fixed number greater than $1/2$: repeating the protocol a constant number of times, the probability of getting the correct answer could be pushed arbitrarily close to 1. (If we want the probability of the correct answer only to be greater than $1/2$, but allow it to tend to $1/2$, we get a different kind of problem. We do not discuss this here, but refer to Alon, Frankl and Rödl (1985) and Babai, Frankl and Simon (1986). For a thorough study of randomized protocols, see also Yao (1983).)

Let us discuss the Encyclopædia Britannica example. We have seen that the best protocol to decide whether two strings of length n are equal is the trivial protocol and requires the communication of n bits. In contrast, we have the following theorem (Freiwalds (1979)):

6.1 Theorem. *There exists a randomized protocol to decide the equality of two strings of length n, using $O(\log n)$ bits.*

Proof. The protocol can be viewed as an extension of "binary check ". Consider the inputs as two natural numbers x and y, $0 \leq x, y \leq 2^n - 1$. Alice selects a prime $p \leq n^2$, computes the remainder x' of x modulo p, and then sends the pair (x', p) to Bob. Now Bob computes the remainder y' of y modulo p, and compares it with x'. If they are distinct, he concludes that $x \neq y$. If they are the same, he concludes that $x = y$.

If the two numbers x and y are equal then, of course, so are x' and y' and so the protocol reaches the right conclusion. If x and y are different then, however, it could happen that x' and y' are the same and the protocol reaches the wrong conclusion. This happens if $p \mid x - y$. Now $|x - y| < 2^n$ and so $x - y$ has fewer than n different prime divisors. On the other hand, Alice had about $\frac{n^2}{2 \log_e n}$ primes to choose from, and so the probability that she chose one of the divisors of $x - y$ tends to 0.

Clearly, this protocol uses at most $4 \log n$ bits. $\square$

In our treatment of deterministic protocols, the general Disjointness problem and the Inner product modulo 2 problem behaved very similarly to the Encyclopædia problem; we just had to replace the identity matrix by some other matrix with obviously high rank. But for randomized protocols, they are substantially more difficult. Before stating this exactly, let us formulate some combinatorial conditions and a general linear algebraic lower bound on the randomized complexity. For simplicity of presentation, we restrict ourselves to $n \times n$ matrices.

Consider a matrix C and any randomized protocol to solve the associated communication problem using at most k bits. For the purposes of lower bounds, we assume that the random number generator is public, i.e., the random bits are available to both players (this only makes the life of Alice and Bob easier, and so our task to give a lower bound more difficult). We may also assume that these random bits are available right at the beginning. Now the players look at this sequence and depending on it, they select a (deterministic) protocol that they follow. This protocol ends up with a partition of the matrix into *terminal submatrices*. This time it will be more convenient to assume that the protocol stops when both players know the answer (both have to have the same answer, which, of course, may be wrong). This means that some of the terminal submatrices are labelled "1" and others are labelled "0", meaning that if for a particular input the protocol ends up with this submatrix, then this is concluding value. Since in this case the conclusion is not necessarily correct, the submatrices labelled "1" may contain 0's and vice versa. (We shall see though that in some average sense, terminal submatrices labelled with "1" must have more 1's than 0's.)

For each submatrix T, let $p(T)$ denote the probability of the event that in the randomly chosen protocol, T is a terminal submatrix labelled "1". Our assumption on the probability of error implies that

(6.1) $$\sum_{\substack{T \\ ij \in T}} p(T) \begin{cases} \geq 2/3, & \text{if } c_{ij} = 1, \\ \leq 1/3, & \text{if } c_{ij} = 0. \end{cases}$$

Moreover, every protocol produces at most 2^k terminal submatrices, and hence

$$\sum_T p(T) \le 2^k.$$

We may consider the linear program

$$\text{minimize} \sum_T p(T)$$

subject to (6.1) and the obvious constraints

(6.2) $$p(T) \ge 0.$$

If $\mu(C)$ denotes the optimum value of this program, then

$$k \ge \log \mu(C).$$

We call this the *linear programming bound* on the randomized communication complexity of the matrix. We can apply linear programming duality and obtain $\mu(C)$ as the maximum of the dual program. This program has variables ϕ_{ij} associated with the entries of the matrix, constraints

(6.3) $$\phi_{ij} \ge 0, \quad \text{if } c_{ij} = 1,$$

(6.4) $$\phi_{ij} \le 0, \quad \text{if } c_{ij} = 0,$$

(6.5) $$\sum_{ij \in T} \phi_{ij} \le 1 \quad \text{for each submatrix } T,$$

and objective function

(6.6) $$\frac{2}{3} \sum_{c_{ij}=1} \phi_{ij} + \frac{1}{3} \sum_{c_{ij}=0} \phi_{ij}.$$

So $\mu(C)$ could also be defined as the maximum of (6.6) subject to (6.3), (6.4) and (6.5). In particular, any feasible solution of (6.3)–(6.5) provides a lower bound on the randomized communication complexity. Of course, if we want the probability of error be at most p, then we can replace (6.6) by the objective function

$$(1-p) \sum_{c_{ij}=1} \phi_{ij} + p \sum_{c_{ij}=0} \phi_{ij}.$$

Note that a feasible solution of (6.3)–(6.5) can be viewed as a matrix Φ. Condition (6.5), which is the most awkward, and contains exponentially many constraints, can be replaced by a somewhat stronger condition using some linear algebra. Let T be any submatrix. Let $a \in \mathbb{R}^n$ be the incidence vector of the set of rows in T and $b \in \mathbb{R}^n$ the incidence vector of the set of columns of T. Then the sum of entries in T is just $a^T \Phi b$, and so (6.5) can be rephrased as

$$a^T \Phi b \le 1$$

whenever a and b are 0-1 vectors. By elementary linear algebra, the left hand side is at most $\| a \| \cdot \| b \| \cdot \| \Phi \|$, where $\| \Phi \|$ is spectral norm of the matrix Φ (recall that this can be expressed as $\Lambda(\Phi^T \Phi)^{1/2}$, where $\Lambda(M)$ denotes the largest eigenvalue of the matrix M). So if we require that

$$(6.7) \qquad\qquad \| \Phi \| \leq \frac{1}{n},$$

then (6.5) is automatically satisfied. Moreover, we can express the objective function as

$$\frac{1}{2} \sum_{i,j} \phi_{ij} + \frac{1}{6} \sum_{i,j} |\phi_{ij}|.$$

Since the first term is at most $1/2$ by (6.5), we can disregard it. The second term involves another norm of the matrix Φ:

$$|\Phi| = \sum_{i,j} |\phi_{ij}|.$$

We can formulate our result as follows:

6.2 Lemma. *For every non-zero matrix Φ satisfying (6.3) and (6.4), we have*

$$\kappa^{\mathrm{rand}}(C) \geq \log \left(\frac{|\Phi|}{n \| \Phi \|} \right) - 3.$$

In particular, we can use here the matrix $2C - J$, where J is the all-1 matrix of size $n \times n$. Clearly $|2C - J| = n^2$, and hence we obtain:

6.3 Corollary. $\kappa^{\mathrm{rand}}(C) \geq \log \left(\frac{n}{\|2C-J\|} \right) - 3.$

As an application, consider the Inner product modulo 2 problem. For this, $2C - J$ is an Hadamard matrix (a ± 1 matrix with any two columns orthogonal) and hence its spectral norm is $\sqrt{n}$. This implies the following result of Chor and Goldreich (1985): *the randomized communication complexity of computing the inner product of two vectors of length m modulo 2 is $\Omega(m)$.*

Note that if Corollary 6.3 gives any valuable result then the matrix C has to be very homogeneous: $\| 2C - J \|$ must be $o(n)$ and hence it follows that each submatrix has to contain almost exactly as many 1's as 0's. But a somewhat weaker condition also works, as shown by Yao (1983):

6.4 Lemma. *Assume that for some $a, b, c > 0$,*

(6.8) *for every submatrix T with $t > an^2$ entries, at least bt entries in T are 0;*
(6.9) *at least cn^2 entries of C are 1.*

Then the randomized communication complexity of the problem is $\Omega \left(\frac{\log(c/a)}{|\log(bc)|} \right)$.

Proof. If there is a randomized protocol using k bits with error probability $1/3$, then repeating this $O(|\log(bc)|)$ times, we get one with error probability less than

$bc/3$. To estimate the complexity of this from below, we construct a feasible solution of (6.3)–(6.5) with objective value $\log(c/a)$; this will prove the lemma. Let

$$\Phi_{ij} = \begin{cases} \frac{1}{an^2}, & \text{if } c_{ij} = 1, \\ \frac{1-b}{abn^2}, & \text{if } c_{ij} = 0. \end{cases}$$

Then a simple computation shows that we have the feasible solution as desired.

□

As an application of this lemma, Babai, Frankl and Simon (1986) show that *the randomized communication complexity of the Disjointness problem for the subsets of an n-element set is $\Omega(\sqrt{n})$.* They restrict the problem to subsets of cardinality $\sqrt{n}$. Then the matrix associated with this restricted problem satisfies hypotheses (6.8) and (6.9) with $a = \text{const}/\sqrt{n}$, b, $c = \text{const}$. The proof is combinatorial and not discussed here.

Let us also mention the paper of Halstenberg and Reischuk (1988), which extends the construction of Duris, Galil and Schnitger (1984), as mentioned earlier, to randomized protocols.

7. VLSI Again: How to Split Information?

Consider the "chip" in Fig. 3, which solves the Equality Problem in a very simple way: It compares x_1 with y_1, etc. and then uses a binary tree to collect this information. It works in $O(\log n)$ time and has area $O(n \log n)$. Doesn't this contradict our Theorem 1.1?

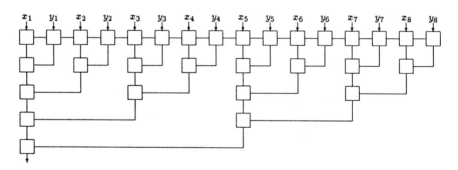

Fig. 3

The trick is of course that the input is coming now in a different arrangement. In general, if we want a chip to compute a Boolean function $f(x_1, \ldots, x_{2n})$ then we can either prescribe where the input bits come in or (depending on the situation) we may choose this ourselves. In this second case (as the above example shows) we can have quite different results. The communication complexity bound works, but we have to use a more involved measure. We can split the variables of f into two sets of n elements, and consider the ordinary communication complexity of the problem of evaluating f if Alice gets one half of the variables and Bob, the other half. This complexity will depend on the partition. We define the *worst-partition communication complexity* of f as the maximum of these complexities. The *best-partition communication complexity* is defined as the minimum over all partitions of the variables into two sets of equal size (we have to add this restriction to avoid putting all variables into one class).

Now the best-partition communication complexity can replace the ordinary communication complexity in lower bounds if we are free to choose the positions of the ports (even inside the chip). (The worst-partition complexity has other uses.) Unfortunately, it is in general more difficult to compute the best-partition complexity. We shall restrict ourselves to an example.

Hajnal, Maass and Turán (1988) settle the following question. Alice and Bob want to decide if a given graph G is connected. Half of the pairs of nodes are supervised by Alice; she knows if these pairs are adjacent in G or not. The other half is similarly supervised by Bob. For a given partition of the edges, this is an ordinary communication problem. We want to find a lower bound valid for all partitions.

Another way to look at this problem is that Alice knows a graph G_A on a given set V of nodes, Bob knows another graph G_B, and they want to decide whether the union is connected. This is now an ordinary communication complexity problem, similar to problem 5.2. What makes different (easier for Alice and Bob, but more difficult for us) is that they both know that Alice's graph must consist of red edges and Bob's graph must consist of blue edges. Now Hajnal, Maass and Turán show that still essentially the same lower bound holds as for a pair of unrestricted graphs:

7.1 Theorem. *The best-partition communication complexity of determining if a graph is connected is $\Omega(n \log n)$.*

Proof. The proof goes by reduction to Corollary 5.11, but this reduction is quite involved, so we only sketch it. Using the powerful Regularity Lemma of Szemerédi (1976) one can prove the following:

7.2 Lemma. *Let $G_1 \cup G_2$ be a partition of the complete graph on the n-element node set V into two subgraphs with $\frac{1}{2}\binom{n}{2}$ edges. Then there exists a set $W \subseteq V$ with $|W| = \Omega(n)$ with the following property: We can associate with each partition P of V two subgraphs $G_1^P \subseteq G_1$ and $G_2^P \subseteq G_2$ such that for every two partitions P and Q of V, the union $G_1^P \cup G_2^Q$ is connected if and only if $P \cup Q$ is the trivial partition of V.*

Consider the matrix C of the communication problem. This lemma says that this contains the matrix C' of the meet problem for the "upside down" partition lattice of W as a submatrix. So $\kappa(C) \geq \kappa(C') \geq \Omega(n \log n)$. □

8. Communication Complexity and Computational Complexity

We discuss some examples where communication complexity seems to have close relationship with certain other complexity issues. The first of these is the work of Yannakakis (1988) about expressing combinatorial optimization problems as linear programs.

a. Most combinatorial optimization problems (Matching, Travelling Salesman, various scheduling problems etc.) can be viewed as the task to find a member of a given set-system $\mathcal{F}$ with maximum value. Furthermore, most often the value is given by associating weights with the elements of the underlying set S, and letting the value of a member of $\mathcal{F}$ be the sum of weights of its elements.

One of the most succesfull approaches to such combinatorial optimization problems is to represent each set $A \in \mathcal{F}$ by its incidence vector $\chi^A \in \mathbb{R}^S$, and consider the convex hull P of these. Our task is then to optimize a linear objective function over P. If we are able to find the system of linear inequalities describing P, then we can apply powerful techniques of linear programming.

There is a basic difficulty with this method: it can easily happen that both the number of vertices and the number of facets of P is exponentially large (in the natural size of the problem, which is a power of the dimension). Hence the system of linear inequalities describing P will be exponentially large. There are ways out of this difficulty; essentially, in some cases one can generate the program as one proceeds with the algorithm (see Grötschel, Lovász and Schrijver (1988), Grötschel and Padberg (1985).)

Another possibility is to reduce the size of the program by introducing new variables. Geometrically, this means to represent P as the projection of another polyhedron Q in higher dimension, but with fewer facets (hopefully only a polynomial number of them). Since projecting a polytope may increase the number of facets, this approach is promising. In fact, if we want to translate into a geometric language simple algorithms consisting of case distinctions and dynamic programming methods, we often end up with introducing new variables.

This leads us to the following, purely geometric question: given a polytope P in $\mathbb{R}^n$, can it be represented as the projection of a polytope Q with "few" (say, $O(n^{\text{const}})$) facets? (Note that the dimension in which Q lives is less than the number of its facets.) If this is the case, and if in addition one can efficiently generate a linear description of Q, then we can optimize any linear objective function over P by lifting it to Q and using any polynomial time linear programming algorithm. This approach was tried (unsuccesfully) to solve the travelling salesman problem in polynomial time (Swart (1986)). It could be expected that no polytope coming from an NP-hard optimization problem can be lifted like this. But even some "nice" polytopes (coming from polynomially solvable problems, like matching or vertex packing in perfect graphs) seem to resist attempts to obtain them as projections of polytopes with a polynomial number of facets.

Yannakakis made an important step in this question by showing that the matching and travelling salesman polytopes of complete graphs cannot be obtained as such projections, if we assume that the natural symmetries of these polytopes can be lifted to Q. He also sketched a possible route to prove such results without this symmetry assumption. This approach uses the following observation. Given a polytope $P \subset \mathbb{R}^n$, consider the following communication complexity problem: Alice is given a vertex v of P, Bob is given a facet F of P, and they have to decide whether $v \in F$. Let us call this the Vertex-Facet problem, and let $C_{VF}(P)$ denote the corresponding matrix (we write a 1 if the corresponding facet contains the corresponding vertex). Then

8.1 Lemma. *If P is the projection of a polytope Q with t facets then $\kappa_0(C_{VF}(P)) \leq \lceil \log t \rceil$.*

Proof. The vertex v is the projection of a vertex w of Q. The facet F is the projection of a face G (usually not a facet) of Q. We can obtain G as the intersection of facets; at least one of these, say G', does not contain w. Now E.T. can announce the name of G' as a proof that $v \notin F$; this takes only $\lceil \log t \rceil$ bits. $\qquad\square$

As a special case, consider the *vertex packing polytope*, i.e., the convex hull of incidence vectors of independent sets of nodes of a graph G. Every (maximal) clique of a graph G defines a facet of the vertex packing polytope. A vertex is not on a face if and only if the corresponding independent set is disjoint from the corresponding clique. Hence by our discussion in Section 4 we obtain:

8.2 Corollary. *If a graph has no polynomial size independent set-clique separating family then its vertex packing polytope cannot be obtained as the projection of a polytope with a polynomial number of facets.*

Unfortunately, we do not know if there is any graph to which this corollary applies. On the other hand, as far as I know there could even be perfect graphs with this property.

b. The second application we discuss is a result of Karchmer and Wigderson (1988). Consider a Boolean function $f : \{0,1\}^n \to \{0,1\}$, and a Boolean circuit evaluating this function. This consists of a directed graph that has $2n$ sources and a single sink. The sources are labelled with the variables $x_1, \ldots, x_n$, and their negations. Moreover, every non-source node has indegree 2, and is labelled either by "AND" or by "OR". To operate the circuit, we feed in the actual values of the variables at the sources, and let each other node compute the conjunction or the disjuncton of the two logical values at the tails of the two incoming edges. The value produced at the sink is the value of the function. The *depth* of the circuit is the longest path from the source to the sink. To prove lower bounds on the depth of a Boolean circuit evaluating various given Boolean functions is one of the hottest topics in theoretical computer science.

With a Boolean function f, we can associate the following communication problem. Alice gets a 0-1 sequence $(a_1, \ldots, a_n)$ and Bob gets a 0-1 sequence $(b_1, \ldots, b_n)$, such that $f(a_1, \ldots, a_n) = 0$ and $f(b_1, \ldots, b_n) = 1$. Their task is to find an index i such that $a_i \neq b_i$. We call this the Difference Problem for f. (Note that this is not a decision problem.)

Karchmer and Wigderson show the following:

8.3 Lemma. *A Boolean function f can be evaluated by a circuit of depth t if and only if the communication complexity of the Difference Problem for f is at most t.*

Proof. We show how to translate a Boolean circuit into a communication protocol; the reverse construction is similar. We suppose that both Alice and Bob know the Boolean circuit (this is part of the protocol). Then Alice can follow the evaluation of $f(a_1, \ldots, a_n)$. This will associate a logical value $\alpha(v)$ with each node v. Similarly, Bob can evaluate $f(b_1, \ldots, b_n)$ and thereby associate a logical value $\beta(v)$ with each node v.

During the protocol, Alice and Bob construct a path "backwards", starting from the sink and ending at one of the sources. The source where the path ends will give the index i that they are looking for. More generally, at each step, they will have a node w such that $\alpha(w) = 0$ and $\beta(w) = 1$. Let uw and $u'w$ be the two edges entering w. If this current node w is labelled "AND" then it is Alice's turn. At least one of $\alpha(u)$ and $\alpha(u')$ is 0; Alice sends a bit to Bob to indicate which one, and they move to that node. (Observe that since $\beta(w) = 1$, it follows that $\beta(u) = \beta(u') = 1$.) If w is labelled "OR" then it is Bob's turn and his message is determined analogously.   □

There is an analogous result if we restrict ourselves to monotone Boolean functions and monotone circuits computing them (i.e., circuits where only the unnegated variables occur at the sources). The corresponding communication problem is only slightly more difficult: Alice and Bob have to find an index i such that $a_i = 0$ and $b_i = 1$ (since the function is monotone, such an index must exist).

Using this transformation of the problem, Karchmer and Wigderson were able to show that every Boolean circuit determining whether a two points in a given graph G can be connected by a path must have depth at least $(\log n)^2$. More recently Karchmer, Newman and Wigderson (unpublished) gave interesting applications of this transformation in the reverse direction: they showed how to design Boolean circuits with small depth for various problems using general results similar to those in Section 3.

c. Hajnal, Maass and Turán (1988) find another connection of this kind. Fix n Boolean variables $x_1, \ldots, x_n$. By a *test tree* we mean a binary tree in which every internal node v has an associated "test function", a Boolean function $g_v : \{0,1\}^n \to \{0,1\}$, and each leaf has value 0 or 1. Such a test tree can be used to compute a Boolean function in n variables: we start from the root and move down to a leave. At each internal node v, compute the test value $g_v(x_1, \ldots, x_n)$. If this is 0, we move right, else we move left. When we arrive at a leaf, we read off the associated logical value. The *depth* of the tree is the length of a longest path from the root to a leaf.

We are interested in finding the minimum depth of a test tree computing a given Boolean function f, where the test functions are restricted to some simple class (if they are restricted to single variables, we get the more usual *decision trees.* The following observation of Hajnal, Maass and Turán gives a lower bound on this depth:

8.4 Lemma. *If a test tree computes a Boolean function f with worst-partition communication complexity k, and every test-function has worst-partition communication complexity l, then its depth is at least k/l.*

One nice set of test functions are disjunctions of variables (i.e., we can test whether a given set of the variables contains a "1"). These have worst-partition communication complexity 2. The lemma implies, in conjunction with Theorem 7.1, that *with such tests one cannot determine the connectivity of a graph in depth less than* $\Omega(n \log n)$.

d. Finally, we sketch an interesting application of communication complexity to random number generation due to Babai, Nisan and Szegedy (1988). Assume that we have a random 0-1 sequence α of length m. We call this the "seed". We want to generate from it a 0-1 sequence β of length $n > m$, which is still "essentially random". By this we mean that it cannot be distinguished from any truly random 0-1 sequence of length n by any machine of some prescribed class; in this case, we consider on-line RAM machines with a sublinear ($O(n^{99})$) bound on the memory. (For this machine, we may even allow to have a true random number generator; note that this machine does not want to generate random numbers; rather, it wants to show that our way of generating them is bad.) The machine is getting the bits of β one by one, and at any time it can occupy only $O(\log n)$ space. Having seen some bits from β, the machine has to compute a "guess" for the next bit. We say that the pseudorandom sequence β fails the test if the probability that this guess is correct is larger than 51%. Babai, Nisan and Szegedy give a construction of a pseudorandom sequence β with $n = m^{c \log m}$ that passes all logspace on-line tests.

Let p, q, r be positive integers with $pqr \leq m$, and consider a $p \times q$ array (v_{ij}), where each $v_{ij} \in \{0,1\}^r$. The bits in all the v_{ij} are truly random bits of the seed. Let $f(u_1, \ldots, u_q) : \{0,1\}^{r \times q} \to \{0,1\}$ be an appropriate function. For every choice of $1 \leq i_v \leq p$ ($v = 1, \ldots, q$), compute the bit $x[i_1, \ldots, i_q] = f(v_{i_1,1}, \ldots, v_{i_q,q})$. Order these bits lexicographically according to the index sequences, to get the sequence β of length $n = p^q$.

Let us show how the pseudorandomness of this sequence depends on communication complexity. Consider the moment when the machine has to guess the bit $x[i_1, \ldots, i_q]$. At this moment, it has already seen all bits coming from lexicographically smaller index sequence. We may split this sequence into q segments as follows: let β_1 be the beginning segment of β consisting of those bits $x[j_1, \ldots, j_q]$ with $j_1 < i_1$; generally, let β_t ($1 \leq t \leq q$) consist of those bits $x[j_1, \ldots, j_q]$ with $j_1 = i_1, j_2 = i_2, \ldots, j_{t-1} = i_{t-1}$ but $j_t < i_t$. Clearly, each β_t is a segment of β and together they make up the portion of β before $x[i_1, \ldots, i_q]$.

Now comes the key observation: the bits in β_t are independent of $v_{i_t,t}$. Let us invite q players and assign the tth player to "supervise" the segment β_t. This means that he can see all bits in β_t. If the machine can make a good guess of $x[i_1, \ldots, i_q]$ then these players can also compute this guess: the first player computes whatever the machine has in its memory when having processed β_1, and sends this to the second player; now he computes from this whatever the machine has in its memory after having processed β_2, etc. The last player will be able to compute the guess of the machine. If the machine uses at most M bits of memory then this protocol involves the communication of not more than Mq bits.

To obtain a lower bound, we make the life of the players even easier: we allow them to "broadcast" all messages (so that all the other players also can learn them), and we tell the tth player the whole seed with the exception of the bits in $v_{i,t}$. So we have the following communication complexity problem. We have q players, and also q vectors $w_1,\ldots,w_q \in \{0,1\}^r$. The tth player knows all vectors except w_t. Their task is to "guess" $f(w_1,\ldots,w_q)$ at the expense of a minimum number of "broadcasted" bits, so that the probability of guessing correctly is at least 51%.

Babai, Nisan and Szegedy consider the function f that gives the parity of the number of positions where each w_t has a 1 (a generalization of inner product) and show by a rather involved extension of the methods in Section 6 that this takes $\Omega(r2^{-q})$ bits. For the choice of $r = m^{.992}$, $q = .001\log m$ and $p = m^{.007}$, this gives a pseudorandom sequence β of length $m^{.001\log m}$, which passes all tests using at most $m^{.99}$ space. It is interesting to remark that this function is logspace computable, so the random number generator works very fast.

References

Aho, A.V., Ullman, J.D., Yannakakis, M. (1983): On notions of informations transfer in VLSI circuits. Proc. 15th ACM STOC, pp. 133–139

Babai, L., Frankl, P., Simon, J. (1986): Complexity classes in communication complexity theory. Proc. 27th IEEE FOCS, pp. 337–347

Babai, L., Nisan, N., Szegedy, M. (1988): Multiparty protocols and logspace-hard pseudo-random sequences. Manuscript

Chor, B., Goldreich, O. (1985): Unbiased bits from sources of weak randomness and probabilistic communication complexity. Proc. 26th IEEE FOCS, pp. 429–442

Duris, P., Galil, Z., Schnitger, G. (1984): Lower bounds on communication complexity. Proc. 16th ACM STOC, pp. 81–89

Freiwalds, R. (1979): Fast probabilistic algorithms. In: Becvar, J. (ed.): Mathematical foundations of computer science 1979. Springer Verlag, Berlin, Heidelberg, pp. 57–69 (Lect. Notes Comput. Sci., Vol. 74)

Grötschel, M., Lovász, L., Schrijver, A. (1988): Geometric algorithms and combinatorial optimization. Springer Verlag, Berlin, Heidelberg (Algorithms Comb., Vol. 2)

Grötschel, M., Padberg, M.W. (1986): Polyhedral computations. In: Lawler, E.L., Lenstra, J.K., Rinnooy Kan, A.H.G., Shmoys, D.B (eds.): The traveling salesman problem. J. Wiley & Sons, New York, NY, pp. 307–360

Hajnal, A., Maass, W., Turán, G. (1988): On the communication complexity of graph properties. Proc. 20th ACM STOC, pp. 186–191

Hall, P. (1936): The Eulerian functions of a group. Q. J. Math., Oxf. II. Ser., 134–151

Halstenberg, B., Reischuk, R. (1988): On different modes of communication. Proc. 20th ACM STOC, pp. 162–172

Karchmer, M., Wigderson, A. (1988): Monotone circuits for connectivity require super-logarithmic depth. Proc. 21st ACM STOC, pp. 539–550

Lipton, R.J., Sedgewick, R. (1981): Lower bounds for VLSI. Proc. 13th ACM STOC, pp. 300–307

Lovász, L. (1979): Combinatorial problems and exercises. North Holland, Amsterdam

Lovász, L., Saks, M. (1988a): Lattices, Möbius functions and communication complexity. Proc. 29th IEEE FOCS, pp. 81–90

Lovász, L., Saks, M. (1988b). (Unpublished)

Mehlhorn, K., Schmidt, E.M. (1982): Las Vegas is better than determinism in VLSI and distributed computing. Proc. 14th ACM STOC, pp. 330–337

Papadimitriou, C.H., Sipser, M. (1983): Communication complexity. Proc. 14th ACM STOC, pp. 196–200

Rota, G.-C. (1964): On the foundations of combinatorial theory I: Theory of Möbius functions. Z. Wahrsch. Verw. Gebiete **2**, 340–368

Stanley, R.P. (1986): Enumerative combinatorics, Vol. 1. Wadsworth & Brooks, Monterey, CA

Swart, E.R. (1986): $P = NP$. Technical Report, University Guelph

Szemerédi, E. (1976): Regular partitions of graphs. In: Bermond, J.-C., Fournier, J.-C., Las Vergnas, M., Sotteau, D. (eds.): Problèmes combinatoires et théorie des graphes. Centre National de la Recherche Scientifique, Paris, pp. 399–401 (Colloq. Int. CNRS, Vol. 260)

Thompson, C.D. (1979): Area-time complexity for VLSI. Proc. 11th ACM STOC, pp. 81–88

Wilf, H.S. (1968): Hadamard determinants, Möbius functions and the chromatic number of a graph. Bull. Am. Math. Soc. **74**, 960–964

Yannakakis, M. (1988): Expressing combinatorial optimization problems by linear programs. Proc. 29th IEEE FOCS, pp. 223–228

Yao, A.C. (1979): Some complexity questions related to distributive computing. Proc. 11th ACM STOC, pp. 209–213

Yao, A.C. (1981): The entropic limitations of VLSI computations. Proc. 13th ACM STOC, pp. 209–213

Yao, A.C. (1983): Lower bounds by probabilistic arguments. Proc. 24th IEEE FOCS, pp. 420–428

An Outline of a Disjoint Paths Algorithm

Neil Robertson and Paul D. Seymour

1. Introduction

For several years now we have been working on what we call the "Graph Minors" project. This is turning into a long series of papers, of which currently sixteen have been written (we shall refer to these as I,II,...,XVI) and about eight more planned. The original goal of the series was Wagner's conjecture that for every infinite set of graphs, one is a minor of another; but it turns out that the theory we developed to prove this conjecture has other applications. One of these (in XIII) is to the disjoint paths problem, and we shall summarize that here, together with the necessary background from other papers of the series.

The disjoint paths problem is, given a graph G and p pairs of vertices of G, to decide if there are p mutually vertex-disjoint paths of G linking the pairs. If p is part of the input of the problem then this is one of Karp's NP-complete problems [5], and it remains NP-complete even if G is constrained to be planar [7]. For p fixed, however, it is more tractable. For $p = 2$ there is a simple algorithm, and we have shown that for any fixed p there is an algorithm with running time $O(|V|^2 \cdot |E|)$. (The problem retains some challenges, however; for instance, no practical algorithm is known even for $p = 3$. Our algorithm for $p = 3$, while polynomially bounded, involves the manipulation of ludicrously large constants and hence is not practical.)

Incidentally, it is crucial that we are working with undirected graphs. For directed graphs, if we ask for vertex-disjoint directed paths linking the pairs, the problem is NP-complete even for $p = 2$ [3]. This is in contrast with the elementary "max-flow" problem, where we ask for p disjoint paths all running from one specified set of vertices to another, for there the directed and undirected problems are virtually identical.

The idea of our algorithm is that if the input is sufficiently rich (more exactly, has high enough "tree-width") then we can find an "irrelevant" vertex, the deletion of which does not change the problem, and we repeat this until the tree-width becomes too small. But then the graph is cut up by an abundance of low order separations, and these can be used to solve the problem directly.

Handling graphs of small tree-width is easy, and the complicated part of the algorithm is finding an irrelevant vertex. Even that is much less complicated than the proof that the algorithm works, that it really does locate an irrelevant vertex, and it is this proof that requires all the machinery from the Graph Minors project.

The paper is organized as follows. After a digression on the $p = 2$ problem, we show how to handle graphs of bounded tree-width. This, with a result of V, will enable us to assume that our input graph contains a large "wall", a subdivision of a piece of the hexagonal lattice. The remainder of the algorithm exploits the presence of this wall to find an irrelevant vertex. It breaks into two parts: one, rather easy, that provides an irrelevant vertex if we have already located a large "clique minor", and the second, harder, part, that uses the wall to produce either an irrelevant vertex or a large clique minor. Then we turn to the proof of correctness. We begin by solving an easier problem, to illustrate the method; we give a proof of correctness for a similar algorithm when the input graph is constrained to lie on a fixed surface. The general proof has a similar step, concerning hypergraphs drawn on a fixed surface; and the remainder of the proof lies in reducing the initial problem about general graphs to this problem about hypergraphs on a fixed surface. This is achieved by applying a theorem of XII, that says roughly that if G contains an enormous wall but no large clique minor, then G can be drawn on a surface of bounded genus up to 3-separations, except for a bounded number of extra vertices and a bounded number of "vortices"-local, highly constrained areas of non-planarity. To prove this, we need a result of IX concerning the existence of large families of disjoint paths "crossing" each other in certain patterns. Thus, our outline of the general proof consists of sketch-proofs of the theorems of IX and XII, and then the application of the theorem of XII to our algorithmic problem.

2. The Two-Paths Problem

When $p = 2$, the disjoint paths problem may be solved by a beautifully simple algorithm, founded on the following theorem. (A *separation* of a graph G is a pair (A, B) of subgraphs with $A \cup B = G$ and $E(A \cap B) = \emptyset$, and its *order* is $|V(A \cap B)|$.)

(2.1) *Let s_1, t_1, s_2, t_2 be distinct vertices of a graph G, such that no separation (A, B) of G of order ≤ 3 has $s_1, t_1, s_2, t_2 \in V(A) \neq V(G)$. Then the following are equivalent:*

    (i) *there do not exist vertex-disjoint paths P_1, P_2 of G such that P_i links s_i and $t_i (i = 1, 2)$*

    (ii) *G can be drawn in a disc with s_1, s_2, t_1, t_2 on the boundary in order.*

This was proved independently by several authors [4, 10, 11, 12], and its proof is non-trivial, probably the simplest proof is given in IX. However, its application to the disjoint paths problem with $p = 2$ is easy. The algorithm proceeds as follows. We are given a graph G and vertices s_1, t_1, s_2, t_2 of it, and we wish to decide if there are disjoint paths linking s_1 with t_1 and s_2 with t_2 respectively. For $k = 0, 1, 2, 3$ in turn, we test if there is a separation (A, B) of our graph of order k with $s_1, t_1, s_2, t_2 \in V(A) \neq V(G)$, and if we find one we reduce G by replacing B by a complete graph with vertex set $V(A \cap B)$; this does not change whether the desired two paths exist, as is easily seen. After making all these reductions we test whether our graph can be drawn as in (2.1)(ii). If so, the two paths do not exist, and otherwise they do. Carefully implemented, this algorithm has running time $O(|V| \cdot |E|)$, where $G = (V, E)$ is the input graph.

A useful generalization of (2.1) is obtained as follows. By a *society* we mean a pair (G, Ω), where G is a graph and Ω is a cyclic permutation of a subset $\overline{\Omega}$ of $V(G)$. A *cross* in (G, Ω) is a pair P_1, P_2 of disjoint paths in G with ends s_1, t_1 and s_2, t_2 respectively, all in $\overline{\Omega}$, such that s_1, s_2, t_1, t_2 occur in Ω in that order (but not necessarily consecutively). (2.1) may be generalized as follows.

(2.2) *Let (G, Ω) be a society such that there is no separation (A, B) of G of order ≤ 3 with $\overline{\Omega} \subseteq V(A) \neq V(G)$. Then the following are equivalent:*

(i)   *there is no cross in (G, Ω)*
(ii) *G can be drawn in a disc with the vertices in $\overline{\Omega}$ drawn on the boundary of the disc in order.*

What if the condition about low order separations fails to hold? Then we can apply the method of the algorithm above to reduce to the case when it does hold, apply (2.2), and reinterpret its statement in terms of the original graph. We obtain a statement which involves drawing hypergraphs in a disc. A *hypergraph* H consists of a set U of *vertices*, a set of F of *edges*, and an incidence relation between them. The vertices incident with an edge are the *ends* of the edge. If we represent the edges of a hypergraph $H = (U, F)$ by closed discs in Σ (where Σ is a disc or, later, a surface) and its vertices by points of Σ, such that the ends of each edge are represented by points of the boundary of the corresponding disc, and every two discs meet only in points representing common ends, we call this a *drawing* of H. From (2.2) we obtain

(2.3) *Let (G, Ω) be a society with no cross. Then there is a hypergraph $H = (U, F)$ with $\overline{\Omega} \subseteq U \subseteq V(G)$, in which every edge has ≤ 3 ends, and a collection $\{A_f : f \in F\}$ of subgraphs of G with union G, mutually edge-disjoint, with $V(A_f \cap A_{f'}) \subseteq U$ for all distinct f, f' and with $V(A_f) \cap U$ equal to the set of ends of f for each $f \in F$, and a drawing of H in a disc with the vertices in $\overline{\Omega}$ drawn on the boundary of the disc in order.*

(Of course, the converse of (2.3) also holds.)

3. Tree-Width

Intuitively, a graph has tree-width $\leq k$ if it can be constructed by piecing together graphs with $\leq k + 1$ vertices in a tree-structure. More precisely, a *tree-decomposition* (T, χ) of G consists of a tree T and a function χ assigning to each vertex t of T a subgraph $\chi(t)$ of G, such that

(i)   the $\chi(t)'s$ are mutually edge-disjoint and have union G
(ii) if $t_1, t_2, t_3 \in V(T)$ and t_2 lies on the path between t_1 and t_3 then $\chi(t_1) \cap \chi(t_3) \subseteq \chi(t_2)$.

Its *width* is the maximum of the $|V(\chi(t))| - 1$, over all $t \in V(T)$; and the *tree-width* of G is the minium width of a tree-decomposition. Thus, trees and forests have tree-width ≤ 1, series-parallel graphs have tree-width ≤ 2, and the complete graph K_n has tree-width $n - 1$.

It is easy to convert an arbitrary tree-decomposition of G into one of no greater width satisfying

(i) T is *ternary*, that is, every vertex of T has degree 1 or 3
(ii) If $t \in V(T)$ has degree 3 then $E(\chi(t)) = \emptyset$
(iii) If $t \in V(T)$ has degree 1 then some vertex or edge of $\chi(t)$ belongs to no other $\chi(t')$.

We call such a tree-decomposition *ternary*. (One advantage of ternary tree-decompositions is that $|V(T)| \leq 2|E(G)|$ if G has no isolated vertices, which is useful for complexity computations.)

Determining the tree-width of an arbitary graph is NP-hard [1], but for fixed k one can decide in polynomial time if a graph has tree-width $\leq k$ [1]. But faster, and just as good for us, is an approximate version of the last – very quickly, one can either decide that G has tree-width $> k$ or decide that G has tree-width $< 4k$. For that we need the following easy result (compare (2.6) of II or (3.1) of XIII):

(3.1) *If G has tree-width $\leq k$ then for every $X \subseteq V(G)$ there is a separation (A_1, A_2) of G of order $\leq k$ such that $|X \cap V(A_1)|, |X \cap V(A_2)| \geq \frac{1}{3}|X|$.*

If (A, B) is a separation of G of order $3k + 1$ and (A_1, A_2) is a separation of A with $A_1, A_2 \neq A$ such that the separations $(A_1, A_2 \cup B), (A_2, A_1 \cup B)$ both have order $3k + 1$, we call $(A_1, A_2 \cup B), (A_2, A_1 \cup B)$ a *successor pair* for (A, B). It follows easily, by applying (3.1) to A with $X = V(A \cap B)$, that if G has tree-width $\leq k$ then every separation (A, B) of order $3k + 1$ with $|V(A)| \geq 3k + 2$ has a successor pair. That yields

(3.2) *For all $k \geq 0$ there is an algorithm with running time $O(|V| \cdot |E|)$ which, for an input graph $G = (V, E)$, decides either that G has tree-width $> k$ or that G has tree-width $< 4k$; and which finds a subgraph G' of G and a ternary tree-decomposition of it of width $< 4k$, such that either $G' = G$ or G has tree-width $\geq k$.*

Proof. We proceed as follows. We choose a separation (A_0, B_0) of order $3k + 1$ with $E(B_0) = \emptyset$ and $A_0 = G$. Now we check that if $|V(A_0)| \geq 3k + 2$ then (A_0, B_0) has a successor pair $(A_1, B_1), (A_2, B_2)$ say; and for each of these, if $|V(A_i)| \geq 3k + 2$ then (A_i, B_i) has a successor pair, and so on. If they all have successor pairs then it is easy to construct from all these separations a ternary tree-decomposition of width $< 4k$, while if one has no successor pair then G has tree-width $> k$. For the second part of the algorithm, we may suppose that we have constructed some (A, B) with no successor pair but with $|V(A)| \geq 3k + 2$, which assures us that G has tree-width $> k$. Now we delete an edge of A from G; the graph we obtain still has tree-width $\geq k$, and we resume looking for successor pairs. Every time we get stuck we delete an edge, and thereby construct a graph G' with the required properties.   □

Now we shall describe a method due to Arnborg and Proskurowski [2] to solve the disjoint paths problem for fixed p when the input graph is provided with a ternary tree-decomposition of bounded width. By a *pairing* of a set X we mean a partition of a subset of X into sets of cardinality 1 or 2. If G is a graph and $P_1, \ldots, P_p$ are vertex-disjoint paths of G, let $X \subseteq V(G)$ include all their ends; then there is an associated pairing of X (sets of cardinality 1 arise from $P_i's$ with no edges and only one vertex.) Such a pairing is said to be *feasible* in G. By the *folio* in G of a subset $X \subseteq V(G)$ we mean the set of all pairings of X which are feasible in G.

(3.3)   *For every $p, k \geq 0$ there is an algorithm with running time $O(|E|)$ which, with input a graph $G = (V, E)$, vertices $s_1, t_1, \ldots, s_p, t_p$ of G, and a ternary tree-decomposition (T, χ) of G of width $\leq k$, determines if there are p vertices-disjoint paths of G linking s_i with $t_i (1 \leq i \geq p)$.*

Proof. We may assume that $s_1, t_1, \ldots, s_p, t_p \in V(\chi(t))$ for every $t \in V(T)$, for we may augment each $\chi(t)$ by adding $s_1, t_1, \ldots, s_p, t_p$ as isolated vertices if they are missing – doing so increases the width of our decomposition by at most $2p$ and so it remains bounded. Choose a vertex $r \in V(T)$ of degree 1. We define $X_r = \{s_1, t_1, \ldots, s_p, t_p\}$ and $H_r = G$. For each $s \in V(T) - \{r\}$, let e be the last edge of the path from r to s and let R, S be the two components of $T \backslash e$, where $s \in V(S)$. Let

$$H_s = \cup(\chi(t) : t \in V(S))$$

and let $X_s = V(\chi(s) \cap \chi(s'))$ where e has ends s, s'. Our objective is to compute for each $t \in v(T)$, the folio of X_t, in H_t. Now if $t \neq r$ and t has degree 1 then the folio of X_t in H_t is easily computed since $|V(\chi(t))|$ is bounded; while if t has degree 3 and neighbours t_1, t_2, t_3 where t_3 is between t and r, then the folio of X_t in H_t may be computed in constant time from a knowledge of the folios of X_{t_i} in H_{t_i} $(i = 1, 2)$, again since $|V(\chi(t))|$ is bounded. (A similar remark applies when $t = r$.) Thus, by starting from the vertices of T furthest from r and working in, we may compute all the desired folios; but the existence of the p paths in question is determined by the folio of X_r in $H_r = G$.   □

By a *subdivision* of a graph H we mean a graph obtained from H by replacing its edges by vertex-disjoint paths (or a graph isomorphic to such a graph). The method above can be adapted, by changing the definition of "folio" suitably, to show

(3.4)   *For any $k \geq 0$ and any graph H there is an algorithm with running time $O(|E|)$ which, with input a graph $G = (V, E)$ and a ternary tree-decomposition of G of width $\leq k$, either finds a subgraph of G which is a subdivision of H or determines that none exists.*

4. Walls and Tree-Width

By a *wall of height h* we mean a portion of the hexagonal lattice as indicated in Figure 1 consisting of h "horizontal" paths and h edges between consecutive pairs of them; or a subdivision of such a graph. By a *brick* of the wall we mean one of the circuits bounding a finite region in the drawing, that is, one of the circuits of length 6 (before subdivision). If a subgraph of G is a wall of height h then G has tree-width $\geq h$, but of more interest is the following partial converse, the main theorem of V.

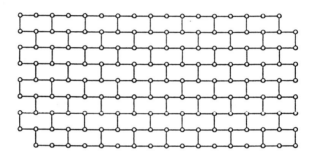

Fig. 1 A wall of height 9

(4.1) *For every $h \geq 0$, every graph of sufficiently high tree-width has a subgraph which is a wall of height h.*

We deduce

(4.2) *For every $h \geq 0$ there is an algorithm with running time $O(|V|^2)$ which, with input a graph $G = (V, E)$, finds a subgraph of G which is a wall of height h, if one exists.*

Proof. Choose k such that every graph of tree-width $\geq k$ contains a wall of height h. The algorithm runs as follows. First we apply (3.2) to the input graph G; we obtain a subgraph G' of G and a ternary tree-decomposition of it of width $< 4k$, such that either $G' = G$ or G' has tree-width $\geq k$. We observe that if G has a wall of height h then so does G' (from our choice of k). Now we use (3.4) applied to G' to locate a wall in G' if one exists. This has running time $O(|V| \cdot |E|)$, but it can be improved to $O(|V|^2)$ by using a theorem of Mader [8] which we omit.   □

We can now present the first step in our general algorithm for the disjoint paths problem, given by the following.

(4.3) *For every $p, h \geq 0$, there is an algorithm with running time $O(|V| \cdot |E|)$ which, with input a graph $G = (V, E)$ and vertices $s_1, t_1, \ldots s_p, t_p$, of G either decides whether there are p disjoint paths of G linking s_i and t_i $(1 \leq i \leq p)$, or locates a wall of height h in G.*

Proof. We first run (4.2) on G. If it gives us a wall we are done, and so we may assume that none exists and hence G has bounded tree-width, by (4.1). We run (3.2) to obtain a ternary tree-decomposition of bounded width, and then run (3.3) to solve the disjoint paths problem. $\qquad\square$

Applying (4.3)(for some suitably large h) will be the first step of our algorithm. It remains to describe how to make use of the wall which we thereby locate. We shall use this wall to find an *irrelevant* vertex, a vertex v such that if the paths exist then they also exist in $G\backslash v$. If we can locate a vertex which we know to be irrelevant, we delete it and go back and apply (4.3) again. Thus, the object of all the rest of the paper is to show how to find an irrelevant vertex if we are given an enormous wall in G.

5. Disjoint Paths on a Surface

The arguments to come are rather complex, and to illustrate our approach in a more digestible fashion we would like to digress at this point, and present an algorithm for the disjoint paths problem when the graph is constrained to lie on a fixed surface. We do not propose this as the best algorithm for the problem, for we need p fixed to ensure polynomiality, while Schrijver [9] has found an algorithm which is polynomial even for variable p, provided that all the s_i's and t_i's lie on a bounded number of regions of the drawing. Our algorithm is very simple, however, and does illustrate the "irrelevant vertex" method, and the proof of its correctness is a rudimentary version of the general proof.

A *surface* is a compact, connected 2-manifold, such as the sphere, torus or projective plane. We do not wish to discuss how a drawing on a surface is presented. If G is drawn on a surface $\sum$, and $W \subseteq G$ is a wall, we call W a *standard* wall if the perimeter of W (the circuit around the outside in Figure 1) bounds a disc $\varDelta$ in $\sum$ and the rest of W is drawn inside this disc. For the moment, let us assume the following result.

(5.1) *For every surface $\sum$ and every $p \geq 0$ there exists $h \geq 0$ such that if a graph G is drawn in $\sum$, and $s_1, t_1, \ldots, s_p, t_p \in V(G)$, and $W \subseteq G$ is standard wall of height h, and none of $s_1, t_1, \ldots, s_p, t_p$ is drawn inside the perimeter of W, then the middle vertex of W is irrelevant.*

This yields an algorithm for the disjoint paths problem for p paths and graphs G drawn on $\sum$, as follows. Choose h to satisfy (5.1), and choose h' somewhat larger than h. We apply (4.3) to G, with h' replacing h, and we may assume that it locates a wall W' of height h' in G. Choose a standard wall $W \subseteq W'$ of height h with none of $s_1, t_1, \ldots, s_p, t_p$ drawn inside its perimeter (there are several ways to do so, if h' is chosen appropriately, and we omit the details). By (5.1) the middle vertex of W is irrelevant. We delete it, and return to apply (4.3) again, and so on.

(5.1) is proved in the next section. Incidentally, (5.1) is inefficient in that we do not really need a large wall surrounding the vertex we proposed to delete, to guarantee its irrelevancy; in fact it suffices that there are a large number of vertex-disjoint circuits, all surrounding discs containing the vertex, and each one inside the next, with all the s_i's and t_i's outside the outermost circuit. (The proof is similar.)

6. Distance on a Surface

Let G be a graph drawn on a surface $\sum$. By a *ring* we mean a non-self-intersecting closed curve F in $\sum$ which meets the drawing only in vertices; its *length* is the number of vertices it passes through.

Now let $W \subseteq G$ be a standard wall. It has *dominance* $\geq h$, if it has height $\geq h$ and each ring of length $< h$ bounds a disc (called its *inside*) in $\sum$ including no horizontal path of W.

By a *degenerate ring* we mean one of the closed curves indicated in Figure 2, again meeting the drawing only in vertices, and its *length* is again the number of vertices passed through as the closed curve is traced (thus, some vertices are counted more than once).

Fig. 2 Degenerate rings

Given a wall $W \subseteq G$ of dominance $\geq h$, a point of $\sum$ is *inside* a degenerate ring F of length $< h$ if either it belongs to F or to the inside of some ring $F' \subseteq F$. We say two points $x, y \in \sum$ have *distance* $\geq h'$ if W has dominance $\geq h'$ and there is no (possibly degenerate) ring of length $< h'$ with x, y both inside it. This defines a metric on $\sum$ with some useful properties, as we shall see.

The following is the main result of VII, modified by some results of XI.

(6.1) *For every surface $\sum$ and integers $p, q \geq 0$ there exists $h > p$ with the following property. Let $\Delta_1, \ldots, \Delta_q \subseteq \sum$ be disjoint closed discs and let $T_1, \ldots T_k$ be trees drawn in $\sum$, mutually disjoint, each meeting each Δ_j only in $bd(\Delta_j)$ and only in a subset of its vertices. Let $V(T_i) \cap (\Delta_1 \cup \ldots \cup \Delta_q) = X_i (1 \leq i \leq k)$ and let $|X_1 \cup \ldots \cup X_k| \leq p$. Let G be a graph drawn in $\sum$, meeting each Δ_j in precisely $\Delta_j \cap (X_i \cup \ldots \cup X_p)$, and all these points being vertices of G. Let W be a standard wall in G of dominance $\geq h$, such that*

(i)   *for $1 \leq j \leq q$ there is no ring of length $< |(X_1 \cup \ldots \cup X_k) \cap \Delta_j|$ with Δ_j inside it*

(ii) *for $1 \leq j < j' \leq q$, Δ_j and $\Delta_{j'}$ have distance $\geq h$.*

   Then there are disjoint subtrees $T'_1, \ldots T'_k$ of G such that $X_i \subseteq V(T'_i)$ for $1 \leq i \leq k$.

The proof of this is long and not very interesting, and we shall not even sketch it. Let us begin directly on its applications.

(6.2) *For every surface $\sum$ and integers $p, q \geq 0$ there exists $h > p$ with the following property. Let $\Delta_1, \ldots, \Delta_q \subseteq \sum$ be disjoint closed discs, and let $X \subseteq bd(\Delta_1) \cup \ldots \cup bd(\Delta_q)$ with $|X| \leq p$. Let G be a graph drawn in $\sum$ meeting $\Delta_1 \cup \ldots \cup \Delta_q$ in precisely X, these all being vertices of G. Let W be a standard wall in G of dominance $\geq h$, such that no vertex of X is drawn inside the perimeter of W. Let v be the middle vertex of W. Then every pairing of X feasible in G is feasible in $G \backslash v$.*

Proof. We proceed by induction on the genus of $\sum$, for fixed $\sum$ on q, and for fixed $\sum, q$ on p. Take a pairing of X which is not feasible in $G \backslash v$; we must show it is not feasible in G. Let W' be a standard wall in $W \backslash v$ of height $h - 1$. We may assume that h is chosen much larger than needed to satisfy (6.1), and so there exists h' with $0 \leq h' \ll h$ such that by (6.1) applied to $G \backslash v$, either W' does not have dominance $\geq h'$ (in $G \backslash v$), or there is a ring (for $G \backslash v$) of length $< |X \cap \Delta_j|$ with Δ_j inside it (for some j with $1 \leq j \leq q$), or Δ_j and Δ'_j have distance $< h'$ in $G \backslash v$. In each case there corresponds a ring or degenerate ring F for $G \backslash v$ of length $< h'$, which therefore meets W only in a narrow annulus around the perimeter. In particular, F goes nowhere near v, and so F is a ring or degenerate ring for G as well. Let us cut $\sum$ along F, splitting all the vertices that F passes through. By doing so we destroy only the outer part of W; most of it survives intact. What we obtain from $\sum$ is a "surface with boundary", and we paste discs on the holes to obtain a surface $\sum'$. In each case it is easy to construct a new set X' (containing some or all of the split vertices and some or all of the old vertices from X) and a pairing of it, and a new graph G', such that if v is irrelevant for this pairing in G' then it is irrelevant for the original pairing in G; and it follows from our inductive hypothesis that v is indeed irrelevant for this pairing, as required. □

We remark that (5.1) follows from (6.2), by choosing a small disc Δ_j for each vertex in $\{s_1, t_1, \ldots, s_p, t_p\}$.

7. Hypergraphs on a Surface

We shall later need a generalization of (5.1) to hypergraphs drawn on a surface. It is proved in much the same way as (5.1), via results generalizing (6.1) and (6.2), but in detail the proof is more complicated (for instance, the 3-fold induction in (6.2) is replaced by something like a 7-fold one) and we shall not attempt to sketch it here. We confine ourselves to a statement of the result.

Let H be a hypergraph. If every edge has two or three ends and we fix a linear order of these ends (the *orientation* of the edge) we obtain what we call a *triadic* hypergraph. We want to define a "minor" of such a hypergraph H, and therefore must define what we mean by "contracting" an edge of H. Edges with two ends are contracted as for graphs (except that such an edge can only be contracted if no other edge is incident with both its ends). For an edge e of size 3, with ends v_1, v_2, v_3 say, we allow four different ways of contracting e; either we identify all of v_1, v_2, v_3 into one new vertex, or we identify some two of them into a new vertex, retaining the third old vertex. In each case, however, if two vertices are to be identified there must be no other edge incident with them both. Let us say that H' is a *minor* of H if H' can be obtained from H by deleting edges and isolated vertices, and contracting edges.

If Λ is a set of "colours" a Λ-*coloured hypergraph* is a triadic hypergraph with a colour from Λ assigned to each edge. A Λ-*coloured minor* of such a hypergraph is a minor as above, with its edges retaining their original colours.

By a *wall* in a triadic hypergraph we mean a wall in its 1-skeleton, the graph obtained by replacing each edge of the hypergraph by a 2-or 3- vertex complete graph. The hypergraph is *free* (relative to a wall) if for each edge e of the hypergraph with k ends say, there is no separation (A, B) of the 1-skeleton of order $< k$ such that A contains all ends of e and B contains a horizontal path of the wall. The extension of (5.1) we require is the following.

(7.1) *For every surface $\sum$, every finite set Λ, and every Λ-coloured hypergraph H which can be drawn in $\sum$ and every $h' \geq 0$, there exists $h \gg h'$ with the following property. Let G be a Λ-coloured hypergraph drawn in $\sum$, and let W be a standard wall in G of height $\geq h$, such that G is free. Suppose that for every edge e of G drawn inside the perimeter of W and every standard subwall W' of W of height $\geq h'$, there is an edge of G drawn inside the perimeter of W' with the same colour, orientation, and number of ends as e. Let v be the middle vertex of W. If G has a Λ-coloured minor isomorphic to H then so does $G\backslash v$.*

In fact, for our application we shall need a generalization of (7.1). By a *hereditary class* C of Λ-coloured hypergraphs, we mean one such that if $G \in C$ and H is isomorphic to a Λ-coloured minor of G then $H \in C$. One example is for a fixed $\sum$ and Λ-coloured H drawable in $\sum$, we take C to be the class of all Λ-coloured hypergraphs drawable in $\sum$ with no minor isomorphic to H. The statement (7.1) can be expressed in terms of this class C instead of explicitly in terms of H, and we would like to know the truth of (7.1) in this form for general hereditary classes of C. More exactly then, we would like

(7.2) *The same as (7.1), except*
  (a) *in the first sentence, change "Λ-coloured hypergraph H" to "hereditary class C of Λ-coloured hypergraphs"*
  (b) *replace the last sentence by "If $G \notin C$ then $G\backslash v \notin C$".*

This is implied by (7.1) if C can be characterized by excluding only finitely many minors, and it follows from our proof of the Wagner conjecture that every such class C can be so characterized. However, this does not constructively yield a value for h satisfying (7.2), and we would like to remain constructive if possible.

It turns out that for certain hereditary classes C, the proof of (7.1) can be transformed directly to a proof of (7.2) without using our "Wagner's conjecture" arguments, and for the application of (7.2) in section 13, the class C concerned is of this special kind. We omit further details of the proof of (7.2).

8. Making Use of a Clique

That concludes our digression; now let us return to the disjoint paths algorithm. As we observed at the end of section 4, it remains to construct an algorithm with running time $O(|V| \cdot |E|)$ which, with input a graph G, vertices $s_1, t_1, \ldots, s_p, t_p$ of G, and an enormous wall in G, locates an irrelevant vertex. (Incidentally, we shall frequently have recourse to words such as large, enormous, small etc. These all refer to constants, independent of the input graph, but dependent on the constant parts of the problem such as the number p of paths we are looking for.) A crucial half-way stage in our method for finding an irrelevant vertex is finding a K_n minor in G, a collection $X_1, \ldots, X_n$ of disjoint connected subgraphs of G, every two of which are joined by an edge of G. Thus, our algorithm is a combination of the following two parts.

(8.1) *For all $p \geq 0$, there is an algorithm with running time $O(|E|)$ which, with input a graph $G = (V, E)$, vertices $s_1, t_1, \ldots s_p, t_p$ of G, and a K_{3p+1} minor of G, finds an irrelevant vertex.*

(8.2) *For all $p \geq 0$ there exists $h \geq 0$ and an algorithm with running time $O(|V| \cdot |E|)$ which, with input a graph $G = (V, E)$, vertices $s_1, t_1, \ldots s_p, t_p$ of G, and a wall in G of height h, finds either an irrelevant vertex or a K_{3p+1} minor of G.*

The first of these is rather straightforward and we shall deal with it now; while the second is proved in the remainder of the paper. We shall need the following theorem, generalizing a result of Larman and Mani [6].

(8.3) *Let $Z \subseteq V(G)$ with $|Z| \leq 2p$, and let $X_1, \ldots, X_{3p}$ be vertex-disjoint subgraphs of G, such that*

(i)   *for each i either X_i is connected or each of its components meets Z*

(ii)   *for $1 \leq i < j \leq 3p$ either some edge of G has an end in X_i and an end in X_j or both X_i and X_j meet Z*

(iii)   *for $1 \leq i \leq 3p$ there is no separation (A, B) of G of order $< |Z|$ with $Z \subseteq V(A)$ and $V(X_i) \subseteq V(B) - V(A)$.*

Then every pairing of Z is feasible in G.

Proof. We proceed by induction on the size of G. If there is a separation (A, B) of G of order $|Z|$ with $Z \subseteq V(A)$ and $V(X_i) \subseteq V(B) - V(A)$ for some i, with $B \neq G$, then the result follows from our inductive hypothesis applied to B and an appropriate pairing in $V(A \cap B)$. We assume then that there is no such separation. Any edge of any X_i may therefore be contracted without violating (i), (ii), or (iii), and so we may assume that each X_i has no edges. Also, each vertex of Z belongs to some X_i, for otherwise some edge of G not in any X_i could be contracted. Then the rest of the proof is easy.   □

(8.4) *Let $Z \subseteq V(G)$ with $|Z| \le 2p$ and let $X_1, \ldots, X_{3p+1}$ be disjoint connected subgraphs of G, every two of which are joined by an edge of G. Choose a separation (A, B) of G such that*

(i)   *$Z \subseteq V(A)$ and $V(X_i) \subseteq V(B) - V(A)$ for some i*
(ii)   *subject to (i), (A, B) has minimum order*
(iii) *subject to (ii), B is minimal.*

Then for every $v \in V(B) - V(A)$, the folios of Z in G and $G \backslash v$ are equal.

Proof. By(8.3), every pairing of $V(A \cap B)$ is feasible in $B \backslash v$.   □

(8.1) is an immediate corollary of (8.4) because it is easy to find (A, B) as in (8.4) by solving $3p + 1$ maximum flow problems.

9. Finding a Flat Wall

It remains to prove (8.2). If W is a wall, it is a subdivision of a graph W' like that in Figure 1. Let us say the *pegs* of W are the vertices of W' on the perimeter. Now let W be a wall in G. It is *flat* if there is a separation (A, B) of G with $W \subseteq A$ and $V(A \cap B)$ the vertices of W on its perimeter, such that (A, Ω) has no cross, where Ω is the set of pegs in their natural cyclic order. Our next objective is:

(9.1) *For every $p, h' \ge 0$ there exists $h \ge 0$ and an algorithm with running time $O(|V| \cdot |E|)$ which, with input a graph G and a wall W in G of height h, finds either a K_{3p+1} minor of G, or a subwall W' of W of height h' and a subset $X \subseteq V(B)$ disjoint from W' with $|X| \le \binom{3p+1}{2}$ such that W' is flat in $G \backslash X$.*

To show (9.1) we need several lemmas which we give in greater generality than needed here, because we shall need them in this more general form later. We recall that H is a *minor* of G if H can be obtained from a subgraph of G by contracting edges.

(9.2) *For every surface Σ and every graph H which can be drawn in Σ there exists $h \ge 0$ with the following property. Let G be a graph drawn in Σ with a standard wall of dominance $\ge h$. Then G has a minor isomorphic to H.*

Proof. We assume H has no isolated vertices. Let $|E(H)| = q$, and choose edges $f_1, \ldots, f_q$ of G pairwise at great distance and such that no f_j lies inside a ring of length ≤ 1. Let us replace each f_j by a small disc Δ_j; we shall apply (6.1) to these discs. Let $|V(H)| = n$, and draw trees T_i $(1 \le i \le n)$ in Σ so that distinct T_i's have disjoint drawings and each T_i meets each Δ_j in a subset of the ends of f_j, in such a way that if we collapse each T_i to a single point, the edges f_j form a graph isomorphic to H. (This is possible since H can be drawn in Σ.) By (6.1),

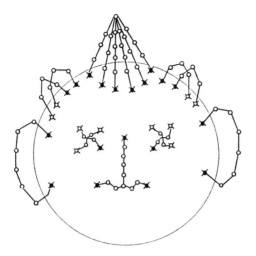

Fig. 3 Eyes, ears, nose, horn and hair

the T_i's can be chosen to be subgraphs of G, and so G has a minor isomorphic to H, as required.   □

For every surface $\sum$ except the sphere, (9.2) can be simplified, for the hypothesis about the existence of the wall can be replaced by the simpler (but essentially equivalent) hypothesis that every non-null-homotopic ring has length $\geq h$. However, we need a version of (9.2) which holds when $\sum$ is the sphere as well, and so for a uniform presentation we keep the wall throughout.

Let K be a subgraph of G. By a K-*path* in G we mean a path in G with distinct ends both in $V(K)$, and with no other vertices or edges in K. If K is connected, an *ear* (for K, in G) is a K-path together with two *roots*, its ends.

(9.3)   *For every surface $\sum$, graph H, and $k \geq 0$, if there are edges $g_1, \ldots, g_k$ of H such that $H \backslash \{g_1, \ldots, g_k\}$ can be drawn in $\sum$, then there is a number $h \geq 0$ with the following property. Let G be a graph, and let K be a connected subgraph of G drawn in $\sum$ with a standard wall of dominance $\geq h$, and let $P_1, \ldots P_k$ be ears, mutually disjoint, with all their roots mutually at distance $\geq h$. Then G has a minor isomorphic to H.*

The proof is similiar to that of (9.2).
Again, let K be a subgraph of G, and now let K be drawn in $\sum$. Let $W \subseteq K$ be a standard wall of dominance ≥ 4. By an *eye* we mean the union of two disjoint K-paths P_1, P_2 with ends s_1, t_1 and s_2, t_2 respectively, such that s_1, s_2, t_1, t_2 all lie on the same region and occur on its border in that order, and there is no ring of length ≤ 3 with $\{s_1, s_2, t_1, t_2\}$ inside it; together with one of s_1, s_2, t_1, t_2, the *root* of the eye. (See Figure 3.)
Again, one can deduce from (6.1) that

(9.4) *For every surface $\sum$, graph H and $k \geq 0$, if H can be drawn in $\sum$ except for k crossings then there is a number $h \geq 4$ with the following property. Let G be a graph and let K be a subgraph of G drawn in $\sum$ with a standard wall of dominance $\geq h$, and let $P_1, \ldots, P_k$ be eyes, mutually disjoint, with all their roots at distance $\geq h$. Then G has a minor isomorphic to H.*

Moreover, (9.2), (9.3), and (9.4) are "effective" in the sense that given a graph satisfying the hypothesis one can construct the minor isomorphic to H. (It will not matter how quickly we can do so, because our algorithm will apply these results only to graphs K which are walls of bounded height and hence are essentially of constant size.)

Let $G, K, \sum, W$ be as in the definition of "eye", with K connected. By a *nose* we mean the union of three paths P_1, P_2, P_3 with distinct ends s_i, t ($i = 1, 2, 3$) respectively, mutually disjoint except for t with no vertex in K except s_1, s_2, s_3. (See Figure 3.) We call t its *tip* and s_1, s_2, s_3 its *roots*. For the rest of this section $\sum$ will be the sphere and $K = W$ will be a wall.

A *mural* in G consists of a wall K drawn in a sphere $\sum$ together with a selection of ears, eyes and noses, mutually disjoint. It has *strength* $\geq h$ if K has height $\geq h$ and all the roots are mutually at distance $\geq h$. If G has no K_{3p+1} minor, then by (9.3) and (9.4) no mural in G with many roots has great strength (for noses may easily be converted into ears). On the other hand, we recall that we are provided with an enormous wall in our input graph, which is a mural of enormous strength but without roots. Our strategy will be to trade strength for roots, and examine the mural where the process stops. We shall see that if we are sufficiently generous in our willingness to trade, then from the "optimal" mural it is easy to locate a wall which is flat in $G\backslash X$, where X is the set of all tips of noses. For let the optimal mural be K together with certain ears, eyes and noses. Let $\Delta \subseteq \sum$ be the disc bounded by the perimeter of K with K drawn on it. Let X be the set of all tips of noses.

(9.5) *Let $W \subseteq K$ be a subwall, standard in Δ, such that all vertices of K enclosed by its perimeter are far from all roots of the mural. Then W is flat in $G\backslash X$.*

Proof. Suppose that there is a K-path P in $G\backslash X$ with one end strictly inside W, and the other strictly outside W. If P meets any ear then we may convert that ear to a nose, a contradiction. If it meets any eye, we may convert that eye to an ear, again a contradiction. If it meets a nose then we may convert the nose into two disjoint ears (for P does not contain the tip of the nose), again a contradiction. Thus, we may assume that P is disjoint from all the ears, eyes and noses. If its end outside of W, t say, is far from all roots, and far from s (its end inside W) then P yields an extra ear, a contradiction. We next suppose that t is close to some root r. If r is the root of an eye, we sacrifice the eye and gain P as an extra ear, a contradiction. Thus, r is a root of an ear or nose. We delete from K a small neighbourhood of r between r and t, and thereby obtain either a nose (in exchange for the ear) or two ears (for the nose), in either case a contradiction. The final possibility is that t is close to s; but since one is inside the perimeter of

W and the other is outside, we can delete a small portion of K between s and t to obtain an extra eye, again a contradiction.

We have proved then that no such path exists, and so there is a separation (A, B) of $G\backslash X$ with $A \cap K = W$ and $V(A \cap B)$ the set of vertices of the perimeter of W. It remains to show that the corresponding society (A, Ω) has no cross. But if it has a cross, then we could create a new eye by sacrificing the part of K inside W (it is here we use the fact that $\overline{\Omega}$ only contains the pegs of W), a contradiction.   □

Our algorithm (9.1) is an immediate consequence of (9.5), for it is easy to find a large wall $W \subseteq K$ far from all roots, and it is easy to test if it is flat (via (2.2)). If it is not flat, we choose a mural with somewhat less strength but with an extra root, and repeat. Either we find a flat wall after a bounded number of iterations, or we find a reasonably strong mural with many roots, when we can locate a K_{3p+1} minor.

10. Finding a Homogeneous Wall

The next, and last, step of the algorithm is to derive from our enormous flat wall a large "homogeneous" wall, which we shall define in a moment. That concludes the algorithm because the middle vertex of a large enough homogeneous wall is irrelevant, and so we have located an irrelevant vertex. Of course, it remains to prove that the middle vertex of a large homogeneous wall is irrelevant, but that is not part of the algorithm.

Let W be a flat wall in $G\backslash X$, and let (A, B) be the corresponding separation of $G\backslash X$. Let Ω be the natural cyclic permutation of the pegs of W, so that (A, Ω) has no cross. By (2.3), there is a hypergraph $H = (U, F)$ with $\overline{\Omega} \subseteq U \subseteq V(A)$ in which every edge has ≤ 3 ends, and a collection $\{A_f : f \in F\}$ of subgraphs of A with union A, mutually edge-disjoint, with $V(A_f \cap A'_f) \subseteq U$ for all distinct f, f', and with $V(A_f) \cap U$ equal to the set of ends of f for each $f \in F$, and a drawing of H in a disc Δ with the vertices in $\overline{\Omega}$ drawn on the boundary of the disc in order. It is possible to choose H and the A_f's with three further properties:

(i)   each $f \in F$ has two or three ends
(ii) for each $f \in F$ with k ends there is no ring (for the drawing of H in Δ) of length $< k$ with the disc representing f inside the ring
(iii) for each $f \in F$ and every two vertices of $V(A_f) \cap U$, there is a path of A_f joining them with no other vertex in U.

Moreover, it is easy to find an algorithm which constructs such an H and family of A_f's (we omit the details). Let us choose subgraphs $C_f(f \in F)$ of G, mutually edge-disjoint and with union $G\backslash(V(B) - V(A))$, so that $X \subseteq V(C_f)$ and $C_f\backslash X = A_f$ for each $f \in F$. We call these C_f's the bags of W. We write $N(C_f) = X \cup (V(A_f) \cap U)$.

We would like to find a large wall, flat in $G\backslash X$ for some small X with $s_1, t_1, \ldots s_p, t_p \in X$. This is easy, given a large wall W flat in $G\backslash Y$ for some Y; for we define $X = Y \cup \{s_1, t_1, \ldots s_p, t_p\}$ and observe that there is a large subwall W' of W disjoint from X and therefore flat in $G \backslash X$.

Next, given a large wall W in $G \backslash X$ where $s_1, t_1, \ldots, s_p, t_p \in X$ and $|X|$ is small, we would like to calculate for each $f \in F$ the folio of $N(C_f)$ in C_f. This, of course, we cannot do in general, for this is a set of disjoint path problems just like our original problem, and we have no control over what lies in C_f. Nevertheless, let us use (4.3) to attempt to calculate all these folios, where h in (4.3) is suitably large. We succeed unless we locate in some C_f a wall of height h, and if this happens we go back and run (9.1) again on this new wall, to obtain a new flat wall of the same height as W. Then we start examining its bags, as before. One can arrange that this process terminates, and thus we can find what we wanted.

In summary

(10.1) *For every $p, h' \geq 0$ there exists $h \geq 0$ such that there is an algorithm with running time $O(|V| \cdot |E|)$ which, with input a graph G, vertices $s_1, t_1, \ldots s_p, t_p$ of G, and a wall of height h in G, either*

   (i)   *finds a K_{3p+1} minor of G, or*
   (ii) *finds a subset $X \subseteq V(G)$ with $|X| \leq \binom{3p+1}{2} + 2p$ and $s_1, t_1, \ldots, s_p, t_p \in X$ and a wall W of height $\geq h'$, flat in $G \backslash X$, and computes the folio of $N(C_f)$ in C_f for each bag C_f.*

Let us assume that we have X, W as in (10.1)(ii), and define $A, B, \Omega, H = (U, F)$ and the C_f's as before. To make H triadic we must choose a linear order of the vertices of each edge; and let us do so such that each edge is oriented in the same sense as Ω. Let us say two edges f, f' of H have the *same colour* if they have the same number of ends, and the natural bijection between their ends maps the folio of C_f to the folio of $C_{f'}$. We see that there are only a bounded number of "colours". We say that W is *homogeneous* if for every subwall W' of W of height h' (where W has height h, and h' is some appropriate number with $1 \ll h' \ll h$) and for every edge e of H there is an edge f of H with the same colour as e drawn inside the perimeter of W'.

If W is not homogeneous, we can find an enormous subwall W', which is automatically flat, such that the corresponding hypergraph H' has fewer colours; and since the number of colours is bounded, if we choose W large enough initially, this process will always terminate with a large homogeneous wall. Then its middle vertex is irrelevant because of the following.

(10.2) *For every p, h there exists $h' \geq 0$ with the following property. Let $\{s_1, t_1, \ldots, s_p, t_p\} \subseteq X \subseteq V(G)$ where $|X| \leq \binom{3p+1}{2} + 2p$, and let W be a flat wall in $G \backslash X$ of height h'. Let $H = (U, F)$ and the C_f's be as before, and suppose that for every subwall W' of W of height h and every edge e of H, some edge f of H drawn inside the perimeter of W' has the same colour. Then the middle vertex of W is irrelevant.*

This concludes the proof of (8.2), except for proving (10.2) which will occupy the remainder of the paper.

11. Disjoint Crossed Paths

Our strategy to prove (10.2) is as follows. We would like to apply (7.2), but that requires all of G (or some derived hypergraph) to be drawn on a surface of bounded genus, and we are presented with a drawing of only part of G on a disc. The problem is, how do we bring the remainder of G under control? It will turn out to be an easy consequence of (8.3) that we can assume that our graph has no large clique minor, and we shall prove a theorem of XII, that every graph with no large clique minor (but with a large wall) can be drawn in a surface of bounded genus (up to 3-separations) except for a bounded number of extra vertices and a bounded number of "vortices". This theorem is the goal of the next two sections, and then we shall apply it to prove (10.2) in section 13.

In this section we shall discuss a theorem of IX, which will be used as a lemma to prove the theorem of XII. Let (G, Ω) be a society (defined in section 2). A *bisection* of Ω is a partition of $\overline{\Omega}$ into two "semicircles", that is, we take distinct $a, b \in \overline{\Omega}$, and let A be all those vertices of $\overline{\Omega}$ after a (or a itself) and before b, and let B be the complementary set, all those which are b or after b and before a. Then (A, B) is a bisection of Ω. By a *transaction* in (G, Ω) we mean a set $\mathcal{P}$ of disjoint paths of G, such for some bisection (A, B) of Ω, each $P \in \mathcal{P}$ has one end in A and the other in B. If (G, Ω) has no transaction of cardinality $> p$ we call it a *vortex of depth* $\leq p$.

(11.1) *If (G, Ω) is a vortex of depth $\leq p$ then for each $v \in \overline{\Omega}$ there is a subgraph G_v of G with the following properties:*

  (i)   *the G_v's are mutually edge-disjoint and have union G*
  (ii) *each G_v intersects the other G_u's only in vertices which belong to $G_{v_1} \cup G_{v_2}$, where v_1, v_2 are consecutive with v in Ω*
  (iii) *if $u, v \in \overline{\Omega}$ are consecutive then $|V(G_u \cap G_v)| \leq p$.*

Proof. For each bisection (A, B) there is a separation (C, D) of G of order $\leq p$ with $A \subseteq V(C)$ and $B \subseteq V(D)$, by Menger's theorem. The result follows easily by "uncrossing" these separations. $\square$

We remark that the converse of (11.1) is false, but if (i),(ii) and (iii) are satisfied then (G, Ω) is a vortex of depth $\leq 2p$.

What really concerns us here is not the structure of societies with no large transaction, but the structure of societies with no large transaction of a certain kind. There are three kinds of transaction of interest to us, indicated in Figure 4.

Ideally, one might hope to understand the structure of societies which do not contain a large crosscap transaction, and similarly for the other two, but we have not been able to do so. What we can do is exclude all three simultaneously; we shall study societies which have no large crosscap, leap or doublecross transaction.

Let $\mathcal{P}$ be a transaction in (G, Ω). If $P \in \mathcal{P}$ and no other member of $\mathcal{P}$ crosses P (in the natural sense) and also P does not lie between two other members of $\mathcal{P}$, we say that P is *peripheral* in $\mathcal{P}$. Thus, P is peripheral if one of the two "segments" of Ω between the ends of P contains no end of any other member

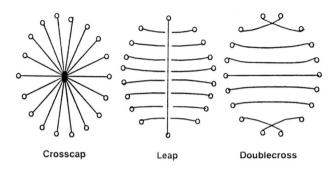

Crosscap   Leap   Doublecross

Fig. 4   Three crooked transactions

of $\mathscr{P}$. It follows that any transaction has at most two peripheral members. A transaction is *crooked* if it has no peripheral members. We see that crosscap, leap and doublecross transactions are all crooked, and most conveniently there is a kind of converse.

(11.2) *Let $t \geq 4$, and let $\mathscr{P}$ be a crooked transaction in (G, Ω) with $|\mathscr{P}| \geq \ (t - 1)(3t - 7)$. Then there exists $\mathscr{P}' \subseteq \mathscr{P}$ with $|\mathscr{P}'| = t$ such that $\mathscr{P}'$ is a crosscap, leap or doublecross transaction.*

The proof is easy and we omit it.

In view of (11.2), we wish to understand the structure of graphs admitting no large crooked transaction. If a society (G, Ω) can be drawn in a disc with Ω on the boundary in order, then it admits no cross and hence certainly no non-null crooked transaction. If it can be drawn in a disc as above except for one crossing, then it may admit a cross but it admits no crooked transaction of cardinality ≥ 3. In contrast, if its drawing has two crossings, sufficiently far apart, then the society may well admit a large doublecross transaction. We shall prove that if a society admits no large crooked transaction then it can be decomposed into a planar annulus (modulo 3-separations) encircling a vortex of bounded depth, and conversely any society decomposible in this way admits no large crooked transaction. By a *planar annulus* we mean a graph G together with two cyclic permutations Ω, Ω' of subsets $\overline{\Omega}, \overline{\Omega}'$ of its vertices, which can be drawn in a sphere Σ in such a way that there are two closed discs, $\Delta, \Delta' \subseteq \Sigma$ with $\Delta \cap \Delta' = bd(\Delta) \cap bd(\Delta')$, the drawing of G meets $\Delta \cup \Delta'$ only in $bd(\Delta) \cup bd(\Delta')$, and the vertices in $\overline{\Omega}$ (respectively $\overline{\Omega}'$) are drawn in $bd(\Delta)$ (respectively, $bd(\Delta')$) in order. We say that a society (G, Ω) is the *composition* of a planar annulus (G_0, Ω, Ω') and a society (G', Ω') if (G_0, G') is a separation of G and $\overline{\Omega}' = V(G_0 \cap G')$. The main result of this section is

(11.3) *Let (G, Ω) be a society admitting no crooked transaction of cardinality $\geq p$. Suppose that there is no separation (A, B) of G of order ≤ 3 with $\overline{\Omega} \subseteq V(A) \neq V(G)$. Then (G, Ω) is the composition of a planar annulus and a vortex of depth $\leq 3p + 7$.*

Proof. We proceed by induction on the size of G, and for fixed G on $|V(G) - \overline{\Omega}|$. We may assume then that no two consecutive vertices in Ω are adjacent, and no vertex in Ω is isolated. Moreover, we may assume that for each $v \in V(G) - \overline{\Omega}$, if we "insert" v into Ω somewhere then the new society we obtain has a crooked transaction of cardinality $\geq p$. We may assume that (G, Ω) has a transaction of cardinality $\geq 3p + 8$, for otherwise (G, Ω) itself is the desired vortex. From these assumptions we shall obtain a contradiction.

We shall use the transaction of cardinality $\geq 3p + 8$ to find one of cardinality $\geq 2p + 4$ with at most one peripheral member; and then use that to find a crooked one of cardinality $\geq p$, a contradiction . The steps are similar (sacrificing $\leq p + 4$ members of our transaction to reduce the number of peripheral members by ≥ 1) and here we only describe the first.

Let $\mathscr{P} = \{P_1, \dots, P_{3p+8}\}$ be a transaction in (G, Ω), with bisection (A, B). Let each P_i have ends $a_i \in A, b_i \in B$, and let $a_{3p+8}, a_{3p+7}, \dots, a_1$ be in order in Ω. We may assume that P_1 is peripheral in $\mathscr{P}$, and P_2 is peripheral in $\mathscr{P} - \{P_1\}$, and so on for the first $p + 5$ paths; and hence $a_{3p+8}, \dots, a_1, b_1, \dots, b_{p+5}$ are in order, and the other b_j's occur after b_{p+5} and before a_{3p+8}. Now we may choose $\mathscr{P}$ so that a_1, b_1 are consecutive in Ω (as is easily seen), and every vertex $v \in V(G) - V(P_1)$ is joined to a vertex of $\overline{\Omega} - \{a_1, b_1\}$ by a path avoiding P_1 (which may be seen with a little more difficulty). Let v be an internal vertex of P_1 (this exists since a_1, b_1 are not adjacent), and insert v into Ω between a_1 and b_1, forming Ω'. Now from our assumption, (G, Ω') has a crooked transaction Q of cardinality $\geq p$, and let us choose it using as few edges not in $P_1 \cup \dots \cup P_{3p+8}$ as possible. One can argue that not all of $P_1, \dots, P_{p+5}$ meet a member of Q, from the minimality of the choice of Q; let P_j be disjoint from every member of Q, where $j \leq p + 5$. Now we may assume that no member of Q crosses P_j (for otherwise we could sacrifice $P_1, \dots, P_{j-1}$ and reduce the number of peripheral members). Hence, since some member of Q has end v, it follows that some cross of Q has all four ends after a_j and before b_j, and (we may assume) is disjoint from $P_{j+1}, \dots P_{3p+8}$. If none of these four ends is v, then we sacrifice $P_1, \dots, P_{j-1}$ and gain the two paths of the cross, again making the desired transaction. It remains to argue that if v is one of the ends of the cross then we can find another cross of which v is not an end, using the connectivity assumptions about (G, Ω) and the properties of P_1, and we omit these details.   □

Like (2.3), there is a version of (11.3) for societies with 3-separations, and we obtain it the same way. There turn out to be some difficulties in the derivation, however, due to triangles which are half in the vortex and half in the annulus, and to rid ourselves of these we increase the depth of the vortex a little. We obtain

(11.4) *Let (G, Ω) be a society admitting no crooked transaction of cardinality $\geq p$. Then there is a hypergraph $H = (U, F)$ with $\overline{\Omega} \subseteq U \subseteq V(G)$, and an edge $f_0 \in F$, such that every other edge has ≤ 3 ends, and a collection $\{A_f : f \in F\}$ of subgraphs of G with union G, mutually edge-disjoint, with $V(A_f \cap A_{f'}) \subseteq U$ for all distinct f, f', and with $V(A_f) \cap U$ equal to the set* *of ends of f for each f ∈ F, and a cyclic permutation Ω' with $\overline{\Omega'}$ the set of ends of f_0, such that (A_{f_0}, Ω') is a vortex of depth ≤ 3p + 9, and a drawing of H in a disc with Ω on the boundary of the disc in order and the ends of f_0 drawn in their order according to Ω'.*

12. Excluding a Minor

Throughout this section, H is a fixed graph and G is a graph with no minor isomorphic to H; and W_0 is a wall in G of enormous height (as a function of H, not of G).We would like to use (11.4) to investigate the structure of G.

Let K be a connected subgraph of G drawn on a surface $\sum$ and let $W \subseteq K$ be a standard wall of dominance ≥ h, where h is a very large function of H. We have already defined what we mean by ears, eyes and noses, in section 9. Now we want two more features, horns and hair. (See Figure 3.) A horn is a generalization of a nose with many paths instead of just three; thus, a *horn* is the union of $k \geq 3$ paths $P_1, \ldots, P_k$, where P_i has distinct ends $s_i, t (1 \leq i \leq k)$, and the P_i's are mutually disjoint except for t, and each P_i has no vertex in K except for s_i. We call t its *tip*, $s_1, \ldots, s_k$ its *roots*, and $P_1, \ldots, P_k$ its *paths*. A *hair* is a K-path with one distinguished end, its *root*, the other end is its *tip*. An *animal* in G consists of $K, W, \sum$ as above together with a selection of *features*, that is, horns, hair and eyes (we do not need ears or noses) mutually disjoint except that one hair can intersect another and hair can intersect eyes. It has *strength* ≥ h if W has dominance ≥ h, each horn has ≥ h roots, all the roots are mutually at distance ≥ h, and for each hair its root and tip are at distance ≥ h. Its *beauty* is the 4-tuple $(\sum, a, b, c)$ where a, b, c are the number of horns, hairs and eyes respectively. We order beauties lexicographically, the first coordinate having greatest influence (and $\sum_1 \leq \sum_2$ if up to homeomorphism $\sum_2$ can be obtained from $\sum_1$ by adding handles or crosscaps).

Now we are given an animal in G of enormous strength and minimal beauty, the wall W_0, and as in section 9, we shall proceed to trade strength for beauty. From (9.2) the first coordinate of beauty will remain bounded (for H can be drawn on all except finitely many surfaces up to homeomorphism). By (9.3) the number of horns remains bounded (because horns contain ears) and by (9.4) so does the number of eyes. What about hair ? This is more difficult, because hairs can intersect one another and their tips need not be far apart. But it turns out that if the number of hairs passes a certain point we can construct a new horn, as follows. We observe first that if some component of the union of the hairs contains many of our hairs, then either we can choose from it many ears, disjoint, with all their roots far apart, contrary to (9.3), or we can choose a new horn in this component. (To see this, choose a minimal subtree connecting all the roots of the hairs. If it has a vertex of high degree we have our horn, and otherwise it is easy to find the ears.) Thus we may assume that each such component contains only a few hairs, and there are many components. We deduce that there are a great many mutually disjoint hairs. By (9.3) not many of them have their tips far apart, and so there are many disjoint hairs with their tips close together. Let us remove from K a small disc containing all these tips; then we find ourselves with a new animal of slightly smaller strength, but with the same number of horns and with many hairs all now lying in one component. Thus we may use the previous argument to obtain the extra horn.

Because of this we can find an "optimal" animal in G, one of enormous strength and limited beauty but which cannot be "locally transformed" to make an animal of somewhat lesser strength and greater beauty. Let $K, W, \sum$ correspond to this animal. ("Locally transformed" means that we only assume the non-existence of more beautiful animals which contain a great deal of W. There may be better animals in other parts of G.)

(12.1) *Let P be a K-path with ends s, t, not passing through any tip of any horn of the animal. If s is far from all roots then P is disjoint from all the features, and t is very close to s, so close that P does not give us a new eye.*

Proof. If P meets a feature, let Q be the subpath of P from s to the first feature it meets. Extend Q through the feature to a root (or close to the root if it is an eye, and extending down the corresponding path if it is a horn). Then we have a new hair (sacrificing the path of the horn if it is a horn) and hence greater beauty, a contradiction.

Thus P meets no features. If t is far from s then P is a new hair. If t is not far from s then we may delete a small part of K to get a new eye unless t is very close, as required.   □

For each root r, let us choose a disc $\Delta_r \subseteq \sum$ centered at r of say twice the radius necessary to contain all $s \in V(K)$ not far from r (in the sense of (12.1)). Choose Δ_r so that $bd(\Delta_r)$ meets the drawing only in vertices, and so that no K-path runs from just inside Δ_r to just outside Δ_r (this is possible by (12.1)). Let N_r be the vertices of K drawn in $bd(\Delta_r)$. Let K_r be the subgraph of K drawn within Δ_r. Let K_0 be the subgraph of K not drawn in the interior of any Δ_r.

(12.2) *For any two roots r_1, r_2 there are only a small number of disjoint K_0-paths in G from N_{r_1} to N_{r_2}.*

Proof. For if we had a large family of such paths, we could find a large subfamily such that the cyclic order of their ends in N_{r_1} and in N_{r_2} coincide (or are completely reversed). Let us delete Δ_{r_1} and Δ_{r_2} from $\sum$ and add a handle with ends the two holes in $\sum$ we have made; then our paths can be drawn on this handle. This yields an animal of greater first coordinate of beauty. We would like to claim that it still has great strength. Unfortunately, although we have many disjoint paths running along the new handle, there may be a lack of connectivity at N_{r_1} or N_{r_2}. To obviate this difficulty we observe that in fact our paths arrive at K_{r_2} far from N_{r_2} (because Δ_{r_2} has twice the radius it needs to capture the ends of such paths). Thus, let us retain an annulus of K_{r_1} of reasonable thickness, and run our paths out through it to N_{r_2} (we have to ensure that once our paths arrive at K_{r_2} they do not leave again, but that can be done). Doing this (and similarly for K_{r_1}) does yield an animal of great strength and increased beauty, a contradiction.   □

(12.3) *If r_1 is the root of a horn or eye, and $r_2 \neq r_1$ is a root, and P is a K-path in G from K_{r_1} to K_{r_2} then P passes through the tip of a horn.*

Proof. Suppose not. We may assume, by diverting P down any other feature that it meets, that P meets no feature except possibly an eye with root r_1 or the path of a horn with root r_1, and the feature of which r_1 is a root. If r_2 is not the root of a horn, then by sacrificing the eye with root r_1 or the path of the horn with root r_1 we gain an extra hair (with root r_1). If r_2 is the root of a horn, we sacrifice both corresponding paths and gain an extra hair. In each case we have a contradiction , as required.   □

In particular, no hair meets an eye. Let Z be the set of tips of all the horns; then $|Z|$ is bounded. By (12.2) and (12.3) there exists $Y \subseteq V(G)$ with $|Y|$ bounded and a subgraph D_r for each root r of the animal such that

(i) $D_r \cap K = K_r$ for each root r

(ii)   the D_r's are edge-disjoint from one another, and their union together with K_0 is G

(iii) if r is a root of an eye or horn then $Z \subseteq V(D_r)$ and $Y \cap V(D_r) = \emptyset$

(iv)   if r is a root of a hair then $Y \cup Z \subseteq V(D_r)$

(v)   for distinct r, r', $V(D_r \cap D'_r) \subseteq Y \cup Z$

(vi)   if F is an eye with root r then $F \subseteq D_r$

(vii) if F is the path of a horn with root r then $F \subseteq D_r$.

For each root r, let Ω_r be the natural cyclic order of N_r.

(12.4) *For each root r, $(D_r \backslash (Y \cup Z, \Omega_r)$ has no large crooked transaction.*

Proof. Suppose that it does; then by (11.2) it has a large crosscap, leap or doublecross transaction. If the former, we delete Δ_r from $\sum$ and add a crosscap on the hole, and obtain an animal of increased beauty, a contradiction. (As for (12.2) we must retain an annulus of K_r to ensure that this animal has reasonable strength. The same difficulty occurs in the arguments below, and we omit those details.) Thus, we have a large leap or doublecross transaction. If r is the root of an eye, we sacrifice the eye by deleting K_r, and adjoin our transaction to what remains of K, thereby gaining a new hair or two new eyes, a contradiction. Similarly, if r is a root of a horn, we remove the corresponding path from the horn, and gain either a new hair or two new eyes. Thus r is the root of a hair F. If we delete Δ_r and add our transaction to what remains of K, a part of F (with a new root but same tip) is still a hair, and its root is far from one of the two "eyes" of the transaction (if our transaction is a doublecross) which yields a new eye for the animal, or far from one end of the leap (if it is a leap) which can be taken as the root of a new hair. We have to worry about other hairs which may intersect our transaction in this case; but those can be shortened, keeping the same root. In all cases, then, we obtain an animal of greater beauty and reasonable strength, a contradiction.   □

In summary

(12.5) *For every graph H there are numbers a, b, c, h with the following property. Let G be a graph with no minor isomorphic to H but with a wall W of height $\geq h$. Then there exists $X \subseteq V(G)$ with $|X| \leq a$ and a hypergraph $J = (U, F)$ in which every edge except $f_1, \ldots f_b$ has two or three ends, with $U \subseteq V(G) - X$, and a drawing of J in some surface in which H cannot be drawn, and a collection $\{A_f : f \in F\}$ of subgraphs of $G \backslash X$, mutually edge-disjoint and with union $G \backslash X$, such that $V(A_f \cap A_{f'}) \subseteq U$ for distinct f, f', and $V(A_f) \cap U$ is the set of ends of f in J for each $f \in F$, if Ω_f is the natural cyclic order of the ends of f in the drawing then (A_f, Ω_f) is a vortex of depth $\leq c$ (for $f = f_1, \ldots, f_b$), and no A_f includes a horizontal path of W.*

13. The Proof of Correctness

Now we use (12.5) to prove (10.2). Let $p, h, h', G, s_1, t_1, \ldots, s_p, t_p, X, W, v, H = (U, F)$ and the C_f's be as in (10.2), where h' is very large compared with p and h. We may assume that the desired p paths exist in G, and for each vertex $u \in V(G) - (V(W) \cup X)$

(i) if $u \in V(G)$ for some C_f then the folio of $N(C_f)$ in $C_f \backslash u$ is different from that of $N(C_f)$ in C_f

(ii) if $u \notin V(C_f)$ for all C_f then the desired p paths do not exist in $G \backslash u$.

(For otherwise the result would follow by induction on the size of G.) Suppose that G has a K_n minor, where we shall specify n in a moment. Let $X_1, \ldots, X_n$ be disjoint connected subgraphs of G every two of which are joined by an edge. If some X_i is a subgraph of $C_f \backslash N(C_f)$ for some f, then all the X_i's meet $V(C_f)$, and $\leq |X| + 3$ of them meet $N(C_f)$; and so $n - |X| - 3$ of them are subgraphs of $C_f \backslash N(C_f)$. By (8.4), if $n - |X| - 3 \geq \frac{3}{2}(|X| + 3)$ then for every vertex u of all but $|X| + 3$ of these X_i's, the folio of $N(C_f)$ in $C_f \backslash u$ is the same as it is in C_f, a contradiction. Thus if $n \geq \frac{5}{2}(|X| + 3)$ then no X_i is a subgraph of any $C_f \backslash N(C_f)$. If we replace each C_f by a complete graph with vertex set $N(C_f)$, we obtain a graph which still has a K_n minor. But if (A, B) is the separation of $G \backslash X$ associated with the flat wall, then under this replacement B is left unchanged and A is changed into a planar graph. Since this planar graph has no K_5 minor, we deduce that at least $n - 4$ of the X_i's meet $X \cup (V(B) - V(A))$, and hence at least $n - |X| - 4$ meet $V(B) - V(A)$. But by (8.4) if $n \geq 3p + 1$ then every vertex of all except $2p$ of the X_i's is irrelevant, and by our assumption every irrelevant vertex belongs to $X \cup V(A)$. If $n - |X| - 4 > 2p$ this is a contradiction. We deduce that G has no K_n minor if

$$n \geq \max[\frac{5}{2}(|X| + 3), |X| + 2p + 5, 3p + 1].$$

Hence we may apply (12.5) to G with $H = K_n$. We deduce that G may be decomposed as in (12.5), and this can be done, moreover, to satisfy conditions like (i), (ii), (iii) at the start of section 10. Now the hypergraph J of (12.5) is not triadic, since it has a bounded number of edges $f_1, \ldots, f_b$ of arbitrary size. However, each of them corresponds to a vortex of bounded depth, and we may apply (11.1) to that vortex, and decompose it into a number of pieces each attached onto the rest by only a bounded number of vertices, arranged in a circular fashion. We would like to make all the overlaps between consecutive pieces the same size, and we can do that by expanding the vortex slightly, allowing it to swallow a small portion of the graph drawn on the surface. Moreover, we would like that for each of these pieces, the overlap with the next piece (of size k say) can be joined to the overlap with the previous piece (also of size k) by k vertex-disjoint paths, through the piece; and this too can be arranged. However, the result of all this "squaring off" is that our pieces now contain two or three vertices of U instead of one. (For the detailed argument, see XIV.) Let us now delete from J the difficult edges $f_1, \ldots, f_b$, and add to J new edges with two or three ends, one for each piece of each vortex, with ends the vertices of U in that piece. We obtain a new hypergraph J' in which every edge has two or three ends. Now let us colour each edge of J' by the folio of the piece of G that it represents, in the natural way. Let C be the class of all such hypergraphs on $\sum$, labelled with these colours, in which if we replace each edge by a graph yielding the right folio, the p paths do not exist. Then C is hereditary, and by (7.2), if $J' \notin C$ then $J' \backslash v \notin C$ (for most of our wall W is undisturbed by this whole process, and so its middle vertex v satisfies the requirements of (7.2)). Hence the p paths exist in $G \backslash v$, that is , v is irrelevant. This proves (10.2).(Actually, we cheated: it is necessary to refine (7.2) further, to make sure that the special edges of J' which arise from vortices behave in the right way. But this gives the idea.)   □

14. Extensions

Although the algorithm just described is cumbersome, it is at least reasonably robust in that it can be adapted to solve several similar problems. For instance, more general than the disjoint paths problems is: given G and disjoint subsets $X_1, \ldots, X_k$ of $V(G)$, to decide if there are disjoint connected subgraphs $G_1, \ldots, G_k$ of G with $X_i \subseteq V(G_i)$ ($1 \le i \le k$). For $\sum |X_i|$ bounded, we can solve this by almost exactly the same algorithm. Secondly, the problem, for a fixed graph H, to decide if the input graph G has a minor isomorphic to H, can be solved similarly (indeed, this one is rather easier – we can do it in $O(|V|^3)$). One can also define minors of directed graphs (meaning a directed graph obtained from a subgraph by contracting edges), and our minor-testing algorithm works for directed graphs too.

By a theorem of XV, every hereditary property of graphs can be characterized by excluding finitely many minors. (A property of graphs is *hereditary* if whenever G has the property and H is isomorphic to a minor of G, H also has the property.) Since we can test for the presence of each excluded minor in $O(|V|^3)$ time, it follows that for every hereditary property there is an $O(|V|^3)$ algorithm to test if a graph has the property. For instance, it follows that for every surface $\sum$ there is an $O(|V|^3)$ algorithm to test if $G = (V, E)$ can be drawn on $\sum$; and there is an $O(|V|^3)$ algorithm to test if $G = (V, E)$ can be embedded in 3-space with no circuit knotted. This is, however, a non-constructive proof of the existence of the algorithm, for in general we have no way to find the excluded minors.

References

[1] Arnborg, S., Corneil, D.G., Proskurowski, A. (1987): Complexity of finding embeddings in a k-tree. SIAM J. Algebraic Discrete Methods **8**, 277–284
[2] Arnborg, S., Proskurowski, A. (1984): Linear time algorithms for NP-hard problems on graphs embedded in k-trees. TRITA-NA-8404, Dept. of Numerical Analysis and Computer Science, The Royal Institute of Technology, Sweden
[3] Fortune, S., Hopcroft, J.E., Wyllie, J. (1980): The directed subgraph homeomorphism problem. J. Theor. Comput. Sci. **10**, 111–121
[4] Jung, H.A. (1970): Eine Verallgemeinerung des n-fachen Zusammenhangs für Graphen. Math. Ann. **187**, 95–103
[5] Karp, R.M. (1975): On the computational complexity of combinatorial problems. Networks **5**, 45–68
[6] Larman, D.G., Mani, P. (1970): On the existence of certain configurations within graphs and the 1-skeletons of polytopes. Proc. Lond. Math. Soc., Ser. III. **20**, 144–160
[7] Lynch, J.F. (1975): The equivalence of theorem proving and the interconnection problem. ACM SIGDA Newslett. **5**, 31–65
[8] Mader, W. (1972): Hinreichende Bedingungen für die Existenz von Teilgraphen, die zu einem vollständigen Graphen homöomorph sind. Math. Nachr. **53**, 145–150
[9] Schrijver, A. (1988): Disjoint circuits of prescribed homotopies in a graph on a compact surface. Report OS-R8812, Mathematical Centre, Amsterdam (J. Comb. Theory, Ser. B, to appear)
[10] Seymour, P.D. (1980): Disjoint paths in graphs. Discrete Math. **29**, 293–309
[11] Shiloach, Y. (1980): A polynomial solution to the undirected two paths problem. J. Assoc. Comput. Mach. **27**, 445–456
[12] Thomassen, C. (1980): 2-linked graphs. Eur. J. Comb. **1**, 371–378

The graph minors series, all by N. Robertson and P.D. Seymour:

Graph minors. I: Excluding a forest. J. Comb. Theory, Ser. B *35*, 39–61 (1983)
Graph minors. II. Algorithmic aspects of tree-width. J. Algorithms 7, 309–322 (1986)
Graph minors. III. Planar tree-width. J. Comb. Theory, Ser. B *36*, 49–64 (1984)
Graph minors. IV. Tree-width and well-quasi-ordering. J. Comb. Theory, Ser. B, to appear
Graph minors. V. Excluding a planar graph. J. Comb. Theory, Ser. B *41*, 92–114 (1986)
Graph minors. VI. Disjoint paths across a disc. J. Comb. Theory, Ser. B *41*, 115–138 (1986)
Graph minors. VII. Disjoint paths on a surface. J. Comb. Theory, Ser. B *45*, 212–254 (1988)
Graph minors. VIII. A Kuratowski theorem for general surfaces. J. Comb. Theory, Ser. B, to appear
Graph minors. IX. Disjoint crossed paths. J. Comb. Theory, Ser. B, to appear
Graph minors. X. Obstructions to tree-decomposition. Manuscript (1984, submitted)
Graph minors. XI. Distance on a surface. Manuscript (1985)
Graph minors. XII. Excluding a non-planar graph. Manuscript (1986)
Graph minors. XIII. The disjoint paths problem. Manuscript (1986)
Graph minors. XIV. Taming a vortex. Manuscript (1987)

Graph minors. XV. Wagner's conjecture. Manuscript (1988)
Graph minors. XVI. Well-quasi-ordering on a surface. Manuscript (1989)

Notes added in proof (1989):
(i) We have improved the running time of the algorithm to $O(|V|^3)$.
(ii) A new proof of (4.1), which is simpler and numerically much improved, appears in Robertson, N., Seymour P.D., Thomas, R. (1989), Quickly excluding a planar graph. Manuscript.

Representativity of Surface Embeddings

Neil Robertson and Richard Vitray

The concept of representativity of a surface embedding was developed by one of the authors and Paul Seymour in [13] in connection with a series of papers on the subject of graph minors. In the graph minor context bounded representativity, for classes of embeddings on a fixed surface other than the sphere, is analogous to bounded grid minor size for embeddings on the sphere. Embeddings of unbounded grid minor size on the sphere generate all sphere embeddings by taking minors. Similarly, embeddings of unbounded representativity on another fixed surface generate all embeddings on that surface by taking minors. In this expository article an elementary theory of this embedding invariant will be developed, some recent developments due to Dan Archdeacon, Scott Randby, Carsten Thomassen, Klaus Truemper and the authors will be outlined, and connections with the graph minor project will be discussed.

1. Notation

Surfaces are understood to be connected compact surfaces with empty boundary. These consist of the *orientable surfaces* Σ_g with *orientable genus* or *handle number* g, homeomorphic to the 2-sphere with $g \geq 0$ handles adjoined, and the *unorientable surfaces* $\tilde{\Sigma}_k$ with *unorientable genus* or *cross-cap number* k, homeomorphic to the 2-sphere with $k \geq 1$ cross-caps adjoined. When S is a topological space denote by $C(S)$ its *set of connected components*. Then a *surface embedding* Ψ is defined to be a triple (Σ, U, V) where Σ is a surface, U is a closed subset of Σ, and V is a finite subset of U such that $C(U - V)$ is a finite set of homeomorphic copies of the open unit interval $(0, 1)$. We denote by $V(\Psi)$ the *vertex set* V of Ψ, by $E(\Psi)$ the *edge-set* $C(U - V)$ of Ψ, and by $F(\Psi)$ the *face set* $C(\Sigma - U)$ of Ψ. Thus the *surface* $\Sigma(\Psi) = \Sigma$ of Ψ is partitioned into *vertices*, *edges* and *faces*, the elements of these respective sets. The *graph* $G(\Psi)$ of Ψ is the graph with the same vertex-set and edge-set as Ψ, in which each edge is incident with the vertices (one or two) in its closure. If $V(\Psi) = \emptyset$ then $E(\Psi) = \emptyset$ and $F(\Psi) = \{\Sigma\}$. Otherwise $V(\Psi) \neq \emptyset$ and $F(\Psi)$ consists of a finite positive number of faces, each homeomorphic to a surface with a finite positive number of points removed. We say two surface embeddings $\Psi_1 = (\Sigma_1, U_1, V_1)$, $\Psi_2 = (\Sigma_2, U_2, V_2)$ are *homeomorphic*, written $\Psi_1 \approx \Psi_2$, when there is a homeomorphism of their surfaces Σ_1, Σ_2 restricting to a homeomorphism of U_1, U_2 and to a bijection between V_1, V_2.

When all the faces of Ψ are homeomorphic to open 2-cells (i.e. open 2-disks) the graph $G(\Psi)$ is nonnull and connected. Such embeddings are called *proper* and are the most common type of embedding considered in topological graph theory. For proper embeddings the *Euler relation* $\chi(\Psi) = V - E + F$ (where we abbreviate the cardinalities of $V(\Psi)$, $E(\Psi)$, $F(\Psi)$ by V, E, F, respectively.) determines the surface $\Sigma(\Psi)$ up to orientability. Then Σ_g satisfies $V - E + F = 2 - 2g$ and $\tilde{\Sigma}_k$ satisfies $V - E + F = 2 - k$. Moreover, $\chi(\Psi)$ is an integer ≤ 2, with $\chi(\Psi) = 2$ only when $\Sigma(\Psi)$ is the 2-sphere. The boundary bd(f) of any $f \in F(\Psi)$ in $\Sigma(\Psi)$ is a closed subset of U containing any edge of Ψ it meets, and thus determines a subgraph of $G(\Psi)$. Following the circular order of the open disk f this subgraph is traced out by a closed walk (possibly a single vertex) in $G(\Psi)$, unique up to rotations and reversal of direction, called the *facial walk* of f (for general faces, each deleted point of the face has such a walk associated). Note that each edge of Ψ appears exactly twice in the facial walks of Ψ and thus has two "sides" along one or two of the faces of Ψ_1 and each vertex of Ψ appears in these facial walks with multiplicity equal to its valency in $G(\Psi)$.

We define an embedding Ψ^* with $\Sigma(\Psi^*) = \Sigma(\Psi)$ and the roles of vertices and faces interchanged as follows. Choose, for each $f \in F(\Psi)$ and $A \in E(\Psi)$ a point $v_f \in f$ and $p(A) \in A$. With each $f \in F(\Psi)$ there is a facial walk having edge-sequence $A_1, A_2, \ldots, A_k$ for some integer $k \geq 0$. When $k = 0$ the open 2-cell face has a single vertex on its boundary and so $\Sigma(\Psi)$ is the 2-sphere and both $G(\Psi)$ and $G(\Psi^*)$ are vertex-graphs. When $k \geq 1$ choose paths meeting only at equal endpoints, joining v_f to $p(A_1), p(A_2), \cdots, p(A_k)$, respectively, in f. Then for each $A \in E(\Psi)$ form a dual edge $A^* \in E(\Psi^*)$ by combining the two paths associated with A in the facial walks and deleting the facial endpoints. Then $V(\Psi^*) = \{v_f : f \in F(\Psi)\}$ and $E(\Psi^*) = \{A^* : A \in E(\Psi)\}$ determines the embedding Ψ^*. Supposing Ψ is proper, one readily sees that Ψ^* is determined up to homeomorphism, is also proper and that Ψ^{**} is homeomorphic to Ψ. We call Ψ and Ψ^* *primal* and *dual* embeddings, respectively, while $G(\Psi)$ and $G(\Psi^*)$ are the *primal* and *dual* graphs of Ψ.

Circuits in a surface Σ are homeomorphic images of the unit circle. There are a limited number of ways to embed a circuit C into a surface Σ. Define $\Sigma \setminus C$ to be the (not necessarily connected) surface-with-boundary formed by cutting Σ along C. Then $\Sigma \setminus C$ has one or two boundary components, which are themselves circuits. In the first case Σ is formed by antipodal identification *(pasting)* of the boundary circuit with itself, and C is called *one-sided* (it goes through a cross-cap, so to speak). In the second case Σ is formed by identifying *(pasting)* the two boundary circuits, and C is called *two sided*. Either $\Sigma \setminus C$ has two connected components and C is called *separating* or $\Sigma \setminus C$ has one connected component and C is called *nonseparating*. If C is separating and one component is homeomorphic to an open 2-cell then C is *trivial* (or *null-homotopic*). All circuits which are not trivial are said to be *essential*. Separating circuits are always two sided as are nonseparating circuits on an orientable surface. However, on an unorientable surface a nonseparating circuit C may be one-sided or two-sided, and $\Sigma \setminus C$ may be orientable or unorientable. We say a nonseparating circuit C is *orienting* or *nonorienting* when $\Sigma \setminus C$ is orientable or unorientable, respectively. When $\chi(\Sigma)$ is even there are no orienting one-sided circuits and when $\chi(\Sigma)$ is odd there are no orienting two-sided circuits. Additionally, there are only trivial circuits when $\chi(\Sigma) = 2$, no nonorienting circuits when $\chi(\Sigma) = 1$, and no nonorienting two-sided circuits when $\chi(\Sigma) = 0$. For orientable surfaces Σ_g there are $1 + \lfloor g/2 \rfloor$ kinds of separating circuits, with $0, 1, 2, \ldots, \lfloor g/2 \rfloor$ handles on the short side. For unorientable surfaces $\Sigma = \tilde{\Sigma}_k$ there are two families of separating circuits, called *orienting* if one component of $\Sigma \setminus C$ is orientable and *nonorienting* when both components of $\Sigma \setminus C$ are unorientable. There are $\lfloor k/2 \rfloor$ kinds of nonorienting separating circuits, with $1, 2, \ldots, \lfloor k/2 \rfloor$ cross-caps on the short side, and $\lceil k/2 \rceil$ orienting separating circuits, with $0, 1, 2, \ldots, \lfloor (k-1)/2 \rfloor$ handles on the orientable side. Thus there are k different kinds of separating circuits in $\tilde{\Sigma}_k$.

In what follows Ψ will always stand for a graph embedding with $\Sigma(\Psi)$ not the sphere. Then the *representativity* of Ψ is defined to be

$$\rho(\Psi) = MIN\{|C \cap G(\Psi)| : C \text{ is an essential circuit of } \Sigma(\Psi)\}.$$

By elementary topology it is enough to use in the definition of $\rho(\Psi)$ essential circuits C which pass through only vertices and faces of Ψ and which use no vertex or face more than once. When the faces are open 2-cells, $\rho(\Psi) \geq 1$, the dual embedding Ψ^* is defined and satisfies $\rho(\Psi) = \rho(\Psi^*) \geq 1$. We remark that references [11] and [17] give other names for $\rho(\Psi)$.

Representativity of an embedding is a measure of how "densely" the graph is embedded onto the surface. A major effect of high representativity is to make the embedding highly "locally planar". In fact, points p on $\Sigma(\Psi)$, reachable from a fixed point p_0 by simple paths which meet $G(\Psi)$ in $< \rho(\Psi)/2$ points, are contained in an open disk on $\Sigma(\Psi)$. Thus the "locally Euclidean" property of the surface is mirrored by the "locally planar" property of the embedded graph. Representativity can also be regarded as a type of connectivity where the "separations" are given by essential circuits in $\Sigma(\Psi)$ rather than by complementary subgraphs. It should be noted however that high representativity does not imply much if anything about the actual degree of connectivity of the graph.

2. Problems

To motivate representativity some basic problems concerning it can be listed.

1. The first problem is to develop a elementary theory of representativity guided by the standard theorems about planar embeddings.
2. A classical problem in graph theory is to show that any 2-connected graph G is isomorphic to $G(\Psi)$ for some embedding Ψ with $\rho(\Psi) \geq 2$. This would imply the *circuit-double-cover conjecture,* that a graph with edge-connectivity ≥ 2 has a family of simple circuits making a 2-fold covering of its edges.
3. What are the best possible lower bounds on $\rho(\Psi)$ required to ensure that an embedding Ψ is of minimum (orientable or unorientable) genus or is uniquely embedded on $\Sigma(\Psi)$? Here *uniquely embedded* means (when $G(\Psi)$ is 3-connected) all homeomorphisms of $G(\Psi)$ extend to homeomorphisms of $\Sigma(\Psi)$ and that no embedding Ψ' exists with $G(\Psi') \approx G(\Psi)$, $\Psi' \not\approx \Psi$ and $\chi(\Psi') \geq \chi(\Psi)$.

4. What can be said about embedding flexibility of graphs G, embeddable on a fixed surface Σ, along the lines of Whitney's theorems from [24] and [25] for planar graphs?

5. It has been conjectured in [1] for any embedding Ψ with $\rho(\Psi) \geq 4$ that $G(\Psi)$ has chromatic number ≤ 5. Similarly, if $\rho(\Psi)$ is sufficiently high $G(\Psi)$ may be conjectured to be Hamiltonian. Under such $\rho(\Psi)$ constraints there may be efficient polynomial-time algorithms to find 5-colorings, 4-colorings, Hamiltonian circuits, etc. As the condition $\rho(\Psi) = \rho(\Psi^*)$ seems to be too strong a requirement for some of these problems it may be more natural to use "edge-width" constraints, as done in [1], where *edge-width* is the minimum girth of essential circuits in $G(\Psi)$.

6. It would be of interest to make more full use of the fundamental group $\pi_1(\Psi) = \pi_1(\Sigma(\Psi))$ by defining

$$\rho(\Psi, C) = MIN\{|C' \cap G(\Psi)| : C' \text{ is a circuit in } \Sigma(\Psi) \text{ homotopic to } C\}$$

and considering optimal families of generators for $\pi_1(\Psi)$ relative to $\rho(\Psi, C)$.

7. There are also the classical algorithmic problems regarding testing minor inclusion or topological inclusion for embeddings, and structural characterizations (refinements of the constraining $\rho(\Psi)$) about excluded minors.

At the time of writing considerable progress has been made by the authors and several others on three more specific versions of the above problems.

A. We conjecture that if $\Sigma(\Psi)$ is the sphere with $k \geq 1$ cross-caps, $G(\Psi)$ is 3-connected, and $\rho(\Psi) \geq k + 3$, then Ψ is unique in the sense that if Ψ' is an embedding with $G(\Psi') \cong G(\Psi)$ and $\Psi' \not\approx \Psi$ then $\chi(\Psi') < \chi(\Psi)$ and $\rho(\Psi') < k + 3$. The orientable case of this with $\rho(\Psi) \geq 2g + 3$, where $\Sigma(\Psi) = \Sigma_g$, is given as Proposition 13.1 here. The condition that $G(\Psi)$ be 3-connected does not restrict the generality of this result in an important way, as Proposition 7.3 makes clear.

B. It seems possible to give a characterization involving generators for $\pi_1(\Psi)$ of the orientable genus of graphs embedded on a fixed unorientable surface, and a polynomial-time algorithm for finding this genus. We hope to show for any minor closed class of graphs that genus testing is either NP-complete or it can be proved to be a polynomial-time problem. References include [6], [15] and an unpublished result of P.D. Seymour concerning graphs of bounded tree-width.

C. There is a structure theory for embeddings Ψ of planar graphs $G(\Psi)$ on nonplanar surfaces $\Sigma(\Psi)$ generalizing the $\rho(\Psi) \leq 2$ result of Proposition 10.1 to generators for $\pi_1(\Sigma)$. The projective plane case and the torus case have been settled in joint work [9] with Bojan Mohar, and it is conjectured that the general case resolves into a "local cross-cap and handle structure" conforming to the projective plane and torus cases. The projective plane case is discussed in Section 5 of this paper. On a more general surface a "local cross-cap structure" would be an expression as in Figure 5.4, but with handles and cross-caps allowed in the triangular patches.

3. Recognizing Embeddings with $\rho(\Psi) \geq 1$, $\rho(\Psi) \geq 2$, and $\rho(\Psi) \geq 3$

When the faces of Ψ are open 2-cells the embedding is called an *open 2-cell embedding* or a *proper embedding*. When the faces of Ψ are open 2-cells bounded by simple circuits in $G(\Psi)$ the embedding is called a *closed 2-cell embedding* or a *simple embedding*. The bounding circuits of the 2-cells in a closed 2-cell embedding are called *facial circuits*. When Ψ is a closed 2-cell embedding and the subgraph of $G(\Psi)$ bounding the faces incident with any vertex is a wheel with ≥ 3 spokes and a possibly subdivided rim, the embedding is called a *wheel-neighborhood embedding*. These types of embeddings are easily recognized, and correspond to 1-connected, 2-connected, and 3-connected graph embeddings, respectively, on the plane. The following Propositions 3.1, 3.2, 3.3 are due to Nagami [11] and Vitray [22].

Proposition 3.1. *An embedding Ψ is an open 2-cell embedding if and only if $\rho(\Psi) \geq 1$ and $G(\Psi)$ is connected.*

Proof. We adopt the convention that the null graph Ω is neither connected nor disconnected. When $G(\Psi) = \Omega$ the graph is not connected and $\Sigma(\Psi)$, the only face, is not an open 2-cell. When $G(\Psi) \neq \Omega$ each face is homeomorphic to a surface with a finite positive number of points removed. Then $\rho(\Psi) = 0$ if and only if some face f contains an essential circuit C of $\Sigma(\Psi)$. An open 2-cell embedding Ψ implies $\rho(\Psi) \geq 1$, and clearly then $G(\Psi)$ is connected, for otherwise a face of one connected component must contain some other component. If Ψ is not an open 2-cell embedding then some face f contains an essential circuit C of f. If C is essential in $\Sigma(\Psi)$ then $\rho(\Psi) = 0$. If C is trivial in $\Sigma(\Psi)$ it bounds a disk D in $\Sigma(\Psi)$ but not in f, whence D contains a component of $G(\Psi)$.

Proposition 3.2. *An embedding Ψ is a closed 2-cell embedding if and only if $\rho(\Psi) \geq 2$ and $G(\Psi)$ is nonseparable.*

Proof. In view of Proposition 3.1 we may assume Ψ is an open 2-cell embedding $\rho(\Psi) \geq 1$ and $G(\Psi)$ is connected. The boundary of an open 2-cell f is a subgraph of $G(\Psi)$; following the natural circular order of the 2-cell a closed walk is traced out in this subgraph (where edges may be repeated at most twice). Then Ψ is a closed 2-cell embedding iff there are no repeated edges or vertices in these closed walks. Suppose there is a repeated vertex x on the facial walk of a face f. Then there is an open 1-cell in f whose endpoints correspond to x and which meets x between the two pairs of consecutive edges of the closed walk meeting x in these two occurences. The closure of this 1-cell is then a circuit, C, meeting $G(\Psi)$ in x and otherwise lying in f. If C is essential then $\rho(\Psi) = 1$, and if C is trivial it bounds a disc in $\Sigma(\Psi)$ containing some but not all of the edges in the boundary of f. The repeated vertex is then a cutvertex of $G(\Psi)$. Finally, if there are no repeated vertices but there is a repeated edge, then $G(\Psi)$ consists of this edge only, which implies $\Sigma(\Psi)$ is the sphere, contrary to assumption. Conversely, it is easy to see that if $\rho(\Psi) = 1$ then there is a repeated vertex on some facial walk, and if $G(\Psi)$ has a cutvertex, the cutvertex is repeated on some facial walk.

Proposition 3.3. *An embedding* Ψ *is a wheel-neighborhood embedding if and only if* $\rho(\Psi) \geq 3$ *and* $G(\Psi)$ *is simple and 3-connected.*

Proof. We may assume, by Proposition 3.2, that Ψ is a closed 2-cell embedding, that $\rho(\Psi) \geq 2$, and $G(\Psi)$ is nonseparable. When Ψ is a wheel-neighborhood embedding, any essential circuit C with $\rho(\Psi) = |G(\Psi) \cap C|$ which passes through vertices and faces of Ψ goes through the center of a wheel-neighborhood and two points on the rim, and so $\rho(\Psi) \geq 3$. Also, after deleting any vertex of $G(\Psi)$ the resulting embedding Ψ' is a closed 2-cell embedding and so $G(\Psi')$ is 2-connected by Proposition 3.2. It follows that $G(\Psi)$ is 3-connected, and by definition of wheel-neighborhood, that $G(\Psi)$ is simple. Conversely, if $\rho(\Psi) \geq 3$ and $G(\Psi)$ is simple and 3-connected, then removing any vertex of $G(\Psi)$ gives an embedding Ψ' that satisfies $\rho(\Psi') \geq 2$ and $G(\Psi')$ is 2-connected. If follows that Ψ' is a closed 2-cell embedding. The face which contains the deleted vertex is thus bounded by a circuit and, adjoining the vertex and its incident edges to the circuit gives the required wheel-neighborhood.

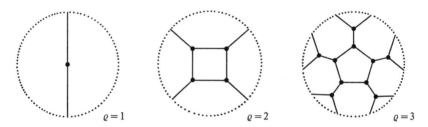

Fig. 3.1 Examples of embedding minimal graphs on the projective plane with representativity 1, 2, and 3

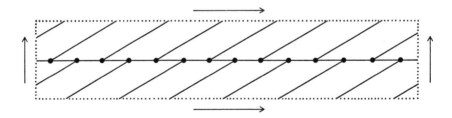

Fig. 3.2 The Heawood graph on the torus with $\rho(\Psi) = 3$

As an illustration of the use of Proposition 3.3 we remark that any embedding Ψ with $G(\Psi)$ a complete graph has $\rho(\Psi) \leq 3$, with $\rho(\Psi) = 3$ if and only if Ψ is a triangulation. Thus if $k \equiv 2(mod\ 3)$ the complete k-graph K_k has $\frac{1}{2}k(k-1) = \frac{1}{2}(3t+2)(3t+1) \equiv 1(mod\ 9)$ edges and hence cannot form a triangulation. In particular $\rho(\Psi) \leq 2$ for any embedding with $G(\Psi) = K_5$.

Figure 3.2 gives an embedding of the Heawood graph in the torus, dual to K_7 in the torus, which has representativity 3.

4. Minor Inclusion for Embeddings

Two embeddings Ψ_1, Ψ_2 are *homeomorphic*, denoted $\Psi_1 \approx \Psi_2$, when there exists a homeomorphism $h : \Sigma(\Psi_1) \to \Sigma(\Psi_2)$ which maps points of vertices, edges and faces, respectively, of Ψ_1 to points of vertices, edges and faces of Ψ_2. Let Ψ be any embedding. Define the automorphism group of Ψ to be the group of homeomorphisms of Ψ onto itself modulo its normal subgroup of homeomorphisms fixing the vertices, edges and faces setwise. Denote this by $A(\Psi)$ and write $A(G)$ and $A_\Psi(G)$ for the (similarly defined) automorphism group of $G(\Psi)$ and its subgroup induced by elements of $A(\Psi)$, respectively.

A fixed graph G may have several embeddings into a fixed surface Σ which are homeomorphic but can be regarded as combinatorially distinct. To formalize the notion of an *embedding of G into Σ*, denote by (Ψ, α) an embedding Ψ such that $\Sigma(\Psi) \approx \Sigma$ and an isomorphism $\alpha : G \to G(\Psi)$. Then two such embeddings (Ψ_1, α_1), (Ψ_2, α_2) are *combinatorially equivalent* when a homeomorphism $h : \Psi_1 \to \Psi_2$ exists such that $\alpha_2 = h\alpha_1$. It follows that for each embedding Ψ with $G(\Psi) \cong G$ and $\Sigma(\Psi) \approx \Sigma$ there are exactly $|A(G)/A\Psi(G)|$ combinatorially distinct embeddings of G into Σ homeomorphic to Ψ. An important theorem of H. Whitney [24] states that embeddings of 3-connected planar graphs into the 2-sphere are combinatorially unique.

To illustrate that Whitney's theorem fails on higher surfaces consider the homeomorphically unique dual pair of the Petersen graph P and the complete graph K_6 on the projective plane $\tilde{\Sigma}_1$. When $G(\Psi) \cong P$ and $\Sigma(\Psi) \approx \tilde{\Sigma}_1$, we have $A(G) \cong S_5$ and $A_\Psi(G) \cong A_5$, where S_5 and A_5 are the symmetric and alternating groups on 5 letters, respectively. The index $[S_5 : A_5] = [A(G) : A_\Psi(G)] = 2$ indicates P has two embeddings in $\tilde{\Sigma}_1$ which are homeomorphic but combinatorially distinct. Note that $A(\Psi) \cong A(\Psi^*)$ and $A_\Psi(G) \cong A_{\Psi^*}(G)$. Thus a similar calculation gives $A(K_6) \cong S_6$, $A_{\Psi^*}(G) \cong A_5$ and $[A(K_6) : A_{\Psi^*}(G)] = 12$ combinatorially distinct embeddings, although again the embedding is unique up to homeomorphism.

We may also wish to distinguish in other ways between homeomorphic embeddings $\Psi_1, \Psi_2 \in \mathscr{E}(\Sigma)$, for a fixed surface Σ, for example those differing by an orientation on Σ when it is orientable, or between the embeddings of two circuits generating a torus. We say Ψ_1 is *homotopic* to Ψ_2, denoted $\Psi_1 \sim \Psi_2$, when there exists a continuous function $F : [0,1] \times \Sigma \to \Sigma$ such that $F(t, x) : \Sigma \to \Sigma$ is a homeomorphism for all $t \in [0,1]$ and $F(0, x)$ is the identity on Σ and $F(1, x)$ is a homeomorphism from Ψ_1 to Ψ_2.

Minors of a graph are formed by a succession of elementary operations, namely *single-edge deletion*, *single-edge contraction*, and *removal of a single isolated vertex*. We introduce these operations for surface embeddings as follows. Let $a \in E(\Psi)$ and denote by Ψ_a' the embedding after a is *deleted* and by Ψ_a'' the embedding after a is *contracted*. When a is deleted its point-set is simply deleted from the point-set of $G(\Psi)$ but not from $\Sigma(\Psi)$. When the edge is a loop Ψ_a''

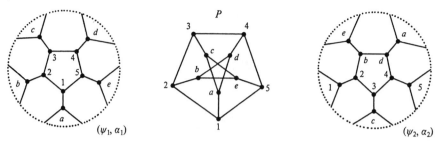

Fig. 4.1 Combinatorially distinct embeddings of P on $\tilde{\Sigma}_1$

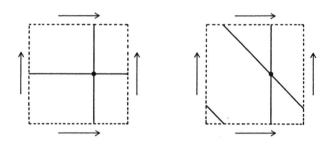

Fig. 4.2 Homotopically different homeomorphic embeddings

becomes an embedding on a pseudo-surface as the image point of a is a pinched point. Thus *we never contract loops*, although they may be deleted, an operation which has the same effect on the graph involved as contraction. Finally, denote by $\Psi - x$ the embedding obtained by *removing an isolated vertex* $x \in V(\Psi)$ from Ψ, or more precisely, from $G(\Psi)$.

If we never contract loops, any embedding Ψ' obtained from Ψ by a succession of the above three operations is called a *minor* of Ψ. Minors in which no contraction is performed are called *restrictions* of Ψ, and in which no deletion is performed are called *contractions* of Ψ. Suppose $\Psi_1 = (\Sigma_1, U_1, V_1)$ and $\Psi_2 = (\Sigma_2, U_2, V_2)$ are embeddings. When $\Sigma_1 = \Sigma_2$, $U_1 \subseteq U_2$ and $V_1 \subseteq V_2$ we call Ψ_1 a *topological minor* of Ψ_2. Then Ψ_1 is obtained from a restriction of Ψ_2 by *suppressing* some vertices of valency 2.

When we restrict ourselves to open 2-cell embeddings we must add the condition that *we cannot delete edges incident with only one face*. In this case the dual embedding is always defined, and there is a strict duality between the operations of deletion and contraction, i.e. $(\Psi_a')^* = (\Psi^*)_a''$ and $(\Psi_a'')^* = (\Psi^*)_a'$. The dual of a restriction is then a contraction and vice-versa. Note that we will abuse notation and use the same symbols to denote vertices, edges and faces in Ψ and their corresponding faces and vertices in Ψ^*.

There are four closely related special minor operations which can be readily defined from the basic operations. Suppose that Ψ is an embedding (perhaps planar) and $x \in V(\Psi), f \in F(\Psi)$. Let $\Psi \setminus x$ denote the result of deleting the edges incident with x and then removing x. We say that $\Psi \setminus x$ is obtained from

Ψ by *deleting* the vertex x. When $\rho(\Psi) \geq 2$ we can write $\Psi/f = (\Psi^* \setminus f)^*$ and say that Ψ/f is obtained from Ψ by *contracting* the face f. Extending this type of notation when $\rho(\Psi) \geq 3$ we write $\Psi \backslash\backslash x = (\Psi \setminus x)/f_x$ where f_x is the face of $\Psi \setminus x$ containing the vertex x, and $\Psi//f = (\Psi/f) \setminus x_f$, where x_f is the vertex of Ψ/f corresponding to the face f. We say $\Psi \backslash\backslash x$ is obtained from Ψ by *removing* the vertex x and $\Psi//f$ is obtained from Ψ by *removing* the face f. We note the obvious duality relationships, and the obvious inequalities concerning the representativity of these embeddings.

Proposition 4.1. *Let Ψ be an embedding and $x \in V(\Psi), f \in F(\Psi)$. Then assuming $\rho(\Psi) \geq 2$ for (a) and (b), and $\rho(\Psi) \geq 3$ for (c) and (d):*

(a) $\Psi \setminus x = (\Psi^*/x)^*$,

(b) $\Psi/f = (\Psi^* \setminus f)^*$,

(c) $\Psi \backslash\backslash x = (\Psi^*//x)^*$, *and*

(d) $\Psi//f = (\Psi^* \backslash\backslash f)^*$.

Proposition 4.2. *Let Ψ be an embedding and $x \in V(\Psi), f \in F(\Psi)$. Then assuming $\rho(\Psi) \geq 2$ for (b) and (d), and $\rho(\Psi) \geq 3$ for (c):*

(a) $\rho(\Psi) - 1 \leq \rho(\Psi \setminus x) \leq \rho(\Psi)$,

(b) $\rho(\Psi) - 1 \leq \rho(\Psi/f) \leq \rho(\Psi)$,

(c) $\rho(\Psi) - 2 \leq \rho(\Psi \backslash\backslash x) \leq \rho(\Psi)$, *and*

(d) $\rho(\Psi) - 2 \leq \rho(\Psi//f) \leq \rho(\Psi)$.

Homeomorphic and homotopic surface embeddings can be used to extend the notion of inclusion of embeddings, just as isomorphism extends the notion of subgraph inclusion. Define *minor inclusion* $\Psi_1 \leq_m \Psi_2$ between two embeddings to mean Ψ_1 is homeomorphic to a minor of Ψ_2. A class $\mathscr{L}$ of embeddings on a fixed surface Σ which is closed (downward) under minor inclusion is called a *minor-closed class* in $\mathscr{E}(\Sigma)$. The set of embeddings not in $\mathscr{L}$ which are *minor-minimal* (i.e. minimal under the minor relation $\leq_m$) form its *obstacle set* $\Omega(\mathscr{L})$. As an example of this, all embeddings on a surface Σ with $\rho(\Psi) < k$ form a minor-closed class $\mathscr{L}_k(\Sigma)$ with obstacle set $\Omega_k(\Sigma)$ composed of the minor-minimal embeddings with $\rho(\Psi) \geq k$. It is a consequence of [14] that obstacle sets for minor closed classes of embeddings are, up to homeomorphism, finite. Example obstacle sets are given in Section 5 for $\rho(\Psi) = 1, 2, 3$ when $\Sigma(\Psi)$ is the projective plane. We also define another form of minor inclusion in $\mathscr{E}(\Sigma)$, where $\Psi_1 \leq_m^s \Psi_2$ means that Ψ_1 is homotopic to a minor of Ψ_2. This will be called *strong minor inclusion*. We note that by using circuits on a standard torus winding around k times, for k any positive integer, an infinite antichain of embeddings under $\leq_m^s$ is obtained. Note also that the various operations and inclusions discussed in this Section do not change the surface involved.

As a general example, in Figure 4.4 we offer $L(K_{3,3})$, the "line-graph" of $K_{3,3}$, which embeds on the projective plane, torus and Klein bottle with different representativities. In Figure 4.5 a circuit C on the torus Σ_1, for the $\rho(\Psi) = 3$ embedding of $L(K_{3,3})$, is given in several homotopically distinct cases. In the first

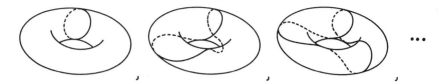

Fig. 4.3 An infinite antichain in $(\mathscr{E}(\Sigma), \leq_m^s)$

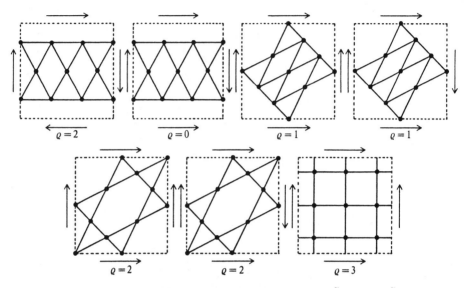

Fig. 4.4 Several different embeddings of $L(K_{3;3})$ in $\tilde{\Sigma}_1, \Sigma_1$ and $\tilde{\Sigma}_2$

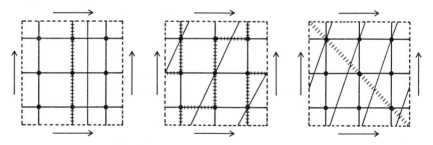

Fig. 4.5 Homotopic embedding of a circuit in Σ_1

two diagrams there is a circuit homotopic to C contained topologically in $G(\Psi)$. Note that in the third diagram the diagonal curve proves there is no such circuit in $G(\Psi)$.

Any curve homotopic to C must meet the diagonal curve in 4 points, but the diagonal curve meets the graph of the embedding in 3 points.

5. Results from Vitray's Thesis on Projective Plane Embeddings

Figure 5.1 displays four typical projective plane embeddings Ψ with circuits in $\Sigma(\Psi)$ meeting their graphs in $0, 1, 2, 3$ points respectively. The obstacle sets for projective plane embeddings Ψ with $\rho(\Psi) \leq 1, 2, 3$ are determined in the thesis [22] and are displayed in Figures 5.2 and 5.3.

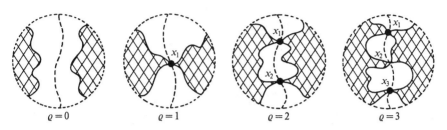

Fig. 5.1 Embeddings Ψ with $\rho(\Psi) = 0, 1, 2, 3$ and $\Sigma(\Psi) = \tilde{\Sigma}_1$

Also established in [22] are the following facts about embeddings on $\tilde{\Sigma}_1$.

1. When $\rho(\Psi) = 0$ or $\rho(\Psi) = 1$ the graph $G(\Psi)$ is planar. Note if $\rho(\Psi) = 1$ and $x \in V(\Psi)$ is the only point of $G(\Psi)$ met by an essential circuit C, then C divides the edges incident with x into two parts. *Twisting* the edges of one part by reversing their order in the embedding transforms the embedding to Ψ' with $\rho(\Psi') = 0$, essentially a planar embedding, i.e. one with the facial circuits of a planar embedding.

2. When $\rho(\Psi) = 2$ the graph $G(\Psi)$ may be planar or nonplanar. Those which are planar fall into two families, as indicated in the Figure 5.4. The shaded triangles represent arbitrary planar subgraphs with three vertices of attachment. The *vertices of attachment* for a subgraph J of G are the vertices of J incident with edges of G not in J. The dual graphs, which are usually non-planar, are indicated alongside. Note that the embedding Ψ_A of this diagram is assumed to have an odd number of shaded triangles. In the case where this number is one we obtain a $\rho(\Psi_A) = 1$ embedding which admits a twist to a $\rho(\Psi'_A) = 0$ embedding. The type Ψ_A embeddings are generalizations of a wheel embedding, while the type Ψ_B embeddings are generalized octahedral embeddings.

It is possible for $G(\Psi)$ and $G(\Psi^*)$ to be planar when $\Sigma(\Psi) = \tilde{\Sigma}_1$ and $\rho(\Psi) = 2$ as shown by Ψ_2 and Ψ_2^* in Figure 5.2. Up to homeomorphism there are three embeddings of the Kuratowski graphs in the projective plane, as given in Figure 5.5 by Γ_1, Γ_2 and Γ_3. We see that Γ_1 and Γ_2 fit into type Ψ_A^* and Ψ_B^* of Figure 5.4, but that Γ_3 has a nonplanar dual $G(\Gamma_3^*)$, which is isomorphic to $K_{3,3}$ plus an edge joining adjacent vertices.

We may note that in a type Ψ_A embedding with $2k + 1$ triangles, one triangle can be re-embedded "on the other side" of the vertical line, and the new

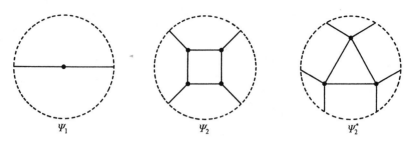

Fig. 5.2 The obstacle sets $\Omega_1(\tilde{\Sigma}_1) = \{\Psi_1\}, \Omega_2(\tilde{\Sigma}_1) = \{\Psi_2, \Psi_2^*\}$

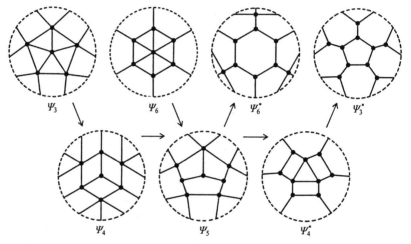

Fig. 5.3 The obstacle set $\Omega_3(\tilde{\Sigma}_1) = \{\Psi_3, \Psi_4, \Psi_5, \Psi_6, \Psi_3^*, \Psi_4^*, \Psi_5^*\}$

embedding is still of type Ψ_A, with two less triangles (combining the three side-by-side triangles into one). As long as $k \geq 2$ the new embedding has $\rho(\Psi) = 2$. When $k = 1$ we obtain the previously mentioned single triangle with $\rho(\Psi_A) = 1$, which can be "twisted" to Ψ_A' with $\rho(\Psi_A') = 0$ or an essentially planar embedding. The Ψ_B embedding can also be "twisted" into Ψ_B' and Ψ_B'' with $\rho(\Psi_B') = 1$ and $\rho(\Psi_B'') = 0$ (here you twist the top and bottom triangular patches into the central face to obtain Ψ_B').

3. When $\rho(\Psi) = 3$ one of the seven embeddings $\Psi_3, \Psi_4, \Psi_5, \Psi_6, \Psi_3^*, \Psi_4^*, \Psi_6^*$ is homeomorphic to a minor of Ψ. An important fact about this class of embeddings is that it is the full class generated on $\tilde{\Sigma}_1$ from Ψ_3 by successive (Y, Δ) or (Δ, Y)—operations (defined in Section 14). As a consequence these embeddings are nonplanar, and remain so when any one vertex is removed. It is also shown there are altogether 15 topologically minimal embeddings with $\rho(\Psi) \geq 3$. The 12 combinatorially distinct embeddings of K_6 in $\tilde{\Sigma}_1$ are the

most possible among the topologically minimal embeddings, and hence for all $\rho(\Psi) \geq 3$ embeddings in which $G(\Psi)$ is 3-connected.

4. When $\rho(\Psi) \geq 4$ and $G(\Psi)$ is 3-connected these embeddings are combinatorially unique on the projective plane.

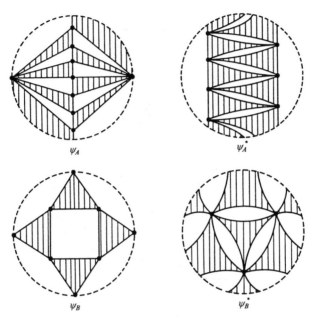

ψ_A $\psi_A^{\bullet}$

ψ_B $\psi_B^{\bullet}$

Fig. 5.4 Embeddings of planar graphs on $\tilde{\Sigma}_1$ and their duals

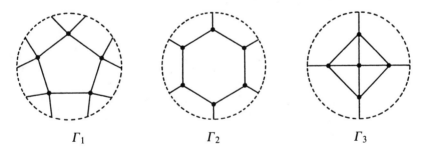

Γ_1 Γ_2 Γ_3

Fig. 5.5 All embeddings of Kuratowski graphs in $\tilde{\Sigma}_1$

6. Reminder About Low Graph Connectivity

Representativity may be thought of as an analogue of graph connectivity appropriate to surface embeddings which, as propositions 3.1, 3.2 and 3.3 make clear, does not imply the usual connectivity. We will discuss some basic facts about low connectivity in order to elaborate our earlier propositions. The three propositions we are leading up to are useful in getting around degeneracies of a trivial type which occur when the graph connectivity is less than the representativity. We also define bridges of a subgraph in order to characterize peripheral circuits in ≥ 3 representative embeddings.

When we can write $G = H \cup K$, for subgraphs H, K of G with no common edges, the pair $\{H, K\}$ is called a *separation* of G. When $|V(H \cap K)| = j$ and $|E(H)| \geq j \leq |E(K)|$, the separation is called a *j—separation* of G. Then G is *connected* if non-null subgraphs H, K do not exist such that $\{H, K\}$ is a 0-separation. Also G is *nonseparable* iff it is connected and no 1-separation $\{H, K\}$ exists. Finally, G is *3-connected* if it is nonseparable and no 2-separations $\{H, K\}$ exists. Thus 3-connected graphs are either rather small (the six nonseparable graphs on ≤ 3 edges) or have no loops or multiple edges and no vertices of valency ≤ 2. We call the ordinary *connected components* (maximal connected subgraphs) of a graph its *1-components* and remark that they are pairwise disjoint, non-null, their union is the whole graph, and each is connected. We call the ordinary *blocks* (maximal nonseparable subgraphs) of a graph its *2-components* and remark they meet pairwise in at most one vertex, they are non-null, their union is the whole graph and each is nonseparable. The 3-components of a graph are a little more complicated, but they can be expressed fairly easily. Their theory is more fully developed in terms of cleavage units in [21] and with slightly different notation in Chapter IV of [22].

We can produce the cleavage units of a nonseparable graph G as follows. If there is a "good" 2-separation $\{H, K\}$ we split G into two graphs H^+, K^+ by adjoining a "new edge" to the vertices of $H \cap K$ in both graphs H, K. This procedure is iterated until no good 2-separation is left among the graphs produced. It can be shown that the graphs obtained, called the *cleavage units* of G, are either circuits, *linkages* (connected loopless graphs on 2 vertices) or 3-connected graphs, and they are uniquely determined apart from the identity of the adjoined edges. *Bad* 2-separations are those admitting decompositions of the type shown in Figure 6.1 below. All other 2-separations are called *good*. Note for bad 2-separations that either H and K are both split as in the first drawing or they both have 1-separations as in the second drawing. Note also that for planar graphs these two drawings have a dual form, and so the dual graph also admits a decomposition where the cleavage units correspond to those of the primal graph under planar duality. Furthermore, if G admits no good 2-separations then, except for circuits and linkages, G admits no bad 2-separations. We call the cleavage units the *3-components* of G.

Another aspect of graph connectivity is embodied in the *bridges* $B_1, B_2,$ $..., B_k$ of a subgraph J of a graph G. They consist either of the *degenerate* bridges which are formed by single edges not in J with their endvertices all in J or the *nondegenerate* bridges which are formed by the connected components of $G - V(J)$, called *kernels*, plus the edges of G joining vertices of the kernels to vertices of J. One example of bridges occurs when Ψ is an embedding and $J \subseteq G(\Psi)$. Then the bridges of J in $G(\Psi)$ are the topological closures in $G(\Psi)$ of the connected components of $G(\Psi) \setminus J$ (where we remove the point-set of J from $G(\Psi)$). For a second example, let K be a 3-component of G and J be the

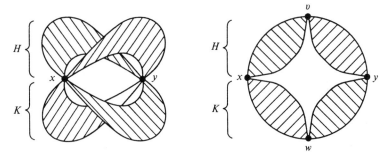

Fig. 6.1 Bad 2-separations of a nonseparable graph G

subgraph of G formed by deleting the new edges from K. Let $A \in E(K) \setminus E(J)$ and B_A be the union of the bridges B of J in G which meet J in the endvertices of A. At least one such bridge exists for each A and all bridges are of this type for some A. Replacing each new edge A by a simple arc in B_A joining the endvertices of A yields a topological embedding of the 3-component K in G. Thus 3-components are topologically contained in G, their vertices are uniquely determined, and their edges are determined up to the choice of simple arcs meeting these vertices only in their endvertices.

7. Essential 1-Components, 2-Components, and 3-Components

We now proceed to the elementary propositions about essential 1-components, 2-components, and 3-components of $G(\Psi)$. A component of $G(\Psi)$ of any of the above types is *trivial* if it contains only trivial circuits (or in the case of 3-components has a topological representation using only trivial circuits). Otherwise the component is called *essential*. The trivial components can be enclosed in disks on the surface; whereas essential components cannot be so enclosed. Given a topological subgraph H of $G(\Psi)$ we denote by Ψ_H the topological minor of Ψ with $H = G(\Psi_H)$. If H is a 1-component, 2-component or 3-component of $G(\Psi)$, then Ψ_H is called a 1-*component*, 2-*component*, or 3-*component*, respectively, of the embedding Ψ. The following Propositions 7.1, 7.2 and 7.3 are also from the thesis [22].

Proposition 7.1. *If $\rho(\Psi) \geq 1$ then $G(\Psi)$ has a unique essential 1-component, say H. Moreover, $\rho(\Psi_H) = \rho(\Psi) \geq 1$.*

Proof. As $\Sigma(\Psi)$ is not the 2-sphere and $\rho(\Psi) \geq 1$ there is an essential circuit C contained in $G(\Psi)$. Let H be the 1-component of $G(\Psi)$ containing C and suppose K is another 1-component of $G(\Psi)$. Then Ψ_H has a face f containing K and there is a circuit C_1 in that face separating H, K in $\Sigma(\Psi)$ and disjoint from $G(\Psi)$. As $\rho(\Psi) \geq 1$ it follows that C_1 is trivial and so bounds a disk in $\Sigma(\Psi)$. The disk contains K as H is essential. Thus K is a trivial 1-component, i.e. H is the only essential 1-component of $G(\Psi)$. Let C_2 be a circuit in $\Sigma(\Psi)$

which meets $G(\Psi_H)$ in exactly $\rho(\Psi_H)$ points. As the other 1-components of $G(\Psi)$ are trivial we can choose C_2 disjoint from them, and so $\rho(\Psi_H) \geq \rho(\Psi)$. Clearly, $\rho(\Psi_H) \leq \rho(\Psi)$ and so $\rho(\Psi_H) = \rho(\Psi)$.

Proposition 7.2. *If $\rho(\Psi) \geq 2$ then $G(\Psi)$ has a unique essential 2-component, say H. Moreover, $\rho(\Psi_H) = \rho(\Psi)$.*

Proof. As $\Sigma(\Psi)$ is not the 2-sphere and $\rho(\Psi) \geq 1$ there is an essential circuit C contained in $G(\Psi)$. Let H be the 2-component of $G(\Psi)$ containing C and let K be any other 2-component of $G(\Psi)$. We may assume K is in the same 1-component of $G(\Psi)$ as H, namely the essential 1-component of $G(\Psi)$, for otherwise K is clearly trivial. Then K is in a unique bridge B of H in $G(\Psi)$, which meets H in a single vertex z, which is a cutvertex of $G(\Psi)$. There is a circuit C_1 in the face of Ψ_H containing B, drawn close to B, meeting $G(\Psi)$ exactly in the vertex z and separating H from B, and hence H from K. As $\rho(\Psi) \geq 2$, the circuit C_1 is trivial and so bounds a disk. This disk contains K as H is essential, and so K is trivial, i.e. H is the only essential 2-component of $G(\Psi)$. As $H \subseteq G(\Psi)$ it follows that $\rho(\Psi_H) \leq \rho(\Psi)$. Let C_2 be a circuit in $\Sigma(\Psi)$ which meets $G(\Psi_H)$ in exactly $\rho(\Psi_H)$ points. As all other 2-components of $G(\Psi)$ are trivial, C_2 can be chosen so that it meets $G(\Psi)$ only in points of H. Thus $\rho(\Psi_H) \geq \rho(\Psi)$, and $\rho(\Psi_H) = \rho(\Psi)$ follows.

Proposition 7.3. *If $\rho(\Psi) \geq 3$ then $G(\Psi)$ has a unique essential 3-component, say H. Moreover, $\rho(\Psi_H) = \rho(\Psi)$.*

Proof. As $\rho(\Psi) \geq 2$ there is a unique essential 2-component, say H_1 of $G(\Psi)$. To obtain a 3-component of H_1 we iterate the following procedure. Take a good 2-separation $\{M_1, N_1\}$ of H_1 and adjoin a new edge X_1 across the two common vertices x_1, y_1 of M_1 and N_1 to form two 2-connected graphs M_1^+ and N_1^+. To find the essential 3-component and guarantee all other 3-components are trivial we must be careful at this stage.

As the separation $\{M_1, N_1\}$ is good, at least one of M_1, N_1 is a bridge of $M_1 \cap N_1$ in H_1. Denote such a bridge by L_1 for reference purposes. Starting from any edge of L_1 incident with x_1, follow the circular sequence in the embedding of edges incident with x_1 in both directions until edges A_1, B_1, respectively, of L_1 are reached followed by edges A_1', B_1' not in L_1. Let f_1, g_1 be the faces of Ψ_H incident with x_1 between $A_1, A_1'; B_1, B_1'$, respectively, in the cyclic order. As A_1', B_1' separate A_1, B_1 and the facial circuits of f_1, g_1 have no repeated vertices it follows that $f_1 \neq g_1$. The facial circuits meet both M_1, N_1 and so they both pass through x_1, y_1. Thus a circuit C_1 exists in $\Sigma(\Psi)$ passing through f_1, g_1 and meeting H_1 exactly in x_1, y_1. Moreover, C_1 can be chosen, as in the latter part of the proof of Proposition 7.2, to not meet any bridge of H_1 in $G(\Psi)$ except in x_1, y_1. As $\rho(\Psi) \geq 3$ it follows that C_1 is not essential and so it separates $\Sigma(\Psi)$ into two parts Δ_1, Σ_1, each a surface-with-boundary, such that $\Sigma(\Psi) = \Delta_1 \cup \Sigma_1$ and $C_1 = \Delta_1 \cap \Sigma_1$, where Δ_1 is a closed disk. In one of Δ_1, Σ_1 the only edges of H_1 meeting x_1 are in L_1, by our construction. As L_1 is a bridge of $M_1 \cap N_1$ it must be entirely in Δ_1 or Σ_1. As H_1 is nonseparable no other bridge of $M_1 \cap N_1$ can lie in that part, and so the 2-separation $\{M_1, N_1\}$ is made by C_1 in $\Sigma(\Psi)$. We may assume $M_1 \subseteq \Sigma_1$ and $N_1 \subseteq \Delta_1$, and may choose X_1 to be a simple arc from x_1 to y_1 through N_1 to form M_1^+. As C_1 meets $G(\Psi)$ only in x_1, y_1 it determines a 2-separation $\{M_1', N_1'\}$ of $G(\Psi)$, where $M_1' = G(\Psi) \cap \Sigma_1$, $N_1' = G(\Psi) \cap \Delta_1$.

Clearly all 3-components of $G(\Psi)$ which are contained in the trivial 2-components of $G(\Psi)$ or in N_1^+ are trivial, and so any essential 3-component is contained in M_1^+. Denote M_1^+ by H_2 and iterate the above procedure until no good 2-separations remain in H_k, at the kth stage. Let H_k^- stand for H_k with its new edges deleted. If X is a new edge of H_k let B_X be the union of the bridges of H_k^- in H_1 attached at the endvertices of X. Each bridge of H_k^- in H_1 is associated with some B_X and each new edge is associated with at least one such bridge in this way. Moreover, for each new edge X of H_k there is a closed disk Δ_X in $\Sigma(\Psi)$ which contains B_X and is otherwise disjoint from H_1. These disks can be chosen so that for distinct new edges X_1, X_2 the $\Delta_{X_1} \Delta_{X_2}$ are disjoint except for a common incident vertex to X_1, X_2 if it exists. Also, the boundary circuit of Δ_X can be chosen to meet $G(\Psi)$ only in the endvertices x, y of X. We take $H = H_k$ to be the essential 3-component of $G(\Psi)$ generated above and topologically contained in $G(\Psi)$. Let C be a circuit in $\Sigma(\Psi)$ meeting H in exactly $\rho(\Psi_H)$ vertices. As the 2-component H_1 can be avoided by C except for points of H, and all other 2-components can be avoided, by earlier arguments, except for points of H, it follows that $\rho(\Psi_H) \geq \rho(\Psi)$. Clearly, $\rho(\Psi_H) \leq \rho(\Psi)$ and so $\rho(\Psi_H) = \rho(\Psi)$. As all topological embeddings of the 3-component H into $G(\Psi)$ are homotopic they are never trivial, and so H is the unique essential 3-component of G. This completes the proof.

The above three propositions allow us to work with embeddings in which the representativity and the connectivity are at the same level $k = 1, 2$ or 3. Then the first three propositions are operative and make some aspects of the embeddings easier to handle. Often in our constructions one embedding is modified to obtain another with a lower bound on the representativity but with no lower bounds on the connectivity. Then the appropriate one of the later three propositions may be invoked, allowing the corresponding one of the former three propositions to be applied. Examples are now given where there are several essential 1-components, 2-components, and 3-components, to illustrate the necessity of the representativity bound. Denote by ϵ_i the number of essential i-components, for $i = 1, 2, 3$.

Remark. Suppose $\rho(\Psi) \geq 3$ for some embedding Ψ. All trivial 3-components, by Proposition 10.1 must be planar, but the essential 3-component must be nonplanar (in fact when $\Sigma(\Psi) \approx \tilde{\Sigma}_1$ it is nonplanar even after deletion of any single vertex) and hence is uniquely determined as a cleavage unit of $G(\Psi)$. Note that circuit and linkage 3-components are never essential. All embeddings of $G(\Psi)$ with essential 3-component H in which Ψ_H is fixed are obtained by re-embeddings (a) of the trivial 1-components in the various faces of Ψ_H, (b) of the bridges of H attached at single nodes of H embedded into the various possible faces of Ψ_H, and (c) of the bridges of H^- in $G(\Psi)$ attached at endvertices of new edges in a disk neighborhood of the new edge. These are all essentially planar re-embeddings. As the planar re-embeddings can be effected by Whitney's 2-isomorphism operations [24] in $G(\Psi^*)$ this aspect of planar embeddings is

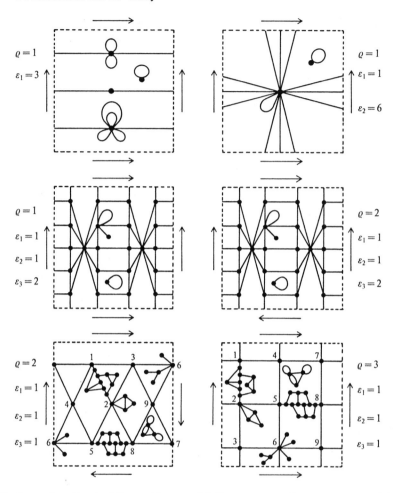

Fig. 7.1 Examples of embeddings with essential 1-components, 2-components and 3-components

carried over to general embeddings. In view of these facts we say that an embedding Ψ of $G(\Psi)$ with $\rho(\Psi) \geq 3$ is *unique up to second order planar flexibility* when for any embedding Ψ' with $G(\Psi') \cong G(\Psi)$ and $\Sigma(\Psi') \approx \Sigma(\Psi)$ we have $\Psi'_H \approx \Psi_H$, where H is the essential 3-component of Ψ.

We can extend results about facial circuits in planar graphs using the concept of a peripheral circuit. There is a general theory of peripheral circuits in [20]. It simplifies matters here to define peripheral circuits in 3-connected graphs where they are circuits with exactly one bridge. Facial circuits in a planar embedding are always peripheral for 3-connected graphs, and by the Jordan Curve Theorem these are exactly all the peripheral circuits. It can be shown [20] that in any 3-connected simple graph each edge is in at least two peripheral circuits. The

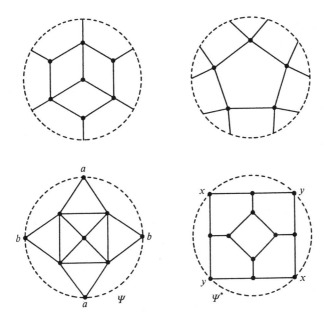

Fig. 7.2 Some embeddings with representativity 2 and their peripheral circuits

following proposition deals with the facial circuits when $\rho(\Psi) \geq 3$, where again all facial circuits of the essential 3-components are shown to be peripheral. The first example in Figure 7.2 shows that all facial circuits in the essential 3-component may be peripheral even though $\rho(\Psi) = 2$. All peripheral circuits in K_5 are its triangles, and so K_5 has no embedding with all its faces peripheral. Thus $\rho(\Psi) \leq 2$ when $G(\Psi) \cong K_5$. The third example below shows that it is possible for all faces of Ψ and of Ψ^* to be peripheral and still to have $\rho(\Psi) = 2$.

Proposition 7.4. *Suppose $\rho(\Psi) \geq 3$ and H is the essential 3-component of $G(\Psi)$. Then all the facial circuits in Ψ_H are peripheral.*

Proof. Let $f \in F(\Psi_H)$ and C_f be its bounding facial circuit. For each $x \in V(C_f)$ the wheel-neighborhood condition at x implies all of the edges of H incident with x and not in C_f (at least one exists) are in the same bridge B of C_f in H. Furthermore, if y is a vertex in C_f adjacent to x then some edge incident with y is also in B. It follows, since H is connected, that C_f has exactly one bridge.

Remark. Suppose $\rho(\Psi) \geq 3$, that H is the essential 3-component of $G(\Psi)$, that $f \in F(\Psi_H)$, and C_f is the corresponding facial circuit in Ψ_H. The bridges of C_f in G consist of an *essential* bridge B_f^+ containing the unique bridge B_f of C_f in H and a number of *trivial* bridges, attached to C_f (if they have vertices of attachment in $G(\Psi)$) on the intervals of C_f between the vertices of attachment of the bridge B_f^+. As all the bridges of C_f in G are trivial except for B_f^+, and all the bridges of H in G are trivial we will often find they do not affect in a serious way whether or not certain embedding properties are true. Thus conclusions reached using C_f, H and the bridge B_f of C_f in H will often extend trivially to $C_f, G(\Psi)$ and the bridge B_f^+ of C_f in $G(\Psi)$.

8. An Algorithm for Computing $\rho(\Psi)$ Running in Polynomial-Time in $|E(\Psi)|$

Proposition 8.1. *There is a polynomial-time algorithm for computing $\rho(\Psi)$ from input Ψ, independent of $\Sigma(\Psi)$.*

Proof. We measure the size of input by $|E(\Psi)|$, assume no isolated vertices exist and that it is easy to check whether or not faces are open disks. For $f \in F(\Psi)$, let $t(f) = Min\{|C_1 \cap G(\Psi)| : C_1$ is an essential circuit of $\Sigma(\Psi)$ meeting $f\}$. We show how to find an essential circuit C in $\Sigma(\Psi)$ meeting $G(\Psi)$ in $\leq t(f)$ points in polynomial-time. Repeating this for all $f \in F(\Psi)$ will give the required polynomial-time computation for $\rho(\Psi)$ since $|F(\Psi)| \leq |E(\Psi)|$, and $Min\{t(f); f \in F(\Psi)\} = \rho(\Psi)$. We assume a presentation of Ψ where (say) a circuit in the dual of the medial graph (the *medial graph* $M(\Psi)$ has vertices in one-to-one correspondence with the edges of Ψ, and edges in one-to-one correspondence with pairs of edges of Ψ with a common incident vertex and a common incident face) can be tested for being essential. This is not hard. If the circuit is not separating (a connectivity test) then it corresponds to an edge of Ψ. If it is separating, by a connectivity test we can find the two parts. Then checking Euler characteristic we can determine if one part is trivial.

First, examine Ψ to see if it is a closed 2-cell embedding. If not then $\rho(\Psi//f) = 1$ and we can trace back a circuit C in $\Sigma(\Psi)$ of the type we want. If yes then $\rho(\Psi) = 0$ or $\rho(\Psi) = 1$ and this is easy to determine. When we have a closed 2-cell embedding Ψ we can assume $G(\Psi)$ is nonseparable by locating the essential block and restricting to it. Again, examining wheel-neighborhoods determines whether $\rho(\Psi) = 2$ or $\rho(\Psi) \geq 3$, and in the latter case we can reduce to where $G(\Psi)$ is 3-connected via polynomial-time methods previously outlined. Now, let $f_1 = f, C_1 = \mathrm{bd}(f_1), \Psi_2 = \Psi//f_1$, and f_2 be the face of $\Psi//f_1$ which contains f_1 (i.e. the "new face"). Repeat the procedure on Ψ_2 checking if f_2 is bounded by a circuit. If not then $\rho(\Psi//f) \geq 2$ and we can test for the wheel-neighborhood property about the new face. If it fails then $\rho(\Psi//f) = 2$ and we can find the circuit C of the type we desire. The procedure terminates at the nth stage when $\rho(\Psi_n) \leq 2$, and produces a circuit C which meets $G(\Psi_n)$ in $t(f_n)$ vertices and hence meets $G(\Psi)$ in at most $t(f_n) + 2(n-1)$ vertices. Observe that $t(f_k) \leq t(f_{k-1}) - 2$, so that $t(f_n) \leq t(f) - 2(n-1)$, and therefore C meets $G(\Psi)$ in at most $t(f)$ vertices.

9. Some Aspects of Representativity in the Graph Minor Papers

In the paper [13] the following main theorem about representativity is proved.

Proposition 9.1. *If Ψ_0 is fixed and Ψ is an arbitrary embedding with $\Sigma(\Psi) = \Sigma(\Psi_0)$, then there is a constant B_0 such that $\rho(\Psi) > B_0$ implies $\Psi_0 \leq_m \Psi$.*

This shows that graphs on nonplanar surfaces with sufficiently high representativity include as surface minors any fixed graph on the surface. Thus the elementary structural consequence of an excluded minor for embeddings on a nonplanar surface is an upper bound on representativity. The planar analogue [12] of Proposition 9.1 states that graphs of sufficiently high "tree-width" contain any fixed planar graph. More concretely, on the plane large "square grid" minors force large tree-width, while on the projective plane large "twisted grid" minors force high representativity (see Figure 14.3).

Embedded on any surface are graphs of *minimum genus* (here *genus* is the orientable handle number or unorientable cross-cap number) in the sense they do not embed on simpler surfaces (surfaces with fewer handles or cross-caps, respectively). Similarly, there are graphs which embed uniquely onto any fixed surface up to second-order flexibilities (analogues discussed in Section 7 of the Whitney twists for non-3-connected graphs in the sphere). These two classes of embeddings are upwardly minor-closed in the sense that if any such embedding is a minor of an embedding then that embedding inherits the property of minimality or uniqueness. We conclude from Proposition 9.1 that for any fixed surface there are bounds on representativity such that if they are exceeded then the corresponding property holds. We show in Section 13 for oriented surfaces that $2g + 3 \leq \rho(\Psi)$ ensures that $G(\Psi)$ is uniquely embedded in this way in $\Sigma(\Psi) = \Sigma_g$, and describe in Section 15 Archdeacon's construction that proves no constant lower bound on $\rho(\Psi)$ suffices for unique embedding.

10. Planar Graphs Embedded on Higher Surfaces

In Figure 10.1 an embedding $\Psi_k (k \geq 1)$ is given of $K_{3,2k+2}$ the complete bipartite graph on 3 plus $2k + 2$ vertices, into $\tilde{\Sigma}_k$. As the graph is bipartite and simple, and all faces are quadrilaterals, a routine Euler characteristic argument shows the cross-cap number k is minimal. Remarkably, these embeddings all have 3-connected planar graph duals, and their representativity is 2. The duals start with the octahedron and can be constructed by successive operations of dividing one triangle into four triangles.

The basic fact about planar graphs embedded on higher surfaces is expressed by the proposition below.

Proposition 10.1. *If $G(\Psi)$ is planar then $\rho(\Psi) \leq 2$.*

Proof. Let Ψ_0 be a plane embedding of G, and Ψ be an embedding of G on another surface, with $\rho(\Psi) \geq 3$. By Propositions 7.3 and 7.4, G has an essential 3-component in Ψ, say H, and all facial circuits of H in Ψ_H are peripheral and so must be facial circuits of H in the embedding $(\Psi_0)_H$. Thus $\Psi_H \approx (\Psi_0)_H$, and so $\Sigma(\Psi_0) \approx \Sigma(\Psi)$, contrary to assumption.

Remark. From the thesis [22], we know that if $\Sigma(\Psi) = \tilde{\Sigma}_1$ and $\rho(\Psi) \geq 3$ then $G(\Psi)$ is not "nearly planar" in the sense that it remains nonplanar after deletion of any vertex and its incident edges. Moreover, all projective planar graphs which are not nearly planar admit only embeddings Ψ with $\rho(\Psi) \geq 3$ on

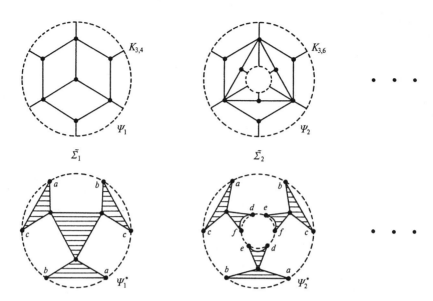

Fig 10.1 Genus embeddings on unoriented surfaces with planar duals

$\tilde{\Sigma}_1$. We may speculate that $\rho(\Psi) \geq 3$ implies $G(\Psi)$ is not nearly planar for any orientable surface $\Sigma(\Psi)$, and that any not nearly planar graph admits a $\rho(\Psi) \geq 3$ embedding on some surface. The former speculation is an open question, while the later speculation is false, as witnessed by complete graphs K_k for $k \equiv 2(mod\ 3)$ and $k > 5$. In fact the complete bipartite graphs $K_{m,n}$ with m, n both odd do not admit $\rho(\Psi) \geq 3$ embeddings, as the nonquadrilateral faces are not peripheral. It seems to be an interesting problem to characterize 3-connected graphs with $\rho(\Psi) \geq 3$ embeddings; though no serious conjecture supported by the evidence seems to be available. We may note that $K_{4,4}$ minus one edge has $V = 8, E = 15$ and can not have only quadrilateral faces, as $4F = 2E$ implies E is even. Thus this minimal not nearly planar graph does not have a $\rho(\Psi) \geq 3$ embedding.

Remark. It is a folklore conjecture concerning surface embeddings that any nonseparable graph containing circuits admits some surface embedding Ψ with a loopless dual or more strongly with $\rho(\Psi) \geq 2$. This is true for planar graphs as any planar embedding shows, and is plausible in general. Trivalent graphs with a 1-factorization are a very general class for which the conjecture is true, as the circuits of edges with 2 colors form a system of facial circuits. In the trivalent case the above conjectures follow from the circuit-double-cover conjecture that any connected graph without an isthmus and with $V \geq 2$ has a system of circuits, exactly two on each edge. The $\rho(\Psi) \geq 2$ conjecture for nonseparable graphs with circuits implies the circuit-double-cover conjecture and so it is of great interest. At any rate the conjecture is obviously fundamental in the context of representativity, and is worth a systematic approach from this viewpoint. Two problems are recommended here which may throw some light on surface embedding methods and not require solving the complete hard problems. The first is to characterize series-parallel networks [4] and [5] which admit $\rho(\Psi) = 2$ embeddings (with $\Sigma(\Psi)$ not the sphere) and the second is to show that planar 3-connected graphs all admit $\rho(\Psi) = 2$ embeddings (with $\Sigma(\Psi)$ not the sphere).

Remark. Another, closely related problem, is to characterize nonplanar embeddings Ψ of planar graphs $G = G(\Psi)$. Recently, Bojan Mohar and the authors [9] have done the $\chi(\Psi) = 0$ case of this, and have prospects of proving a general conjecture. As a further problem along this line we can ask for the structure of graphs admitting minimum genus embeddings (of either type) which have planar duals.

11. Projective Graphs Embedded in Higher Surfaces

The two theorems of this Section strengthen the uniqueness result of [22] on embeddings Ψ with $\rho(\Psi) \geq 4$ and $\Sigma(\Psi) = \tilde{\Sigma}_1$. These theorems say that under these hypothesis if H is the essential 3-component of Ψ then any embedding Ψ', with $G(\Psi) \cong G(\Psi')$ and $\Psi_H \not\approx \Psi'_H$, satisfies $\rho(\Psi') \leq 3$, and that examples of embeddings Ψ, Ψ' exist with $\Sigma(\Psi) \approx \tilde{\Sigma}_1$ and $\rho(\Psi)$ arbitrarily large, having $\rho(\Psi') = 3$ and $\Sigma(\Psi')$ of arbitrarily high genus. Note that the second proposition is modelled on the result of Carsten Thomassen regarding torus embeddings discussed in Section 12.

Proposition 11.1. *Let Ψ_0, Ψ be embeddings with $G(\Psi_0) = G(\Psi)$, where $\Sigma(\Psi_0) \approx \tilde{\Sigma}_1$. Then $\rho(\Psi) \geq 4$ implies $\Psi_H \approx (\Psi_0)_H$, where H is the essential 3-component of Ψ.*

Proof. Note that $\chi(\Psi) \leq \chi(\Psi_0) = 1$. Assume $\Psi_H \not\approx (\Psi_0)_H$. Then there is an $f \in F(\Psi_H)$ with facial circuit C_f which is not a facial circuit of $(\Psi_0)_H$. As C_f has only one bridge in H it must be an essential circuit in $\Sigma(\Psi_0)$. By Proposition 4.2, $\rho(\Psi_H//f) \geq 2$. Let K be the essential 2-component of $\Psi_H//f$. Denote by C the facial circuit of the face f' of Ψ_K containing f. Note that $\rho((\Psi_0)_K) = 0$, whence $G((\Psi_0)_K)$ embeds trivially in $\Sigma(\Psi_0) \approx \tilde{\Sigma}_1$. All facial circuits of Ψ_K, except possibly C, are peripheral in Ψ_H and, being disjoint from C_f, are peripheral in $(\Psi_0)_K$, and hence are facial circuits in $(\Psi_0)_K$. Adjoining to $(\Psi_0)_K$ an open 2-disk f'' with boundary C in place of f' gives a planar embedding Ψ' of K homeomorphic to Ψ_K, as $\rho(\Psi_K) \geq 2$. Thus $\Sigma(\Psi') \approx \Sigma(\Psi_K)$, contrary to assumptions.

Proposition 11.2. *For each integer $m \geq 3$ there is an embedding Ψ_0 with $\Sigma(\Psi_0) = \tilde{\Sigma}_1$ and $\rho(\Psi_0) \geq m$, and an embedding Ψ with $G(\Psi) = G(\Psi_0)$ and Euler characteristic $\chi(\Psi) \leq 2 - (m-1)(2m-3)/2$, and $\rho(\Psi) = 3$.*

Proof. We choose on $\tilde{\Sigma}_1$ exactly $2m$ essential circuits in general position and define Ψ_0 to be the embedding formed with $G(\Psi_0)$ the union of the $2m$ circuits. A simple induction argument shows that Ψ_0 is facially 2-colorable (add two circuits at a time and change parity of colors in one of the two faces determined by the two circuits), and $\rho(\Psi) \geq m$ (as any essential circuit must meet all $2m$ of the circuits and no point meets more than 2 such circuits). We can easily count for Ψ_0 that $V_0 = m(2m-1)$, $E_0 = 2m(2m-1)$ and $F_0 = 2m^2 - m + 1$ (the latter follows as $V_0 - E_0 + F_0 = 1$ here).

Define Ψ with $G(\Psi) = G(\Psi_0)$ and with facial circuits given by the $2m$ essential circuits defining Ψ_0 plus the facial circuits in Ψ_0 of one color class of faces under the facial 2-coloring of Ψ_0. Now this is easily checked to be a well-defined embedding, and clearly $V = V_0, E = E_0$ and WLOG, $F \leq 2m + (2m^2 - m + 1)/2$. We have

$$\chi(\Psi) \leq -m(2m-1) + 2m + (2m^2 - m + 1)/2$$
$$= (-2m^2 + 5m + 1)/2$$
$$= 2 - (2m-3)(m-1)/2.$$

When $m \geq 3$, we have $\chi(\Psi) \leq -1$, and hence by Proposition 11.1, $\rho(\Psi) \leq 3$. Since the intersection of the boundaries of two faces of Ψ consists of either the empty set, a single vertex, or an edge and its endvertices, it follows that $\rho(\Psi) \geq 3$. Thus $\rho(\Psi) = 3$.

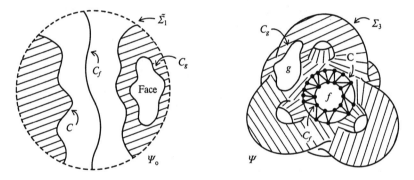

Fig. 11.1 Illustration of the proof of Proposition 11.1

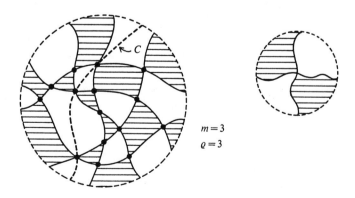

Fig 11.2 Illustration of the proof of Proposition 11.2

12. Toroidal Graphs Embedded in Higher Surfaces

At one time the authors conjectured that $\rho(\Psi_0) \geq 4$ implies $\rho(\Psi) \leq 3$ and $\chi(\Psi) \leq \chi(\Psi_0)$ for all embeddings $\Psi \not\approx \Psi_0$ with $G(\Psi) \cong G(\Psi_0)$. This has turned out to be quite wrong, and Carsten Thomassen's example [17] was the first to demonstrate this. Recently, Dan Archdeacon has produced an example [2] that shows no fixed lower bound on $\rho(\Psi_0)$ will guarantee the embedding has optimal Euler characteristic or that the embedding is unique for $\Sigma(\Psi_0)$. We have mentioned earlier that if $\Sigma = \Sigma(\Psi_0)$ is fixed then such a bound does exist. We will come back to Archdeacon's theorem later.

Proposition 12.1 (Robertson and Vitray). *Let Ψ_0, Ψ be embeddings with $G(\Psi_0) = G(\Psi)$, where $\Sigma(\Psi_0) \cong \Sigma_1$ and $\chi(\Psi) \leq \chi(\Psi_0) \leq 1$. Then $\rho(\Psi) \geq 5$ implies $\Psi_H \approx (\Psi_0)_H$, where H is the essential 3-component of Ψ.*

Proof. Assume $\Psi_H \not\approx (\Psi_0)_H$. Then there is a face $f \in F(\Psi_H)$ whose facial circuit C_f is not a facial circuit of $(\Psi_0)_H$. As C_f has only one bridge in $G(\Psi)$, by Proposition 7.4, it must be an essential circuit in $\Sigma(\Psi_0)$. By Proposition 4.2, $\rho(\Psi_H//f) \geq 3$, which implies $G(\Psi_H//f)$ is nonplanar, by Proposition 10.1. However, $G(\Psi_H//f)$ has an essentially planar embedding via Ψ on the torus $\Sigma(\Psi)$. This contradiction proves the theorem.

Proposition 12.2 (Thomassen). *For all integers $m, n \geq 2$ there is an embedding Ψ_0 with $\Sigma(\Psi_0) = \Sigma_1$ and $\rho(\Psi_0) \geq min(2m, 2n)$, and an embedding Ψ with $G(\Psi) = G(\Psi_0), \rho(\Psi) = 4$ and Euler characteristic $\chi(\Psi) = 2(m + n - mn)$.*

Proof. Construct, for positive integers m, n, an embedding $\Psi_0 = \Psi_0(m, n)$ on the torus Σ_1 as follows. Choose a family of $2m$ disjoint essential circuits and a family of $2n$ disjoint essential circuits so that any two circuits from different families meet in a single point, where they cross. This forms a 4-regular toroidal lattice of dimensions $2m \times 2n$ and such a lattice is facially 2-colorable, in colors red and white say, so that each edge meets different colored faces. Now split each vertex into an edge forming its two incident red quadrilaterals so that the white faces become octagonal, and all vertices are trivalent. There are still $2n$ pairwise disjoint "vertical" essential circuits and $2m$ pairwise disjoint "horizontal" essential circuits, and so $\rho(\Psi_0) = min(2m, 2n)$. Define another embedding Ψ with $G(\Psi) = G(\Psi_0)$ and with facial circuits given by the $2m$ horizontal, $2n$ vertical essential circuits of Ψ_0, and the $2mn$ quadrilaterals. As $G(\Psi_0)$ is trivalent this automatically gives a surface, and its Euler characteristic is $\chi(\Psi) = 8mn - 12mn + 2mn + 2m + 2n = 2m + 2n - 2mn$, as required. The dual graph of Ψ is simple and 3-colorable, and so chains of faces including essential circuits in Ψ cannot be of length ≤ 2 and are of length 3 only if all three kinds of faces are used, which occurs only at adjacent triples, forming a topological disk, contrary to the essential circuit. Hence $\rho(\Psi) \geq 4$. By the previous theorem if $\Psi \not\approx \Psi_0$ then $\rho(\Psi) \leq 4$ and so $\rho(\Psi) = 4$ (this is easily seen directly). We note that $m, n \geq 2$ implies $\chi(\Psi) \leq 0$ and that when $m = n = 2$ we have $\Psi \approx \Psi_0$, and so this theorem gives new embeddings when $m + n \geq 5$ and $\chi(\Psi) < 0$.

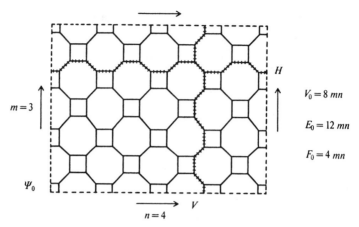

$$V_0 = 8\,mn$$

$$E_0 = 12\,mn$$

$$F_0 = 4\,mn$$

Fig. 12.1 Illustration of the proof of Proposition 12.1

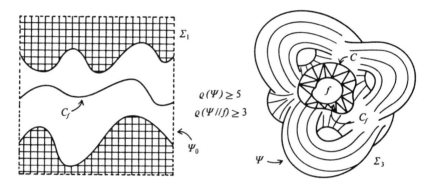

$$\varrho(\Psi) \geq 5$$

$$\varrho(\Psi /\!/ f) \geq 3$$

Fig. 12.2 A toroidal lattice which embeds on higher surfaces with representativity 4

13. Unique Embeddings for Orientable Surfaces

The general orientable surface analogue of Propositions 11.1 and 12.1 is not much harder to prove and so we include it here. At time of writing it is not known that this is best possible, even for Σ_2.

Proposition 13.1. *Let* Ψ_0, Ψ *be embeddings with* $G(\Psi_0) = G(\Psi)$, *where* $\Sigma(\Psi_0) \cong \Sigma_g$ *for some positive integer g, and* $\chi(\Psi) \leq \chi(\Psi_0)$. *Then* $\rho(\Psi) \geq 2g + 3$ *implies* $\Psi_H \approx (\Psi_0)_H$, *where H is the essential 3-component of* Ψ.

Proof. We remark that Proposition 10.1 guarantees that if $g = 0$ (so that $G(\Psi)$ is planar) then $\Sigma(\Psi)$ cannot be nonplanar with $\rho(\Psi) \geq 3$, and so the embedding Ψ must also be planar. The planar embeddings differ by a sequence of Whitney "twists" and so Ψ and Ψ_0 are equivalent in this sense (which coincides with the 3-component equivalence on higher surfaces).

If $g = 1$ then Proposition 13.1 is the same as Proposition 12.1. Thus $g \geq 2$ can be assumed. Let H be the essential 3-component of Ψ and assume $\Psi_H \not\approx (\Psi_0)_H$. Then some facial circuit C_f, for $f \in F(\Psi_H)$, of Ψ_H is not a facial circuit of $(\Psi_0)_H$. We have $\rho(\Psi_H) \geq 2g + 3$, which implies that C_f has a single bridge B in H. Suppose that C_f separates $\Sigma(\Psi_0)$. The bridge B must be contained in one side of the separation, and so the other side must be disjoint from H. The side of the separation disjoint from H cannot be an open 2-cell since C_f is not a facial circuit of $(\Psi_0)_H$, therefore replacing it with an open 2-cell gives an embedding $(\Psi_1)_H$ of H on a surface $\Sigma_{g'}$, where $g' < g$. By induction $\Psi_H \approx (\Psi_1)_H$ which implies, by Proposition 7.3, that $\chi(\Psi) = g' < g$, contrary to assumption. Thus we may assume C_f is nonseparating in $\Sigma(\Psi_0)$. But now $\rho((\Psi_0)_{H//f}) = 0$ and $\rho(\Psi_{H//f}) \geq 2g + 1$ so that we can construct Ψ_0' with $G(\Psi_0') = H//f$ and $\Sigma(\Psi_0') \cong \Sigma_{g-1}$ and note that $\rho(\Psi_{H//f}) \geq 2(g-1) + 3$. By induction, as $g - 1 \geq 1$, we have $\Psi_0' \approx \Psi_{H//f}$, once again contradicting $\chi(\Psi) \leq \chi(\Psi_0)$. It follows that $\Psi_H \approx (\Psi_0)_H$, as required.

14. Relevance of the (Y, Δ)-Transformations

A theorem of Dirac [4] and Duffin [5] shows that the class of graphs not topologically containing K_4 (the so-called series-parallel graphs) can be generated by iterating the following elementary operations.

1. Series-contraction of edges and subdivision of edges.
2. Parallel deletion of edges and parallel addition of edges.
3. Deletion or addition of pendant edges; either links or loops.
4. Removal or creation of isolated vertices.

In fact the whole class can be produced from the null graph Ω by iterations of the positive operations in these four pairs of inverse operations. When these operations are carried out on graphs in embeddings the usual restrictions to respecting the embedding apply. In particular, no "series-contractions" of loops is allowed and "parallel deletion or addition" is only allowed involving pairs of edges bounding a face. Similarly, pendant loops must bound a face (though, strictly speaking it is enough to add pendant links, unless preserving duality is important, because loops can be produced via the other operations from pendant links).

5. A (Y, Δ)-*transformation* replaces a trivalent vertex and its three incident edges (not two parallel) by a triangle on its three adjacent vertices. A (Δ, Y)-*transformation* replaces the three edges of a triangle by three edges from a new vertex to the three vertices of the triangle.

An old result from chapter 15 of [7] says that if (Y, Δ)-transformations and their inverses, (Δ, Y)-transformations, are allowed then all planar graphs can be produced from these operations. In fact, if series edges and parallel edges are suppressed as soon as they appear, all 3-connected planar graphs can be reduced to the simple K_4 via these two operations, keeping the graphs involved 3-connected (after suppressing the series and parallel edges).

Note that on a surface in a (Y, Δ)-transformation the triangle must bound the face containing the old trivalent vertex, and that a (Δ, Y)-transformation can only be applied to a triangle bounding a face, with the new trivalent vertex chosen inside the face. We state an interesting basic theorem of Klaus Truemper [19].

Proposition 14.1. *The class $\mathscr{L}$ of graphs produced from $\{\Omega\}$ by the above five pairs of inverse operations is minor-closed.*

Indeed, the closure $\overline{\mathscr{M}}$ of any minor-closed class $\mathscr{M}$ of graphs under these operations is itself minor-closed. Truemper applies this theorem to prove all planar graphs can be reduced to Ω via the above operations.

Proposition 14.2. *The class of $\mathscr{L}$ of planar graphs can be produced from $\{\Omega\}$ by iterating the five pairs of operations (1),(2),(3),(4),(5) restricted to the sphere.*

Proof. We first show that (a) for all positive integers n the $n \times n$ grid can be reduced to Ω by these operations, and (b) any planar graph G is a minor of the $n \times n$ grid for sufficiently large n. Then Proposition 14.2 is an immediate consequence of Proposition 14.1 noting that it remains valid when (5) is restricted to the 2-sphere. We need five more operations easily established from the basic ones to work with grids; these are given in the Figure 14.1.

The method of reducing an $n \times n$ grid is to remove one row at a time until a $1 \times n$ grid is produced, then to apply the fourth operation reducing this to a link-graph and then to Ω. An example is given in Figure 14.2 for the 3×3 grid which is typical of the general case.

To show that a fixed graph G can be found as a minor of an $n \times n$ grid for large enough n we explain a method suggested by Bruce Reed. It is easy to see that some subdivision of G can be drawn in a 2-page "book" with all its vertices on the spine and all its edges confined to one page or the other. For example, draw a simple closed curve on the plane through all the vertices of G, passing perhaps through some of the edges of G in a finite number of simple crossings. This gives an ordering of the vertices of G and the crossing vertices to be placed on the spine. The interior and the exterior of the simple closed curve become the two pages of the "book". Let N be the number of edges in the subdivision of G found above and find G as a minor of the $2N \times 2N$-grid as follows. Place the $2N$ edge-ends in their order along the spine of the book in one-to-one correspondence with the vertices along the $(N + 1)st$ row of the grid. Then for each vertex in the spine the corresponding incident edge-ends will be consecutive along the $(N+1)st$ row. These edge-ends can then be contracted along this row to form the image of that vertex in the minor of the grid isomomorphic to the subdivision of G. Suppose two vertices in the $(N + 1)st$ row correspond to the ends of an edge in the book. We embed all the edges from the page containing the furthest apart such pair in the top half of the grid, and all the edges of the other page in the bottom half. If the two edge ends are distance k apart in the grid we join them by vertical paths of lengths $\lceil k/2 \rceil$ appropriately above or below the two edge ends and a horizontal path of length k connecting the upper or lower, respectively, endpoints of these paths. Clearly, such edges fit in the pages and do not cross

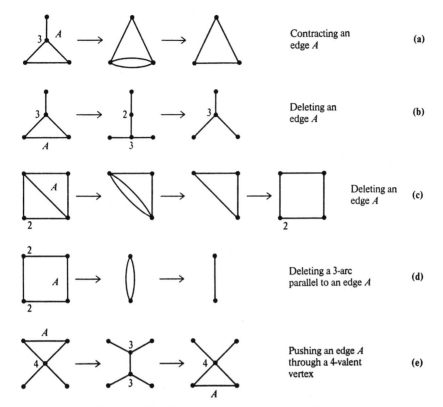

Fig. 14.1   Useful operations for working with grids

each other because of their natural planar nesting. Also, clearly G is a minor of the graph now drawn on the grid. This completes the proof of Proposition 14.2.

We now state Scott Randby's theorem, which simplifies the theorems of [22] reported in Section 5 about minor-minimal embeddings Ψ on $\tilde{\Sigma}_1$ in the cases $\rho(\Psi) = 1, 2, 3$ and extends them to general $\rho(\Psi)$. Projective grids are depicted in Figure 14.3. Note that Ψ_6^* of Figure 5.3 is homeomorphic to Ψ here.

Proposition 14.3. *The class $\mathscr{L}$ of all embeddings Ψ on the projective plane with $\rho(\Psi) = k$ can be generated from a unique embedding, the projective grid of size k, by the five pairs of operations in this Section, for all positive integers k.*

This remarkable theorem does not extend unscathed to higher surfaces, because already on the torus and Klein bottle there are graphs in which all vertices and all faces have valency 4 (≥ 4 on higher surfaces). Such graphs do not admit any of the operations of type (5) in this Section and so remain embedded in any graph produced from them by the five pairs of operations.

It is worthwhile remarking that the operations of this Section restricted to the surface do not effect the representativity of a surface embedding. This leads

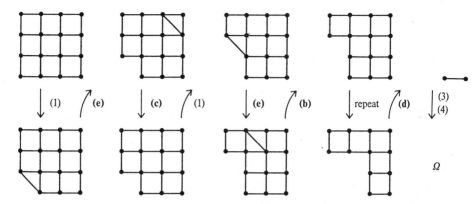

Fig. 14.2 Reducing a 3×3 grid to Ω

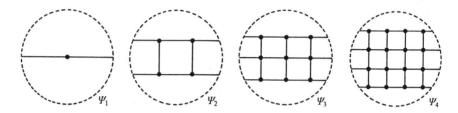

Fig. 14.3 The projective $k \times k$ Grids Ψ_k

to a somewhat more general observation which requires more notation. Let Σ be a (nonspherical) surface and C be an essential circuit on Σ. Define, for any embedding Ψ with $\Sigma(\Psi) = \Sigma$, the parameters

$\rho(\Psi, C) = \min\{|C_1 \cap G(\Psi)| : C_1$ is a circuit of $\Sigma(\Psi)$ homotopic to $C\}$

$\delta(\Psi, C) = \max\{|\mathscr{P}| : \mathscr{P}$ is a set of pairwise disjoint circuits of $G(\Psi)$ homotopic to $C\}$.

These parameters also are invariant under the operations of this Section.

Proposition 14.4. *Let Ψ_1, Ψ_2 be embeddings with $\Sigma(\Psi_1) = \Sigma(\Psi_2)$ which are equivalent under the operations (1),(2),(3),(4),(5) of this section. Then for any essential circuit C on Σ we have*

$$\rho(\Psi_1, C) = \rho(\Psi_2, C) \text{ and } \delta(\Psi_1, C) = \delta(\Psi_2, C).$$

Proof. Note that the operations of this Section come in inverse pairs, and so it is only necessary to show these parameters are monotone under the operations. Also, the trivial operations (3) and (4) obviously have no effect on these parameters and so can be ignored. It is also evident that the series operations (1) and the parallel operations (2) carried out to respect the surface do not effect the parameters. To check the operations (5), let C_1 be a circuit in $\Sigma(\Psi)$ homotopic to C for which $|C_1 \cap G(\Psi)| = \rho(\Psi, C)$. As usual, C_1 can be chosen to meet $G(\Psi)$ only in vertices and to pass through no face twice. In Figure 14.4 the effect of operations (5) on C_1 is shown with a segment of C_1 given by a dashed curve. Let $t \in V(\Psi)$ be a trivalent vertex adjacent to 3 distinct vertices x, y, z in $G(\Psi)$. If $t \notin C_1$ then a (Y, Δ) transformation at t produces a new triangular face at x, y, z surrounding t which does not meet C_1. Thus ρ is monotone here. If $t \in C_1$, we can deform C_1 homotopically to C_1' avoiding t and going through one of x, y, z with $|C_1' \cap G(\Psi)| = \rho(\Psi, C)$ and the earlier case obtains. To check the (Δ, Y)-transformations, let x, y, z be the vertices of a triangular face of Ψ and chose t in that face. Again, if the face is not used by C_1 the parameter is monotone under the (Δ, Y)-transformation. If the face is used then C_1 goes through x, y (say) and can be pushed homotopically out of the triangle to give the previous case. Regarding the parameter $\delta(\Psi, C)$, to show essential circuits in $G(\Psi)$ map to homotopic essential circuits under the transformations (5) and that the pairwise disjoint condition is uneffected is quite routine, and so this parameter is also monotone. It follows that both parameters are in fact invariant under these operations.

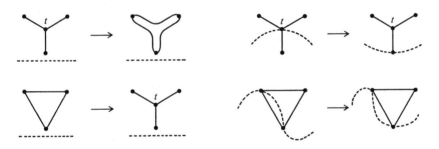

Fig. 14.4   Effect of (Y, Δ)- and (Δ, Y)-transformations on $\rho(\Psi, C)$

Problems of Leo Moser [10] and John Conway, which have been partially solved in [3] and [8], involve characterizing:

1. the minor closed class $\mathscr{C}_1$ of graphs embeddable in 3-space with no circuits knotted and no set of disjoint circuits linked, and
2. the minor-closed class $\mathscr{C}_2$ of graphs embeddable in 3-space with no circuits knotted.

Define a 3-space embedding to be *unknotted* iff every circuit bounds a 2-disk in 3-space; and to be *unknotted and unlinked* iff every circuit bounds a 2-disk in 3-space disjoint from the embedded graph. The unlinkedness condition certainly implies sets of disjoint circuits are unlinked. If follows that the classes $\mathscr{C}_1$ and $\mathscr{C}_2$ are closed under the operations of this Section. Their obstacle sets can thus be simplified by identifying equivalent graphs under these operations (essentially operations (5)). We note that fairly extensive classes can be defined which admit these embeddings. Thus $\mathscr{C}_1$ contains all graphs embeddable in the sphere after removing one of their vertices, and $\mathscr{C}_2$ contains all graphs embeddable in the sphere after removing two of their vertices. The following proposition appears in [22].

Proposition 14.5. *If Ψ satisfies $\Sigma(\Psi) = \tilde{\Sigma}_1$ then $G(\Psi) \in \mathscr{C}_1$ if and only if $\rho(\Psi) \leq 2$.*

Proof. If $\rho(\Psi) \leq 2$ then there exists $x \in V(\Psi)$ such that $G(\Psi) - \{x\}$ is planar (as $\rho(\Psi \setminus x) \leq 1$) and so $G(\Psi) \in \mathscr{C}_1$. If $\rho(\Psi) \geq 3$ then $G(\Psi)$ contains as a minor a graph in the obstacle set $\Omega_3(\tilde{\Sigma}_1)$, which is known to be not in $\mathscr{C}_1$.

Remark. It would be interesting if $\Omega(\mathscr{C}_1)$ was generated by K_6, and $\Omega(\mathscr{C}_2)$ was generated by K_7. This could be approached by elaborating the interior structure of $\mathscr{C}_1$ and of $\mathscr{C}_2$ and relating it to the exclusion of the obstacles. In fact, the first of these assertions has been conjectured by Sachs [16], and recently has been claimed true by Motwani, Raghanathan and Saran [8]. We note that graphs embeddable in the sphere after deleting one vertex cannot all be obtained from Ω by our operations. A very general class of bipartite graphs embeddable in the sphere exists, with one class of vertices trivalent and the other of valency ≥ 4. Joining all the trivalent vertices on the sphere to a single vertex off the sphere produces a class of graphs with girth ≥ 4 and valency ≥ 4. Such graphs cannot be reduced to Ω by the operations of this Section.

15. Dan Archdeacon's Example of Embeddings Ψ_1, Ψ_2 with $G(\Psi_1) \cong G(\Psi_2)$, $\Sigma(\Psi_1) \not\approx \Sigma(\Psi_2)$ and $\rho(\Psi_1), \rho(\Psi_2) \geq B$

This example shows there exist graphs embeddable in different surfaces both with very high representativity. His construction is in two parts.

Part 1. Construct embeddings Ψ with arbitrarily large representativity such that $G(\Psi)$ and $G(\Psi^*)$ are both bipartite and have arbitrarily high girth.

Part 2. Construct embeddings Ψ_1', Ψ_2' of the medial graph of Ψ with $\Sigma(\Psi_1') \not\approx \Sigma(\Psi_2')$ which can be extended to embeddings Ψ_1, Ψ_2 on $\Sigma(\Psi_1'), \Sigma(\Psi_2')$, respectively, where $G(\Psi_1) = G(\Psi_2)$ and both embeddings have arbitrarily large representativity.

As Part 2 is the simplest to describe we will start with it. Consider the embedding Ψ constructed in Part 1. The medial graph $M(\Psi)$ defined in Section 8 is a sort of surface "line-graph" which combines the graphs $G(\Psi)$ and $G(\Psi^*)$ on a unified basis. An example is given in Figure 15.1 of the medial graph of an embedding Ψ where both $G(\Psi)$ and $G(\Psi^*)$ are bipartite.

Note that $M(\Psi)$ is regular of valency 4 and can be regarded as the graph $G(\Gamma)$ of an embedding Γ with $\Sigma(\Gamma) = \Sigma(\Psi)$. Now $G(\Gamma)$ naturally decomposes into circuits which are essential and "cross through" each other at alternate edges at each vertex. Such a decomposition is certainly unique, if it exists. Existence hinges

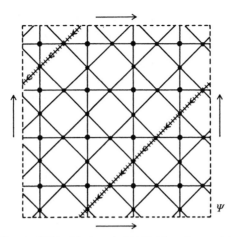

Fig. 15.1 Embedding Ψ With $G(\Psi)$ and $G(\Psi^*)$ bipartite, and its medial graph

on whether or not a path P in $G(\Gamma)$, with successive edges non-consecutive at each vertex, can double back on itself via a new edge at a vertex it has already crossed through. This does not happen because of the double bipartiteness. Suppose the faces of Γ corresponding to the vertices of Ψ divide under their bipartition into type (A) and type (B), and the faces of Γ corresponding to the faces of Ψ divide under their bipartition into type (X) and type (Y). Then we can easily show that a circuit is always formed by the path P. Start say with an edge meeting faces of type $(A), (X)$. The next edge meets faces of type $(Y), (B)$ and then the two types of edges alternate along P so that the path cannot cross itself internally, as this would require a type $(A), (Y)$ or type $(B), (X)$ edge. An Euler characteristic argument then shows that these circuits are essential. Suppose such a circuit bounds a disk in $\Sigma(\Gamma)$. The subgraph of $G(\Gamma)$ on the closed disk has $2m$ vertices of valency 3 on the circuit and n vertices inside. Then $V = 2m + n, E = 3m + 2n$, and $F = m + n + 2$, using $V - E + F = 2$. We may assume that all vertices and faces of Ψ have valency ≥ 4 (as they are arbitrarily large by the construction of Part 1). Counting the valency $2m$ of the outer face, $2m + 4(m + n + 1) \leq 2E = 6m + 4n$, whence $4 \leq 0$, the required contradiction.

The two new embeddings Ψ_1, Ψ_2 are constructed as follows. First embeddings Ψ_1', Ψ_2' with graph $G(\Gamma)$ are defined, with *new* facial circuits corresponding to the circuits decomposing $G(\Gamma)$ in common, and the *old* facial circuits corresponding, respectively, to the faces of Ψ in Ψ_1' and to the vertices of Ψ in Ψ_2'. It is easy to see these facial circuits define a surface embedding of $G(\Gamma)$ in each case, and that $\Sigma(\Psi_1') \not\approx \Sigma(\Psi_2')$ when $|V(\Psi)| \neq |F(\Psi)|$, which is usually the case. Essential circuits in $\Sigma(\Psi_1')$ and $\Sigma(\Psi_2')$ through the old faces still meet $G(\Gamma)$ in $\geq \rho(\Psi)$ vertices. However shorter essential circuits can occur using new faces. To prevent this we fill in the new faces, common to both embeddings, with dense planar graphs so that essential circuits on the surface meeting the extended graph G' in a minimum number of vertices pass only through the old faces. Thus embeddings Ψ_1, Ψ_2 of high representativity of the same graph on different surfaces are constructed.

Part 1 is not so straightforward. It begins with an embedding Ω_1 of the complete bipartite graph $K_{2m,2m}$, dual to a circuit of length $2m$ with each edge multiplied in parallel $2m$ times. This has the special edge-connectivity condition that each pair of vertices in the graphs are contained in m edge-disjoint nonseparating circuits of the graphs. Then a covering space construction is made which raises the girth and representativity by at least 2 and preserves bipartiteness and the edge-connectivity condition. This is iterated until the required embedding Ψ is constructed.

16. Relationship Between Genus and Representativity

We have shown earlier that if $\Sigma(\Psi) \approx \Sigma_g$ and $\rho(\Psi) \geq 2g + 3$ then the essential 3-component H of $G(\Psi)$ has no other embedding of characteristic $\geq 2 - 2g$ but Ψ_H, and all embeddings of characteristic $< 2 - 2g$ have representativity $< 2g + 3$. We can also prove a similar result with $\Sigma(\Psi) \approx \tilde{\Sigma}_k$ and $\rho(\Psi) \geq k + 3$, when $k = 1, 2, 3, 4$ (the case $k = 1$ is given here) but the arguments are much harder. Also, when $g \geq 2$ or $k \geq 2$, we cannot yet show the bound on representativity is best possible to obtain these conclusions.

Recently, Thomassen [18] has shown how to reduce the problem "does the graph G have $\leq k$ vertices meeting all its edges" to the problem for an associated graph G_1 with $\leq cV^3$ vertices "does the graph G_1 have an orientable embedding of genus $\leq E - V + k$". This settles the order of complexity for testing graph genus as being NP-complete. It does leave open the question of finding interesting classes of graphs where Thomassen's argument cannot be made and of settling the order of complexity there. Two such classes strongly related to graph minors are (1) graphs of tree-width $\leq w$ for some fixed w (recently settled to be in polynomial time by P.D. Seymour), and (2) graphs embedded on a fixed nonorientable surface $\tilde{\Sigma}_k$. In the first case Thomassen's reduction still applies, but the vertex-covering of edges problem is linear-time on the class. In the second case the reduction fails to maintain an unoriented genus bound, and there is good evidence the problem is polynomial-time from Proposition 16.1 and [15]. Solution of these two problems may allow a solution of a more general problem, namely do the minor-closed classes of graphs into which Thomassen's reduction of an NP-complete problem cannot be made admit polynomial-time algorithms to test orientable genus? The following proposition from [6], conjectured by Joe Fiedler, shows how to test genus for graphs on the projective plane. A similar result for graphs embeddable in the Klein bottle with representativity ≥ 4 is given in [15].

Proposition 16.1. *Let* Ψ *be an embedding with* $\Sigma(\Psi) \approx \tilde{\Sigma}_1$ *and* $G(\Psi)$ *nonplanar. Then* $g = \lfloor \rho(\Psi)/2 \rfloor$ *is the minimum genus for an orientable embedding of* $G(\Psi)$.

Corollary 16.2. *There is a polynomial-time algorithm for finding the minimum orientable genus of graphs embeddable in the projective plane.*

Proof of 16.1. Firstly, by surgery we obtain from Ψ the embedding Ψ_1 with $\Sigma(\Psi_1) \approx \Sigma_g$ for $g = \lfloor \rho(\Psi)/2 \rfloor$. Let C be a circuit in $\rho(\Psi)$ with $|C \cap G(\Psi)| =$

$\rho(\Psi)$ and enumerate in circular order the points of $C \cap \rho(\Psi)$ by $z_1, z_2, \ldots, z_p$, where $p = 2g$ or $p = 2g + 1$. Split the embedding at C for $z_1, z_2, \ldots, z_{2g}$ and twist the embedding at z_{2g+1} if necessary to obtain a planar embedding of the split graph, say Ψ_2. In $\Sigma(\Psi_2)$ cut $2g$ small holes very close to the pairs $z_1, z_2; z_3, z_4; \ldots; z_{2g-1}, z_{2g}$ one for each of the two copies of these vertex pairs. Then adjoin a handle between the holes next to the two copies z_1, z_2 and repeat this up to the two copies of z_{2g-1}, z_{2g}. Then the split can be repaired along these handles, embedding $G(\Psi)$ on Σ_g with embedding Ψ_1.

Secondly, we show by induction that a better embedding is impossible. In general as $G(\Psi)$ is nonplanar and we can embed $G(\Psi)$ on the torus if $\rho(\Psi) \leq 3$ by the procedure in the above paragraph the theorem is true in this range (clearly $2 \leq \rho(\Psi)$ when $\Sigma(\Psi) \approx \tilde{\Sigma}_1$ and $G(\Psi)$ is nonplanar). The difficulties are at the $\rho(\Psi) = 4$ level. Assuming $\rho(\Psi) \geq 4$, we let Ψ' be an embedding of the essential 3-component H of Ψ onto an orientable surface of minimum genus h. As $\Sigma(\Psi_H) \not\approx \Sigma(\Psi')$ some facial circuit C of Ψ_H is not a facial circuit of Ψ'. As facial circuits of H are peripheral and h is the minimum genus, C is a nonseparating essential circuit of Ψ'. But then $(\Psi_H)//f$ (where f is the face bounded by C) satisfies $\rho(\Psi) - 2 \leq \rho((\Psi_H)//f)$, and $(\Psi_H)//f$ embeds on a surface of genus $h - 1$. The induction hypothesis holds provided $G((\Psi_H)//f)$ is nonplanar, which is true by Proposition 10.1, if $\rho(\Psi) \geq 5$. However, this is also true when $\rho(\Psi) = 4$, because $\rho((\Psi_H)/f) \geq 3$, $\Sigma(\Psi) \approx \tilde{\Sigma}_1$, and $(\Psi_H)//f$ is obtainable by deleting the vertex of $(\Psi_H)/f$ corresponding to f. Such graphs are nonplanar by the theorem mentioned in (3) of Section 5, from [22], concerning minor-minimal embeddings on the projective plane with representativity ≥ 3. This completes the proof.

Proof of 16.2. This clearly follows from any polynomial-time algorithms for embedding graphs and computing $\rho(\Psi)$ on the projective plane.

References

[1] Albertson, M.O., Stromquist, W.R. (1982): Locally planar toroidal graphs are 5-colorable. Proc. Am. Math. Soc. **84**, 449–456

[2] Archdeacon, D. (1988): Densely embedded graphs. Manuscript

[3] Conway, J.H., Gordon, C.McA. (1983): Knots and links in spatial graphs. J. Graph Theory 7, 445–453

[4] Dirac, G.A. (1952): A property of 4-chromatic graphs and remarks on critical graphs. J. Lond. Math. Soc. **27**, 85–92

[5] Duffin, R.J. (1965): Topology of series-parallel networks. J. Math. Anal. Appl. **10**, 303–318

[6] Fiedler, J.R., Huneke, J.P., Richter, R.B., Robertson, N. (1989): Computing the orientable genus of projective graphs. Manuscript

[7] Grünbaum, B. (1967): Convex polytopes. J. Wiley & Sons, New York, N.Y. (Pure Appl. Math., Vol. 16)

[8] Motwani, R., Raghunathan, A., Saran, H. (1988): Constructive results from graph minors: linkless embeddings. Proc. 29th IEEE FOCS, pp. 398–407

[9] Mohar, B., Robertson, N., Vitray, R. (1989): Planar graphs on non-planar surfaces I. Manuscript

[10] Moser, L. (1966): Research problems in discrete geometry. Report 48-1, McGill University, Montreal

[11] Negami, S. (1988): Re-embedding of projective planar graphs. J. Comb. Theory, Ser. B **44**, 276–299

[12] Robertson, N., Seymour, P.D. (1986): Graph minors. V. Excluding a planar graph. J. Comb. Theory, Ser. B **41**, 92–114

[13] Robertson, N., Seymour, P.D. (1988): Graph minors. VII. Disjoint paths on a surface. J. Comb. Theory, Ser. B **45**, 212–254

[14] Robertson, N., Seymour, P.D.: Graph minors. VIII. A Kuratowski theorem for general surfaces. J. Comb. Theory, Ser. B, to appear

[15] Robertson, N., Thomas, R. (1989): On the orientable genus of graphs embedded in the Klein bottle. Manuscript

[16] Sachs, H. (1984): On spatial representation of finite graphs. In: Hajnal, A., Lovász, L., Sós, V. (eds.): Finite and infinite sets II. North Holland, Amsterdam (Colloq. Math. Soc. János Bolyai, Vol. 37), 649-662

[17] Thomassen, C. (1988): Embeddings of graphs with no short non-contractible cycle. Manuscript

[18] Thomassen, C. (1988): The graph genus problem is NP-complete. Manuscript

[19] Truemper, K. (1989): On the delta-wye reduction for planar graphs. J. Graph Theory **13**, 141–148

[20] Tutte, W.T. (1963): How to draw a graph. Proc. Lond. Math. Soc., III. Ser. **13**, 743–768

[21] Tutte, W.T. (1966): Connectivity in Graphs. Univ. of Toronto Press, Toronto, Ontario; Oxford Univ. Press, London (Mathematical Expositions, No 15)

[22] Tutte, W.T. (1984): Graph Theory. Addison-Wesley, Menlo Park, CA (Encycl. Math. Appl., Vol. 21)

[23] Vitray, R. (1987): Representativity and flexibility of drawings of graphs on the projective plane. Ph. D. Thesis, Ohio State University

[24] Whitney, H. (1933): On the classification of graphs. Am. J. Math. **55**, 245–254

[25] Whitney, H. (1933): 2-isomorphic graphs. Am. J. Math. **55**, 236–244

Homotopic Routing Methods

Alexander Schrijver

1. Introduction

The problem:

(1) Given: — a graph $G = (V, E)$,
     — pairs $r_1, s_1, \ldots, r_k, s_k$ of vertices of G;
     find: — pairwise disjoint paths $P_1, \ldots, P_k$ in G, where
     P_i connects r_i and s_i (for $i = 1, \ldots, k$),

is NP-complete, even for planar graphs, both in the vertex-disjoint and in the edge-disjoint case (Lynch [26]). In some special cases, however, there is a polynomial-time method for (1). These cases usually also give rise to a theorem characterizing the existence of a solution as required.

Moreover, if G is planar, one can design a heuristic or enumerative approach based on the topology of the plane. It amounts to selecting a, possibly small, set of faces $I_1, \ldots, I_p$ of G so that each of the vertices $r_1, s_1, \ldots, r_k, s_k$ is incident with at least one of these faces. Next we choose (or enumerate) for each pair r_i, s_i a curve C_i in $\mathbb{R}^2 \backslash (I_1 \cup \ldots \cup I_p)$ connecting r_i and s_i. Our problem then is to find pairwise disjoint paths $P_1, \ldots, P_k$ so that P_i is homotopic to C_i in the space $\mathbb{R}^2 \backslash (I_1 \cup \ldots \cup I_p)$. If such P_i are found, we have solved our original problem. Otherwise, we choose other curves C_i (i.e., representing other homotopies), and try again.

So in this approach (proposed by Pinter [31]) we must solve the following problem:

(2) Given: — a planar graph $G = (V, E)$, embedded in $\mathbb{R}^2$,
     — faces $I_1, \ldots, I_p$,
     — curves $C_1, \ldots, C_k$ with end points on the boundary of
     $I_1 \cup \ldots \cup I_p$,
     find: — pairwise disjoint simple paths $P_1, \ldots, P_k$ in G where
     P_i is homotopic to C_i in $\mathbb{R}^2 \backslash (I_1 \cup \ldots \cup I_p)$, for
     $i = 1, \ldots, k$.

This problem also emerges from Robertson and Seymour's work on graph minors [34]. It turns out that this problem can be solved in polynomial time in the vertex-disjoint case. The edge-disjoint case appears to be more difficult (due to the fact that the curves $C_1, \ldots, C_k$ then can be quite wild). In fact, Kaufmann and Maley

[17] recently showed that the edge-disjoint version is NP-complete. In some special cases, a polynomial-time algorithm for the edge-disjoint case has been found.

In this paper we give a survey of the results and methods for problems (1) and (2). We moreover describe some links with problems on disjoint circuits in graphs on compact surfaces, and on disjoint trees of given homotopies.

Some Conventions and Terminology

By an *embedding* of a graph $G = (V, E)$ in the plane or any other surface, we mean an embedding without intersecting edges. When speaking of a planar graph, we implicitly assume it to be embedded in the plane. We identify an embedded graph with its topological image. Edges are considered as open curves, and faces as open regions. By bd (...) we denote the boundary of ... . An $r - s$-*path* is a path from r to s.

2. Vertex-Disjoint Paths and Trees

As mentioned, the problem:

(3)   Given: — a graph $G = (V, E)$,
          — pairs $r_1, s_1, \ldots, r_k, s_k$ of vertices of G,
      find: — pairwise vertex-disjoint paths $P_1, \ldots, P_k$ in G, where
              P_i connects r_i and s_i (for $i = 1, \ldots, k$),

is NP-complete (Lynch [26]). On the other hand, Robertson and Seymour [35] showed:

Theorem 1. *For each fixed k, there is a polynomial-time algorithm for* (3).

In fact, the algorithm has running time $O(|V|^2 \cdot |E|)$, but the constant depends heavily on k. For details, see also Robertson and Seymour [36].

For the special case of planar graphs, there are some further polynomial-time methods. Clearly, a necessary condition for planar G is:

(4)   (*Cut condition*) for each closed curve D in $\mathbb{R}^2$, the number of intersections with G is at least the number of pairs r_i, s_i separated by D.

Here D *separates* r_i, s_i if each curve connecting r_i and s_i intersects D. Obviously, in (4) we may restrict D to closed curves intersecting G only in vertices of G, and not in edges.

Robertson and Seymour [33] observed that there is an easy algorithm for (3) in the case where G is planar, and $r_1, s_1, \ldots, r_k, s_k$ all lie on the boundary of one face I. In that case a necessary condition is:

(5)   (*Cross-freedom condition*) no two pairs r_i, s_i and r_j, s_j are crossing.

Here r_i, s_i and r_j, s_j are said to *cross* if r_i, s_i, r_j, s_j are all distinct and r_i, r_j, s_i, s_j occur cyclically (clockwise or anti-clockwise) around the boundary of I:

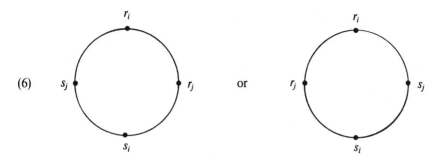

(6)

Theorem 2. *If G is planar and $r_1, s_1, \ldots, r_k, s_k$ all are on the boundary of one face I, problem* (3) *is solvable in polynomial time.*

Proof. Without loss of generality, $r_i \neq s_i$ for all i. We first check if $r_1, s_1, \ldots, r_k, s_k$ are all distinct and if the cross-freedom condition holds. The cross-freedom condition implies that there exists a pair r_i, s_i so that at least one of the two $r_i - s_i$ paths along the boundary of I does not contain any r_j or $s_j (j \neq i)$. Without loss of generality, $i = 1$. Let Q_1 be this path. Now if (3) has a solution, there is one with $P_1 = Q_1$ (as in any solution of (3), path P_1 can be "pushed" against the boundary of I). Leaving out the vertices in Q_1 from G, together with all edges incident to them, we obtain a graph G'. We next solve problem (3) for $G', r_2, s_2, \ldots, r_k, s_k$. If we find paths $P_2, \ldots, P_k$ then $P_1, P_2, \ldots, P_k$ form a solution to the original problem. Otherwise, (3) has no solution.   □

In fact this algorithm also easily implies the following theorem:

Theorem 3. *Let G be planar, so that $r_1, s_1, \ldots, r_k, s_k$ are all on the boundary of one face. Then* (3) *has a solution if and only if the cut condition and the cross-freedom condition holds.*

In fact, the following generalization of Theorem 2 follows from the homotopic approach to be described in Section 5 below (see Theorem 34):

Theorem 4. *For each fixed p there exists a polynomial-time algorithm for problem* (3), *whenever G is planar so that $r_1, s_1, \ldots, r_k, s_k$ can be covered by the boundaries of at most p faces.*

We conjecture that also the following holds:

Conjecture. Problem (3) is solvable in polynomial time whenever the graph $H := (V, E \cup \{\{r_1, s_1\}, \ldots, \{r_k, s_k\}\})$ is planar.

So here the pairs $\{r_1, s_1\}, \ldots, \{r_k, s_k\}$ are now edges in H, which edges we may assume to form a matching in H.

Extension to Disjoint Trees

There is a direct extension of the above results to trees instead of paths. Consider the problem:

(7)   Given: — a graph $G = (V, E)$,
              — subsets $W_1, \ldots, W_k$ of V,
         find: — pairwise vertex-disjoint trees $T_1, \ldots, T_k$ where T_i
                  covers W_i (for $i = 1, \ldots, k$).

Again this problem is NP-complete (as it generalizes (3)). For planar graphs we can proceed similarly to above. Again a necessary condition is:

(8)   (*Cut condition*) for each closed curve D in $\mathbb{R}^2$, the number of intersections with G is at least the number of W_i separated by D.

Here D *separates* W_i if D separates at least two points in W_i.

   If all points in $W_1 \cup \ldots \cup W_k$ are on the boundary of one face I, there is the following necessary condition:

(9)   (*Cross-freedom condition*) no two sets W_i and W_j are crossing.

Here W_i and W_j are said to *cross* if W_i contains two points r', s' and W_j contains two points r'', s'' so that the pairs r', s' and r'', s'' cross.

   Now the following two theorems extend Theorems 2 and 3:

Theorem 5. *If G is planar, and all vertices in $W_1 \cup \ldots \cup W_k$ are on the boundary of one face I, then problem (7) can be solved in polynomial time.*

Theorem 6. *Let G be planar, so that all points in $W_1 \cup \ldots \cup W_k$ are on the boundary of one face. Then problem (7) has a solution if and only if the cut condition and the cross-freedom condition hold.*

   Again, the following generalization of Theorem 5 follows from the homotopic approach to be described in Section 5 below:

Theorem 7. *For each fixed p there exists a polynomial-time algorithm for problem (7), whenever G is planar so that the vertices in $W_1 \cup \ldots \cup W_k$ can be covered by the boundaries of at most p faces.*

   The conjecture above can be extended as follows. Let $u_1, \ldots, u_k$ be new (abstract) vertices. Let F be the set of all pairs $\{u_i, w\}$ where $i \in \{1, \ldots, k\}$ and $w \in W_i$. We conjecture:

Conjecture. Problem (7) is solvable in polynomial time whenever the graph $H := (V \cup \{u_1, \ldots, u_k\}, E \cup F)$ is planar.

3. Edge-Disjoint Paths and Multicommodity Flows

We now turn to the edge-disjoint case. Consider the problem:

(10)   Given — a graph $G = (V, E)$,
              — pairs $r_1, s_1, \ldots, r_k, s_k$ of vertices of G,
         find: — pairwise edge-disjoint paths $P_1, \ldots, P_k$ in G where
                  P_i connects r_i and s_i (for $i = 1, \ldots, k$).

It is not difficult to see that Robertson and Seymour's theorem (Theorem 1 above) implies:

Theorem 8. *For each fixed k, there exists a polynomial-time algorithm for problem* (10).

This follows by considering the line-graph of G. In general however, problem (10) is NP-complete, even for planar G.

Again, a necessary condition for (10) is:

(11) *(Cut condition)* for each $X \subseteq V : |\delta(X)| \geq |\rho(X)|$.

(12)

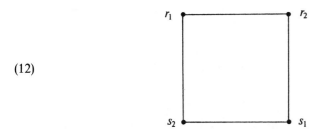

Here $\delta(X)$ denotes the set of edges with exactly one end point in X. By $\rho(X)$ we denote the set of those $i \in \{1, \ldots, k\}$ for which exactly one of r_i and s_i belongs to X.

As is well-known, Menger's theorem [27] states that the cut condition is also sufficient if $r_1 = \ldots = r_k$ and $s_1 = \ldots = s_k$. We leave it as an exercise to derive from this that the cut condition is sufficient if we require only $r_1 = \ldots = r_k$.

However, in the general case it is not a sufficient condition, as is shown by the simple example of (12) above.

So one may not hope for many more interesting cases where the cut condition suffices.

It turns out however that one more condition (which is clearly *not* a necessary condition) is quite powerful:

(13) *(Parity condition)* for each vertex v of G, the number

$$|\delta(\{v\})| + |\rho(\{v\})|$$

is even.

In particular, every vertex not in $\{r_1, s_1, \ldots, r_k, s_k\}$ should have even degree. This is why cases satisfying (13) sometimes are called *eulerian*.

The following is a theorem of Lomonosov [22, 23, 24] (extending earlier results of Hu [11], Rothschild and Whinston [37, 38], Dinits [1], Papernov [30] and Seymour [53] (cf. Lovász [25], Seymour [52])):

Theorem 9. *The cut condition implies that problem* (10) *has a solution, in case the parity condition holds and* $\{r_1, s_1\}, \ldots, \{r_k, s_k\}$ *do not contain four pairs forming one of the following configurations:*

(14)

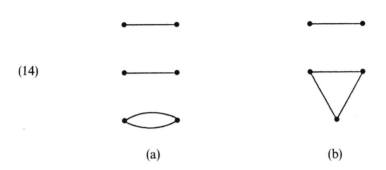

(a)   (b)

For a proof, see also Frank [6].

It is not difficult to see that excluding (14)(a) and (b) is equivalent to the condition that the graph on $\{r_1, s_1, \ldots, r_k, s_k\}$ with edges $\{r_1, s_1\}, \ldots, \{r_k, s_k\}$ (possibly parallel) is

(15) either (i) the complete graph K_4 (possibly with parallel edges), or (ii) the circuit C_5 (possibly with parallel edges), or (iii) the union of two stars (possibly with parallel edges), or (iv) a graph consisting of three disjoint edges.

The following examples show that the condition in Theorem 9 is in a sense tight:

(16)

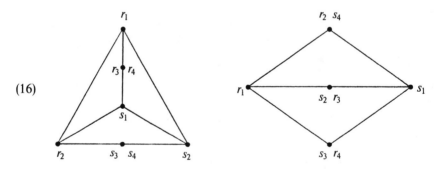

Theorem 9 has the following implication for multicommodity flows. For any "demand" function $d : \{1, \ldots, k\} \longrightarrow \mathbb{Q}_+$ and any "capacity" function $c : E \longrightarrow \mathbb{Q}_+$, let a *multicommodity flow* be a system of paths $P_{11}, \ldots, P_{1t_1}, P_{21}, \ldots, P_{2t_2}, \ldots, P_{k1}, \ldots, P_{kt_k}$, together with a system of rationals $\lambda_{11}, \ldots, \lambda_{1t_1}, \lambda_{21}, \ldots, \lambda_{2t_2}, \ldots, \lambda_{k1}, \ldots, \lambda_{kt_k} \geq 0$ satisfying:

(17) (i) $\displaystyle\sum_{j=1}^{t_i} \lambda_{ij} = d_i$ $(i = 1, \ldots, k)$,

(ii) $\displaystyle\sum_{i=1}^{k}\sum_{j=1}^{t_i}\lambda_{ij}\mathscr{X}^{P_{ij}}(e) \leq c(e)$ $(e \in E)$.

Here $\mathscr{X}^P(e)$ denotes the number of times P passes e.

If the λ_{ij} are integral, we say that the multicommodity flow is *integral*. If the λ_{ij} are half-integral, we say that the multicommodity is *half-integral*. If $d_i = 1$ for all i and $c(e) = 1$ for all e, we call a multicommodity flow a *fractional* solution to problem (10). Indeed, an integral multicommodity flow then corresponds to a solution to (10).

Again we have a cut condition necessary for the existence of a multicommodity flow (given a demand function d and a capacity function c):

(18)   (*Cut condition*) for each $X \subseteq V : \sum_{e\in\delta(X)} c(e) \geq \sum_{i\in\rho(X)} d_i$.

Note that there are the following implications:

(19)   ∃ integral multicommodity flow $\Longrightarrow$
         ∃ half-integral multicommodity flow $\Longrightarrow$
         ∃ multicommodity flow $\Longrightarrow$
         cut condition.

Now Theorem 9 implies that in some cases we can reverse the implications, as was shown by Papernov [30] (forming an extension of Ford and Fulkerson's max-flow min-cut theorem [5]). Consider the property

(20) $\{r_1, s_1\}, \ldots, \{r_k, s_k\}$ do not contain one of the following configurations:

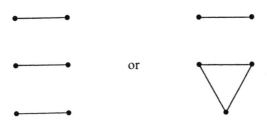

Theorem 10. *If d and c are integral-valued, and condition (20) is satisfied, then the cut condition (18) is equivalent to the existence of a half-integral multicommodity flow.*

This can be derived from Theorem 9 by replacing each edge e of G by $2c(e)$ parallel edges, and each pair $\{r_i, s_i\}$ by $2d_i$ "parallel" pairs.

Theorem 11. *If (20) is satisfied, then the cut condition is equivalent to the existence of a multicommodity flow.*

This can be seen by multiplying d and c by some natural number K so that Kd and Kc are integral, and next by applying Theorem 10.

This theorem is tight in the sense that if (20) is not satisfied, there exists a graph G, a demand function d and a capacity function c for which the cut condition is satisfied, but no multicommodity flow as required exists - this can be derived directly from the examples (16).

If $d \equiv 1$ and $c \equiv 1$, Theorem 10 reduces to:

Theorem 12. *If (20) is satisfied, then the cut condition (11) is equivalent to the existence of a half-integral solution to problem (10).*

Karzanov [16] gave an extension of part of this result. Consider the property:

(21) $\{r_1, s_1\}, \ldots, \{r_k, s_k\}$ do not contain one of the following configurations:

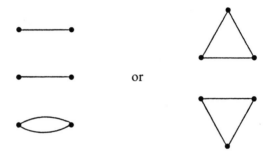

It can be checked easily that this means that the graph on $\{r_1, s_1, \ldots, r_k, s_k\}$ with edges $\{r_1, s_1\}, \ldots, \{r_k, s_k\}$ is:

(22)   either (i) the complete graph K_5 (possibly with parallel edges), or (ii) the union of a triangle and a star (possibly with parallel edges), or (iii) the union of two stars (possibly with parallel edges), or (iv) the graph consisting of three disjoint edges.

Karzanov showed:

Theorem 13. *If (21) holds and the parity condition holds, then the existence of a fractional solution to (10) implies the existence of a solution to (10).*

Again this implies:

Theorem 14. *If (21) holds, then (10) has a half-integral solution if and only if (10) has a fractional solution.*

Example (23) shows that it is necessary to exclude the second configuration in (21).

In this example a fractional solution exists, but no integral solution. It is not known to me if also the first configuration in (21) must be excluded. In

[24], Lomonosov gives an example showing that it is necessary to require that $\{r_1, s_1\}, \ldots, \{r_k, s_h\}$ do not contain 38 pairs, covering 6 points, so that they fall apart in three sets of parallel edges, of sizes 2, 18 and 18, respectively.

(23)

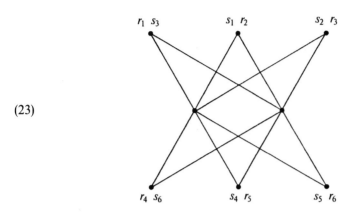

Duality

Some of the results above have a "dual" counterpart, in terms of packing of cuts as was noticed by Karzanov [14] and Seymour [51]. Consider the convex cone K in $\mathbb{R}^k \times \mathbb{R}^E$ consisting of all vectors $(d; c)$ for which (17) has a solution $\lambda_{ij} \geq 0$. So K is the convex cone generated by all vectors

(24)   (i) $(\varepsilon_i; \mathscr{X}^P)$ $(i = 1, \ldots, k; P$ is an $r_i - s_i$ path),

         (ii) $(0; \varepsilon_e)$ $(e \in E)$.

Here ε_i denotes the i-th unit basis vector in $\mathbb{R}^k$, and ε_e denotes the e-th unit basis vector in $\mathbb{R}^E$. By χ^P we denote the function in $\mathbb{R}^E$ given by $\chi^P(e) :=$ the number of times P passes e.

Now the content of Theorem 11 is that, if (20) is satisfied, then K is exactly the cone of all vectors which have nonnegative inner product with all vectors:

(25)   (i) $(-\mathscr{X}^{\rho(X)}; \mathscr{X}^{\delta(X)})$ $(X \subseteq V)$,

         (ii) $(\varepsilon_i; 0)$ $(i = 1, \ldots, k)$,

         (iii) $(0; \varepsilon_e)$ $(e \in E)$.

Here $\mathscr{X}^{\rho(X)}$ and $\mathscr{X}^{\delta(X)}$ denote the incidence vectors of $\rho(X)$ and $\delta(X)$, respectively.

Now by duality (Farkas' lemma), the convex cone generated by the vectors (25) is exactly equal to the set of vectors having nonnegative inner product with all vectors (24) (if (20) is satisfied). In fact, it is equivalent to the following:

Theorem 15. *Let* (20) *be satisfied. Then there exist cuts* $\delta(X_1), \ldots, \delta(X_t)$ *and rationals* $\mu_1, \ldots, \mu_t \geq 0$ *so that:*

(26)   (i) $dist_G(r_i, s_i) = \sum (\mu_j \mid i \in \rho(X_j))$ (for each $i = 1, \ldots, k$),

      (ii) $\sum_{j=1}^{t} \mu_j \mathcal{X}^{\delta(X_j)}(e) \leq 1$ (for each $e \in E$).

Here $dist_G(r, s)$ denotes the distance between r and s in G. To derive Theorem 15 note that the vector

(27) $(-dist_G(r_1, s_1), \ldots, -dist_G(r_k, s_k); 1, \ldots, 1)$

has nonnegative inner product with all vectors in (24). Hence it can be written as a nonnegative linear combination of vectors in (25), yielding cuts $\delta(X_j)$ and rationals μ_j as required.

Now Karzanov [15] showed that if G is bipartite, we can take the μ_j integral. That means:

Theorem 16. *Let G be bipartite, and $r_1, s_1, \ldots, r_k, s_k$ be vertices of G so that (20) is satisfied. Then there exist pairwise disjoint cuts $\delta(X_1), \ldots, \delta(X_t)$ so that for each $i = 1, \ldots, k$:*

(28) $dist_G(r_i, s_i) = $ the number of cuts $\delta(X_j)$ separating r_i and s_i.

Here $\delta(X)$ is said to *separate* r and s if X contains exactly one of r and s. Theorem 16 extends theorems of Hu [12] and Seymour [50] for the case $k = 2$.

Theorem 16 implies:

Theorem 17. *The μ_j in Theorem 15 can be taken from $\{\frac{1}{2}, 1\}$.*

This follows by replacing each edge of G by two edges in series, thus making a bipartite graph.

For a short proof of some of the results in this section, see [47].

4. Edge-Disjoint Paths in Planar Graphs

Although the forbidden configurations given in Section 3 are "tight", there are more cases where the cut condition suffices, if we restrict G to planar graphs. Again, we consider problem (10). So we have a graph $G = (V, E)$ and pairs $r_1, s_1, \ldots, r_k, s_k$ of vertices, and we ask for pairwise edge-disjoint paths $P_1, \ldots, P_k$, where P_i connects r_i and s_i ($i = 1, \ldots, k$).

A basic result due to Okamura and Seymour [29] requires the following property for planar G :

(29) G has a face I so that $r_1, s_1, \ldots, r_k, s_k$ all belong to the boundary of I.

Theorem 18. *Let G be planar so that (29) is satisfied. Moreover, let the parity condition (13) hold. Then (10) has a solution if and only if the cut condition holds.*

For a proof we refer to Frank [6].

In fact, Okamura [28] showed that condition (29) can be weakened to:

(30) G has faces I_1, I_2 so that for each $i = 1, \ldots, k : r_i, s_i \in bd(I_1)$ or $r_i, s_i \in bd(I_2)$.

Theorem 19. *Let G be planar so that (30) is satisfied. Moreover, let the parity condition (13) hold. Then (10) has a solution if and only if the cut condition holds.*

Also a proof of this theorem is given in Frank [6].

One may not allow in Okamura's theorem "mixed pairs", i.e. pairs r_i, s_i with $r_i \in bd(I_1)$ and $s_i \in bd(I_2)$. Neither can one extend the theorem to more than two faces. These facts are shown by the following example:

(31)

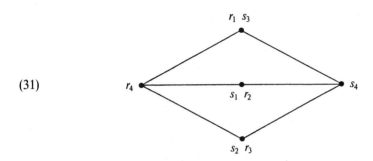

In fact, in this example not even a fractional solution exists. The following example (Hurkens, Schrijver and Tardos [13]), with mixed pairs, satisfying the parity condition, has a fractional solution, but no integral solution:

(32)

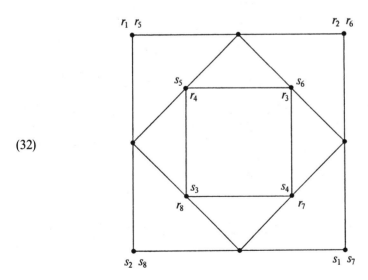

In [46] we showed that in a particular case of mixed pairs the cut condition suffices. Let I_1 and I_2 be two faces of G, where I_1 is (without loss of generality)

the unbounded face. Let $r_1, s_1, \ldots, r_k, s_k$ be vertices so that:

(33) $r_1, \ldots, r_k$ are on bd (I_1) in clockwise order,

              $s_1, \ldots, s_k$ are on bd (I_2) in anti- clockwise order.

Theorem 20. *Let G be planar so that* (33) *is satisfied. Moreover, let the parity condition* (13) *hold. Then* (10) *has a solution if and only if the cut condition holds.*

Example (31) also shows that we cannot allow $r_1, \ldots, r_k$ and $s_1, \ldots, s_k$ to occur both in clockwise order on $bd(I_1)$ and $bd(I_2)$, respectively.

Seymour [54] considered the following property:

(34)   the graph $(V, E \cup \{\{r_1, s_1\}, \ldots, \{r_k, s_k\}\})$ is planar.

Theorem 21. *Let* (34) *and the parity condition* (13) *be satisfied. Then* (10) *has a solution if and only if the cut condition is satisfied.*

Again, for a proof see Frank [6].
In fact, the proofs of Theorems 18, 19, 20 and 21 all yield polynomial-time algorithms for finding the required paths. These theorems also imply the following:

Theorem 22. *Let G be planar, and let* (30), (33) *or* (34) *be satisfied. Then problem* (10) *has a half-integral solution if and only if the cut condition is satisfied.*

More generally,

Theorem 23. *Let G be planar, and let* (30), (33) *or* (34) *be satisfied. Let $d \in \mathbb{Z}_+^k$ and $c \in \mathbb{Z}_+^E$. Then there exists a half-integral multicommodity flow if and only if the cut condition* (18) *is satisfied.*

Duality

Similar to the cut packing results in Section 3 dual to Theorem 9, there are theorems dual to Theorems 18, 19, 20 and 21.

The following result (Hurkens, Schrijver and Tardos [13]) is dual to the Okamura-Seymour theorem (Theorem 18):

Theorem 24. *Let G be a planar bipartite graph. Then there exist pairwise disjoint cuts $\delta(X_1), \ldots, \delta(X_t)$ so that for each pair of vertices u, v on the outer boundary:*

(35) $dist_G(u, v) =$ number of cuts $\delta(X_j)$ separating u, v.

In fact, this can be derived from the Okamura-Seymour theorem, as we will show now. Let

(36) $(v_0, e_1, v_1, \ldots, v_{k-1}, e_k, v_k)$

be the vertices and edges on the outer boundary of G, where $v_k = v_0$. Define for each pair e_i, e_j :

(37) $r(e_i, e_j) := \frac{1}{2}(dist_G(v_{i-1}, v_{j-1}) + dist_G(v_i, v_j) - dist_G(v_{i-1}, v_j) - dist_G(v_i, v_{j-1}))$.

It is not difficult to see that this number is 0 or 1 (as G is bipartite and planar). Let Q be the set of pairs $\{e_i, e_j\}$ with $r(e_i, e_j) = 1$. Now for each v_g, v_h one has:

(38) $dist_G(v_g, v_h)$ = number of pairs $\{e_i, e_j\} \in Q$ crossing $\{v_g, v_h\}$.

Here $\{e_i, e_j\}$ crosses $\{v_g, v_h\}$ if v_g and v_h belong to different components of the circuit (36) after deleting e_i and e_j. Equality (38) follows from (assuming without loss of generality $0 = g < h < k$) :

(39)   number of pairs $\{e_i, e_j\} \in Q$ crossing $\{v_g, v_h\} = \sum_{i=1}^{h} \sum_{j=h+1}^{k} r(e_i, e_j) =$
$\frac{1}{2}\sum_{i=1}^{h} \sum_{j=h+1}^{k}(dist_G(v_{i-1}, v_{j-1}) + dist_G(v_i, v_j) - dist_G(v_{i-1}, v_j) - dist_G(v_i, v_{j-1}))$
$= dist_G(v_0, v_h)$

(by cancellation).

Now we can apply the Okamura-Seymour theorem to a slight modification of the dual graph of G, so that (38) implies that for each $\{e_i, e_j\} \in Q$ there exists a cut $\delta(X_{ij})$ containing e_i and e_j, in such a way that the $\delta(X_{ij})$ are pairwise disjoint. By (38) again, these cuts have the required property (35).

In [49] it is shown that the more general dual to Okamura's theorem (Theorem 19) also holds:

Theorem 25. *Let G be a planar bipartite graph, and let I_1 and I_2 be two of its faces. Then there exist pairwise disjoint cuts $\delta(X_1), \ldots, \delta(X_t)$ so that (35) holds for each pair of vertices u, v with $u, v \in bd(I_1)$ or $u, v \in bd(I_2)$.*

We do not see a direct way of deriving this from Okamura's theorem. Similar results hold for the duals of Theorem 19 and 20:

Theorem 26. *Let G be a planar bipartite graph, and let $r_1, s_1, \ldots, r_k, s_k$ be pairs of vertices so that (33) or (34) is satisfied. Then there exist pairwise disjoint cuts $\delta(X_1), \ldots, \delta(X_t)$ so that for each $i = 1, \ldots, k$:*

(40) $dist_G(r_i, s_i)$ = number of cuts $\delta(X_j)$ separating r_i and s_i.

With respect to (33) this follows from the results in [46]. For (34), this follows from the "sums of circuits" theorem of Seymour [51], as was communicated to me by A.V. Karzanov: Let $H = (W, F)$ be a planar graph, and let $g : F \longrightarrow \mathbb{Z}_+$ be so that $\sum_{e \ni v} g(e)$ is even for each vertex v; Seymour's theorem says that g is a nonnegative integral combination of incidence vectors of circuits in H, if and only if

(41) $g(e') \leq \sum_{e \in D \setminus e'} g(e)$

for each cut D and each $e' \in D$. Theorem 26 is derived by applying Seymour's theorem to the graph H dual to $(V, E \cup \{\{r_1, s_1\}, \ldots, \{r_k, s_k\}\})$, with $g(e) := 1$ for edge e of H dual to an edge in E, and $g(e) := dist_G(r_i, s_i)$ for edge e of H dual to $\{r_i, s_i\}(i = 1, \ldots, k)$.

The Projective Plane and the Klein Bottle

Some of the results have an analogue in terms of compact surfaces. First consider the projective plane S. It arises from the disk

(42)

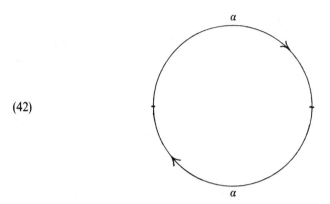

by identifying opposite points. There are two types of simple closed curves on S : the homotopically trivial closed curves, and the homotopically nontrivial closed curves (which form one homotopy class).

The homotopically trivial closed curves are those closed curves C whose removal disconnects S. The homotopically nontrivial closed curves are not disconnecting.

The homotopically trivial closed curves are also those closed curves C which are *orientation-preserving*, i.e., after one turn of C the meaning of "left" and "right" is not changed. The homotopically nontrivial closed curves are those closed curves C which are *orientation-reversing*, i.e., after one turn of C the meaning of "left" and "right" is exchanged.

Now Lins [21] proved:

Theorem 27. *Let $G = (V, E)$ be an eulerian graph embedded on the projective plane S. Then the maximum number of pairwise edge-disjoint homotopically nontrivial circuits in G is equal to the minimum number of edges intersecting all homotopically nontrivial circuits.*

This theorem can be derived from the Okamura-Seymour theorem (Theorem 18) as follows. Let $F \subseteq E$ be a minimum set of edges intersecting all homotopically nontrivial circuits in G. It is not difficult to see that there exists a homotopically nontrivial simple closed curve D in S so that F is the set of edges intersected by D. Removing D from S gives a disk, on which $G' := (V, E \backslash F)$ is embedded. Let $\{r_1, s_1\}, \ldots, \{r_k, s_k\}$ be the collection of pairs of end points of the edges in F (so $k = |F|$). The fact that F has minimum size implies that the cut condition (11) is satisfied with respect to $G', r_1, s_1, \ldots, r_k, s_k$. Hence by the Okamura-Seymour theorem there exist pairwise edge-disjoint paths $P_1, \ldots, P_k$ in G' connecting $r_1, s_1; \ldots; r_k, s_k$, respectively. Extending these paths with the edges in F gives a set of k circuits as required.

In fact, by a construction of Lins [21], one can derive conversely the Okamura-Seymour theorem from Lins' theorem (Theorem 27). Indeed, in the Okamura-Seymour theorem one may assume without loss of generality that all pairs r_i, s_i and r_j, s_j are crossing (with respect to the unbounded face). If they are not, there are two pairs r_i, s_i and r_j, s_j so that (may be after interchanging r_i and s_i) r_i, r_j, s_j, s_i are in this order on the boundary of the unbounded face (clockwise, say), and so that the path Q from r_i to r_j along this boundary (clockwise) does not contain any other vertices from $r_1, s_1, \ldots, r_k, s_k$:

(43)

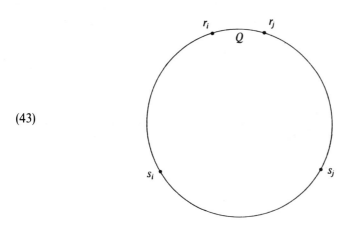

Now extend G, in the unbounded face, as follows:

(44)

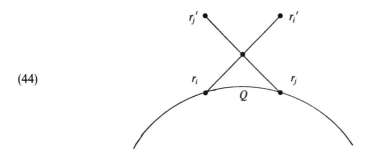

Replace r_i by r_i' and r_j by r_j'. It is not difficult to see that both the conditions and the conclusion of the Okamura- Seymour theorem are invariant under this modification.

After a finite number of such modifications we obtain a situation where $r_1, s_1, \ldots, r_k, s_k$ are pairwise crossing. After that we can embed the graph $(V, E \cup \{\{r_1, s_1\}, \ldots, \{r_k, s_k\}\})$ in the projective plane, in such a way that a circuit is orientation-reversing if and only if it contains an odd number of edges from $\{r_1, s_1\}, \ldots, \{r_k, s_k\}$. If the cut condition (11) is satisfied, the minimum size of an edge set intersecting all orientation-reversing circuits is k. Hence by Lins' theorem, there exist k pairwise edge-disjoint orientation-reversing circuits, each of which cannot contain more than one edge from $\{r_1, s_1\}, \ldots, \{r_k, s_k\}$. Hence each contains exactly one such edge. It gives in the original graph G paths as required.

By passing over to the surface dual, Theorem 27 gives:

Theorem 28. *Let $G = (V, E)$ be a bipartite graph embedded on the projective plane S. Then the minimum length of an orientation-reversing circuit is equal to the maximum number of pairwise disjoint edge sets, each intersecting all orientation-reversing circuits.*

Theorems 27 and 28 on the projective plane, parallel to the Okamura-Seymour theorem, can be extended as follows to the Klein bottle, extending Okamura's theorem (Theorem 19) and Theorem 20.

Note that the Klein bottle can be constructed from the cylinder in two possible ways. First, we can identify opposite points on one boundary, and similarly identify opposite points on the other boundary:

(45)

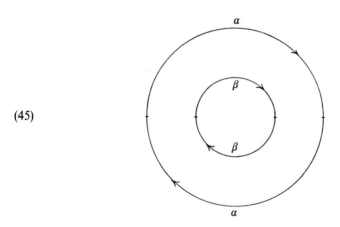

A second representation also comes from the cylinder. Now we identify one boundary in clockwise orientation with the other boundary in anti-clockwise orientation:

(46)

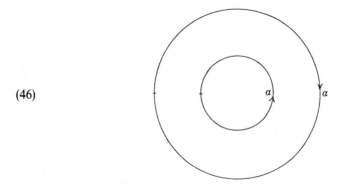

This is the usual representation of the Klein bottle.

Now in [46] we showed:

Theorem 29. *Let G be an eulerian graph embedded on the Klein bottle. Then the maximum number of pairwise edge-disjoint orientation-reversing circuits in G is equal to the minimum number of edges intersecting all orientation-reversing circuits.*

This can be derived from Theorems 19 and 20, in a similar way as Lins' theorem is derived from Theorem 18. In fact Theorems 19 and 20 correspond to the two representations of the Klein bottle described above. It is not difficult to see (by adding a "cross-cap") that Theorem 29 implies Lins' theorem.

Similarly, from Theorem 25 one can derive an extension of Theorem 28:

Theorem 30. *Let G be a bipartite graph embedded on the Klein bottle. Then the minimum length of an orientation-reversing circuit in G is equal to the maximum number of pairwise disjoint edge sets, each intersecting all orientation-reversing circuits.*

5. Vertex-Disjoint Homotopic Paths and Trees

The problem:

(47)   Given: — a planar graph $G = (V, E)$,
            — pairs $r_1, s_1, \ldots, r_k, s_k$ of vertices,
    find: — pairwise vertex-disjoint paths $P_1, \ldots, P_k$ in G,
            where P_i connects r_i and s_i (for $i = 1, \ldots, k$),

is NP-complete. So in order to solve this problem, one seemingly is bound to nonpolynomial or suboptimal methods, like enumeration and heuristics.

Pinter [31] proposed to make use of the topology of the plane, and to classify the possible solutions after their homotopy with respect to certain "holes" in the plane.

That is, select a number of faces $I_1, \ldots, I_p$ (including the unbounded face), such that $r_1, s_1, \ldots, r_k, s_k$ all are on the boundary of $I_1 \cup \ldots \cup I_p$. Two curves C, C' : $[0, 1] \longrightarrow \mathbb{R}^2 \backslash (I_1 \cup \ldots \cup I_p)$ are called *homotopic* (in the space $\mathbb{R}^2 \backslash (I_1 \cup \ldots \cup I_p)$) if there exists a continuous function $\Phi : [0, 1] \times [0, 1] \longrightarrow \mathbb{R}^2 \backslash (I_1 \cup \ldots \cup I_p)$ such that

(48) $$\Phi(x, 0) = C(x), \Phi(x, 1) = C'(x),$$
$$\Phi(0, x) = C(0), \Phi(1, x) = C(1)$$

for all $x \in [0, 1]$. Note that this implies that $C(0) = C'(0)$ and $C(1) = C'(1)$.

So C and C' are homotopic if C can be shifted continuously over $\mathbb{R}^2 \backslash (I_1 \cup \ldots \cup I_p)$ to C', without changing the beginning point or the end point of the curve.

Homotopy determines an equivalence relation between curves. Curves C, C' beeing homotopic is denoted by $C \sim C'$.

Since each path in G can be considered as a curve in $\mathbb{R}^2 \backslash (I_1 \cup \ldots \cup I_p)$, it also belongs to some homotopy class. So one approach to solve problem (48) is first to choose for each pair r_i, s_i a homotopy class of curves connecting r_i and s_i (represented by one curve C_i), and next to find paths $P_1, \ldots, P_k$ so that $P_i \sim C_i$ (for $i = 1, \ldots, k$).

This approach can be done in an enumerative way, by enumerating all possible choices of homotopy classes (there are some direct ways of ensuring finiteness of this enumeration, by excluding trivially infeasible choices), or alternatively in a heuristic way, by guessing a choice of homotopy classes, and locally improving it in case it turns out infeasible.

This approach asks for solving the following problem:

(49)   Given: — a planar graph $G = (V, E)$, embedded in $\mathbb{R}^2$,
         — faces $I_1, \ldots, I_p$ of G (including the unbounded face),
         — curves $C_1, \ldots, C_k$ with end points on $bd(I_1 \cup \ldots \cup I_p)$,
       find: — pairwise vertex-disjoint simple paths $P_1, \ldots, P_k$ where
         P_i is homotopic to C_i in $\mathbb{R}^2 \backslash (I_1 \cup \ldots \cup I_p)$ (for
         $i = 1, \ldots, k$).

The following was shown in [7]:

Theorem 31. *Problem* (49) *is solvable in polynomial time.*

The proof in [7] used the ellipsoid method. Below we shall give a sketch of the method based on [42]. In [48] we give an algorithm with running time $0(|V|^2 \cdot \log^2 |V|)$. Earlier, a polynomial-time algorithm for (49) was given by Leiserson and Maley [20] in case G is a "grid" graph - an important case for VLSI-design. Moreover, Robertson and Seymour [33] gave a polynomial-time algorithm for (49) if $p = 1$ (which is Theorem 2) and if $p = 2$.

In fact, in [44] we gave an polynomial-time algorithm for a problem more general than (49), viz. where one wants to connect sets of points by trees instead of paths - see below.

Sketch of the Algorithm for (49)
We give a sketch of the algorithm of [42] for problem (49), leaving out many details. It consists of four basic steps:

(50)   I.   Uncrossing $C_1, \ldots, C_k$,
       II.   Determining the system $Ax \leq b$,
       III. Solving the system $Ax \leq b$ in integers,
       IV.   Shifting the curves.

I. Uncrossing $C_1, \ldots, C_k$

We first "uncross" $C_1, \ldots, C_k$ so as to make them simple and pairwise disjoint. That is, if C_i and C_j have a crossing x, they should have a second crossing y so that the parts of C_i and C_j in between of x and y are homotopic:

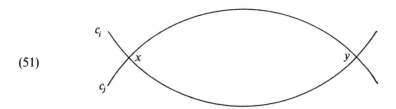

(51)

So roughly speaking, none of the faces $I_1, \ldots, I_p$ is contained in the region enclosed. If C_i and C_j have a crossing x, and they would not have a second crossing y with this property, then problem (49) has no solution.

Now replace (51) by:

(52)

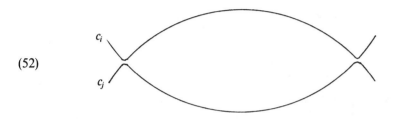

Now the new C_i and C_j are homotopic to the original C_i and C_j. In a similar way we can uncross self-crossings of any C_i. Repeating this we will end up with

(53) curves $\tilde{C}_1, \ldots, \tilde{C}_k$ in $\mathbb{R}^2 \backslash (I_1 \cup \ldots \cup I_p)$ being simple and pairwise disjoint, so that $\tilde{C}_i \sim C_i$ for $i = 1, \ldots, k$

(or curves with this property do not exist at all, in which case (49) trivially has no solution). Without loss of generality, $\tilde{C}_i = C_i$ for all i.

II. Determining the System $Ax \leq b$

We next determine a system $Ax \leq b$ of linear inequalities (where A is a matrix and b is a column vector).

First "blow up" the graph G slightly. That is, each vertex v of G becomes a disk D_v, and each edge e becomes a "channel":

(54)

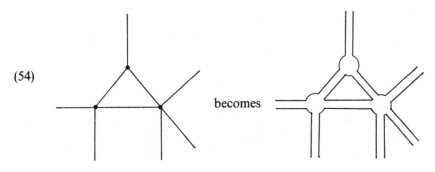

becomes

Let H be the blown-up "graph". Each face F of G corresponds naturally to a "face" F' of H. We may assume (by shifting slightly), that $I_1' = I_1,\ldots,I_p' = I_p$.

We can "push" the C_i so that they are simple and pairwise disjoint and they are in the interior of H. So we get, e.g.,

(55)

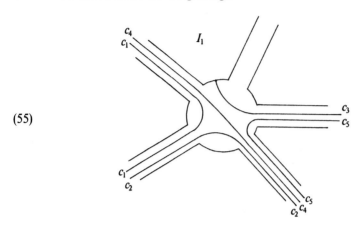

Consider now any disk D_v together with all C_i passing D_v :

(56)

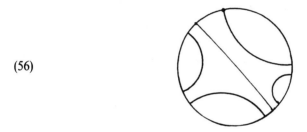

Each time a curve C_i passes D_v, we introduce a small line segment crossing C_i in D_v :

(57)

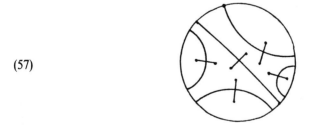

We do this for each vertex v of G. This gives us a set $\mathscr{L}$ of pairwise disjoint line segments. Let U be the set of end points of line segments in $\mathscr{L}$. So $|U| = 2|\mathscr{L}|$. We call $u,u' \in U$ *mates* if they are the two end points of one line segment in $\mathscr{L}$.

Next introduce a variable x_u for each $u \in U$. The value of this variable x_u will stand for the distance over which we will shift the corresponding curve C_i to obtain a path P_i in G as required.

We give four classes of linear inequalities in the $x_u (u \in U)$. First:

$$(58) \qquad x_u + x_{u'} = 0 \quad \text{if } u \text{ and } u' \text{ are mates.}$$

Second consider $u, u' \in U$ so that u is end point of a line segment crossing C_i and u' is end point of a line segment crossing C_j with $j \neq i$, so that u and u' belong to the same component of:

$$(59) \qquad \mathbb{R}^2 \backslash (I_1 \cup \ldots \cup I_p \cup C_1[0,1] \cup \ldots \cup C_k[0,1]).$$

Let for any curve D in $\mathbb{R}^2$:

$$(60) \qquad \varphi(D) := \text{the number of faces } F' \text{ of } H \text{ passed by } D$$
$$\text{(counting multiplicities).}$$

Define:

$$(61) \quad \beta_{u,u'} := \min \{\varphi(D) \mid D \text{ is homotopic in } \mathbb{R}^2 \backslash (I_1 \cup \ldots \cup I_p) \text{ to some}$$
$$\text{curve in (59) connecting } u \text{ and } u'\}.$$

Now we require:

$$(62) \qquad x_u + x_{u'} \leq \beta_{u,u'} - 1.$$

It means that if curve C_i is shifted at u over a distance x_u (in the direction of u), then curve C_j should be shifted over a distance of at least $x_u + 1 - \beta_{u,u'}$, in the negative direction. In particular, (62) gives that if u and u' belong to the same disk D_v as in:

(63)

then $x_u + x_{u'} \leq -1$.

Third, consider $u, u' \in U$ so that u is end point of a line segment in $\mathscr{L}$ crossing C_i, and u' is end point of a line segment in $\mathscr{L}$ also crossing C_i. Moreover, let there exist a curve D satisfying:

(64)   (i) D is a curve in (59) connecting u and u',

(ii) D is not homotopic to any curve in $C_i[0,1] \cup \bigcup_{\ell \in \mathscr{L}} \ell$

connecting u and u'.

Now let

(65) $\beta_{u,u'} := \min\{\varphi(\tilde{D}) \mid \tilde{D}$ is homotopic to some

curve D satisfying (64)$\}$.

We require:

(66) $$x_u + x_{u'} \leq \beta_{u,u'} - 1.$$

This includes the case $u = u'$, where (64) (ii) means that D is homotopically nontrivial, and where (66) becomes $2x_u \leq \beta_{u,u} - 1$.

Finally, for any $u \in U$ let

(67) $\beta_u := \min\{\varphi(D) \mid D$ is homotopic in $\mathbb{R}^2 \setminus (I_1 \cup \ldots \cup I_p)$ to some

curve in (59) connecting u and $bd(I_1 \cup \ldots \cup I_p)\}$.

We require:

(68) $$x_u \leq \beta_u.$$

It means that we should not shift curve C_i over one of the "holes" $I_1, \ldots, I_p$.

By $Ax \leq b$ we denote the system of linear inequalities made up by (58), (62), (66) and (68). It can be shown that the right hand sides in these inequalities can be calculated in polynomial time.

III. Solving the System $Ax \leq b$ in Integers

In general, solving a system of linear inequalities in integers is NP-complete. However, our matrix A is of a special type. It satisfies:

(69) $$\sum_{j=1}^{n} |a_{ij}| \leq 2 \quad \text{for}\ i = 1, \ldots, m,$$

where $A = (a_{ij})$ has order $m \times n$, say. In that case, $Ax \leq b$ can be solved in integers, e.g., with the "Fourier-Motzkin elimination method" (cf. [39]).

This method eliminates variables one by one. Order the inequalities in $Ax \leq b$ as:

(70) $$2x_1 \leq \gamma_1$$
$$-2x_1 \leq \gamma_2$$
$$x_1 + a_1 x' \leq \delta_1$$
$$\vdots \quad \vdots \ \vdots$$
$$x_1 + a_{m'} x' \leq \delta_{m'}$$

$$-x_1 + a_{m'+1}x' \le \delta_{m'+1}$$

$$\vdots \quad \vdots \quad \vdots$$

$$-x_1 + a_{m''}x' \le \delta_{m''}$$

$$a_{m''+1}x' \le \delta_{m''+1}$$

$$\vdots \quad \vdots$$

$$a_m x' \le \delta_m$$

where $a_1,\ldots,a_m$ are row vectors of dimension $n-1$, where $x' := (x_2,\ldots,x_n)^T$, taking possibly $\gamma_1 = \infty$ or $\gamma_2 = \infty$.

Now if $\gamma_1 + \gamma_2 < 0$ then (70) clearly has no solution. If $\gamma_1 + \gamma_2 = 0$ and $\gamma_1 = -\gamma_2$ is odd, then (70) has no integer solution. So we may assume:

$$(71) \qquad \lceil -\tfrac{1}{2}\gamma_2 \rceil \le \lfloor \tfrac{1}{2}\gamma_1 \rfloor.$$

We can put (70) in another form:

$$(72) \qquad -\tfrac{1}{2}\gamma_2 \le x_1 \le \tfrac{1}{2}\gamma_1,$$
$$a_j x' - \delta_j \le x_1 \le \delta_i - a_i x' \quad i = 1,\ldots,m'; j = m'+1,\ldots,m'';$$
$$a_i x' \le \delta_i \qquad i = m''+1,\ldots,m.$$

Eliminating x_1 gives:

$$(73) \qquad (a_i + a_j)x' \le \delta_i + \delta_j \qquad i = 1,\ldots,m'; j = m'+1,\ldots,m'';$$
$$a_i x' \le \tfrac{1}{2}\gamma_2 + \delta_i \qquad i = 1,\ldots,m';$$
$$a_j x' \le \tfrac{1}{2}\gamma_1 + \delta_j \qquad j = m'+1,\ldots,m'';$$
$$a_i x' \le \delta_i \qquad i = m''+1,\ldots,m.$$

If (70) has an integer solution, also (73) must have an integer solution. System (73) is again of the same type as the original system; i.e., the corresponding matrix satiesfies (69) again. So we can solve (73) recursively. Let x' be an integral solution to (73). Hence, using (71), we have:

$$(74) \qquad \lceil \max\{\tfrac{-1}{2}\gamma_2, \max_{m'+1 \le j \le m''}(a_j x' - \delta_j)\} \rceil \le \lfloor \min\{\tfrac{1}{2}\gamma_1, \min_{1 \le i \le m'}(\delta_i - a_i x')\} \rfloor.$$

This implies that we can find an *integer* x_1 satisfying (72). Thus we have found an integer solution to (70).

The polynomial running time bound of this method follows from the fact that the system (73) can be reduced to a system with $O(n^2)$ inequalities: for any set of inequalities with equal left hand side, we consider only that one with lowest right hand side.

IV. Shifting the Curves

We call the integers x_u found by solving $Ax \leq b$ the *shift numbers*. They determine the distance and direction of shifting of the curves C_i. We carry out this shifting in small steps. Roughly it works as follows.

If all x_u are equal to 0, then no two distinct curves C_i pass the same disk D_v. Indeed, if two different curves C_i and C_j would pass disk D_v, then there are two different curves C_i and C_j passing D_v in such a way that they are incident to the same component of

$$(75) \qquad D_v \backslash (C_1[0,1] \cup \ldots \cup C_k[0,1]),$$

like in:

$$(76)$$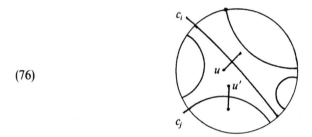

Let u and u' be as indicated. By (62), $x_u + x_{u'} \leq -1$, contradicting $x_u = x_{u'} = 0$.

Moreover, if one curve C_i passes a disk D_v more than once, we can similarly derive from (66) that the "loop" in between of the two passes of C_i through D_v is homotopic to some curve in D_v. So we can shortcut C_i. Repeating this, we obtain $C_1, \ldots, C_k$ so that each D_v is passed at most once in total. Shrinking H to G the curves $C_1, \ldots, C_k$ transform to pairwise disjoint simple paths in G as required.

If not all x_u are 0, select one with $x_u = M > 0$ as large as possible. Let u belong to component K of

$$(77) \qquad D_v \backslash (C_1[0,1] \cup \ldots \cup C_k[0,1]).$$

Suppose there is another point $u' \in U$ in K:

$$(78)$$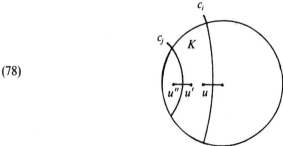

If $j \neq i$, then by (62), $x_{u'} + x_u \leq -1$, and hence $x_{u''} = -x_{u'} \geq x_u + 1 = M + 1$, contradicting the maximality of x_u. If $j = i$, then the loop in between of two passes of C_i through D_v is homotopic to some curve in D_v, so we can shortcut C_i.

So we may assume that no such u' exists. It means that K is "on the border" of D_v. Consider a longest subcurve of C_i so that a consecutive series of line segments has end point u with $x_u = M$:

(79)

with $x_{u_0} < M, x_{u_1} = x_{u_2} = \ldots = x_{u_{s-1}} = x_{u_s} = M$ and $x_{u_{s+1}} < M$. (Such a longest path exists, as at the beginning and end of C_i we have $x_u = 0$ (by (58) and (68)).)

Consider a neighbourhood of H at the same side of C_i as $u_1, \ldots, u_s$:

(80)

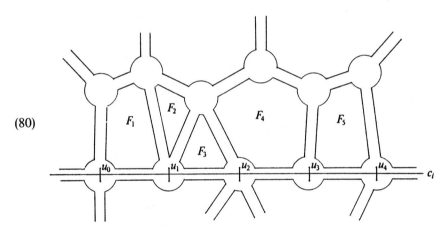

By (68), none of the faces $F_1, \ldots, F_5$ (in this example) belongs to $I_1, \ldots, I_p$. So we can shift C_i as:

(81)

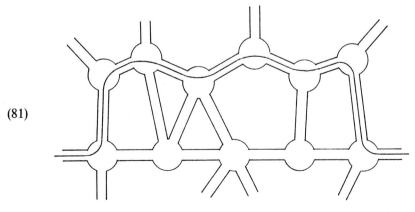

We introduce new line segments $\ell_1,\ldots,\ell_s$, crossing the new part of C_i in the corresponding disks. Let $u'_1,\ldots,u'_{s'}$ be the end points at the lower side in (81). So we replace $u_1,\ldots,u_s$ and their mates by $u'_1,\ldots,u'_{s'},u''_{1'},\ldots,u''_{s'}$. The same we do for the variables. The new variables we set:

$$(82) \qquad \begin{aligned} x_{u'_1} &:= x_{u'_2} := \ldots := x_{u'_{s'}} := M - 1, \\ x_{u''_1} &:= x_{u''_2} := \ldots := x_{u''_{s'}} := -M + 1, \end{aligned}$$

leaving the remaining variables invariant. It is not difficult to see that (generally) we obtain in this way an integral solution for the system of linear inequalities corresponding to the modified system.

Repeating this "local" shifting, we obtain after a polynomial number of steps a system with all x_u equal to 0, in which case the C_i give paths in G as required.

On the Correctness of the Method

The correctness of the method follows from the following fact:

(83)   problem (49) has a solution $\Leftrightarrow$ system $A \leq b$ has an integral solution.

The implication $\Leftarrow$ is proved by showing the correctness of the above shifting process. The implication $\Rightarrow$ is proved by deriving shift numbers x_u from any solution of (49).

The implication $\Rightarrow$ can also derived in the following way. Let $A = (a_{ij})$ be any integral $m \times n$-matrix satisfying

$$(84) \qquad \sum_{j=1}^{n} |a_{ij}| \leq 2 \quad \text{for each} \quad i = 1,\ldots,m.$$

Let b be an integral column vector of dimension m. In characterizing the solvability of $Ax \leq b$ in integers, consider first the case that each row of A contains one $+1$ and one -1. Then A is the incidence matrix of some directed graph. We can consider b as a length function on the edges of this directed graph. Then a solution x of $Ax \leq b$ is called a *potential*. It satisfies:

$$(85) \qquad x_w - x_v \leq b_{vw} \quad \text{for any edge} \quad vw.$$

As is well-known, such an integral potential exists if and only if each directed cycle has nonnegative length.

The general case can be studied in terms of *bidirected graphs*. We can in fact identify the matrix A with a bidirected graph. The vertices are identified with the columns (or column indices) of A, and the edges with the rows (or row indices) of A. An edge connects v and w if $a_{ev} \neq 0$ and $a_{ew} \neq 0$. So we have $++$ edges, $+-$ edges, and $-$ edges, indicated as

(86)

A row with a ± 2 can be seen as a loop. There are two types: $++$ loops and $--$ loops, indicated as:

(87)

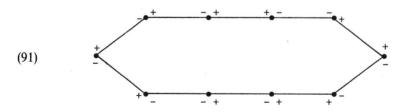

We call a row with only one ± 1 an *end*, at the corresponding vertex v. They can be indicated as:

(88)

Call a sequence

$$(89) \qquad (v_0, e_1, v_1, \ldots, e_d, v_d)$$

a *bidirected cycle* if:

(90)   (i) $v_0 = v_d$;

      (ii) e_i is an edge or loop connecting v_{i-1} and v_i $(i = 1, \ldots, d)$;

      (iii) $a_{e_i v_i} \cdot a_{e_{i+1} v_i} < 0$ (for $i = 1, \ldots, d-1$), and $a_{e_1 v_0} \cdot a_{e_d v_0} < 0$

(the vertices $v_1, \ldots, v_d$ need not all be distinct). An example is:

(91)

A first necessary condition for the existence of a solution of $Ax \le b$ is:

(92)   each bidirected cycle has nonnegative length

(where the *length* of cycle (89) is $\sum_{j=1}^{d} b_{e_j}$). This follows from:

$$(93) \qquad \sum_{j=1}^{d} b_{e_j} \ge \sum_{j=1}^{d} (a_{e_j v_{j-1}} x_{v_{j-1}} + a_{e_j v_j} x_{v_j}) = \sum_{j=1}^{d} (a_{e_j v_j} + a_{e_{j+1} v_j}) x_{v_j} = 0$$

(taking $e_{d+1} := e_1$, and assuming for simplicity that no e_j is a loop - the general case is left as an exercise).

Call a sequence

(94) $$(e_1, v_1, e_2, v_2, \ldots, v_{d-2}, e_{d-1}, v_{d-1}, e_d)$$

a *link* if

(95)   (i) e_1 is an end at v_1 and e_d is an end at v_{d-1};

     (ii) e_i is an edge or loop connecting v_{i-1} and $v_i (i = 2, \ldots, d-1)$;

     (iii) $a_{e_i v_i} \cdot a_{e_{i+1} v_i} < 0$ (for $i = 1, \ldots, d-1$).

As a second necessary condition for the existence of a solution of $Ax \le b$ we have:

(96)   each link has nonnegative length.

It can be shown that (92) and (96) together are sufficient for the existence of a rational solution of $Ax \le b$. However, for an integral solution we need one further condition. Call a cycle (89) *doubly-odd* if there exists a t with $0 < t < d$ so that:

(97)   (i) $v_0 = v_t = v_d$;

     (ii) $a_{e_1 v_0} \cdot a_{e_t v_t} > 0$ and $a_{e_{t+1} v_t} \cdot a_{e_d v_d} > 0$;

     (iii) $\displaystyle\sum_{j=1}^{t} b_{e_j}$ is odd and $\displaystyle\sum_{j=t+1}^{d} b_{e_j}$ is odd.

An example of a cycle satisfying (i) and (ii) is:

(98)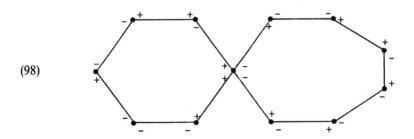

Now a necessary condition for the existence of an integral solution of $Ax \le b$ is:

(99)   each doubly-odd cycle has positive length.

This follows from (assuming again for simplicity that no e_j is a loop):

$$(100) \qquad \sum_{j=1}^{t} b_{e_j} \geq \sum_{j=1}^{t} (a_{e_j v_{j-1}} x_{v_{j-1}} + a_{e_j v_j} x_{v_j}) =$$

$$= a_{e_1 v_0} v_0 + \sum_{j=1}^{t-1} (a_{e_j v_j} + a_{e_{j+1} v_j}) x_{v_j} + a_{e_t v_t} x_{v_t} = 2 x_{v_0}.$$

Since the left hand side is odd, we should have strict inequality if x is integral. Hence we have strict inequality in (93).

Now conditions (92), (96) and (99) are sufficient for the existence of an integral solution of $Ax \leq b$:

Theorem 32. *Let A be an integral matrix satisfying (84), and let b be an integral column vector. Then $Ax \leq b$ has an integral solution x, if and only if*:

(101) (i)   each bidirected cycle has nonnegative length;
     (ii) each link has nonnegative length;
     (iii) each doubly-odd cycle has positive length.

It is not difficult to derive a proof of this theorem with the help of the Fourier-Motzkin elimination method described above.

From Theorem 32 one can derive the following theorem [42,43]:

Theorem 33. *Problem (49) has a solution if and only if*:

(102) (i) there exist pairwise disjoint simple curves $C'_1, \ldots, C'_k$ in $\mathbb{R}^2 \setminus (I_1 \cup \ldots \cup I_p)$ so that $C'_i \sim C_i$ (for $i = 1, \ldots, k$);
     (ii) for each curve $D : [0,1] \longrightarrow \mathbb{R}^2 \setminus (I_1 \cup \ldots \cup I_p)$ with end points on bd $(I_1 \cup \ldots \cup I_p)$ one has:

$$cr \, (G, D) \geq \sum_{i=1}^{k} mincr \, (C_i, D);$$

     (iii) for each doubly-odd closed curve

$$D : S_1 \longrightarrow \mathbb{R}^2 \setminus (I_1 \cup \ldots \cup I_p)$$

not passing obligatory points one has:

$$cr \, (G, D) > \sum_{i=1}^{k} mincr \, (C_i, D).$$

Here we use the following notation and terminology. We denote:

$$(103) \qquad cr \, (G, D) := |\{x \in [0,1] \mid D(x) \in G\}|,$$

$$cr \, (C, D) := |\{(x, y) \in [0,1] \times [0,1] \mid C(x) = D(y)\}| ;$$

$$mincr \, (C, D) := min \, \{cr(\tilde{C}, \tilde{D}) \mid \tilde{C} \sim C, \tilde{D} \sim D\}.$$

So $cr \, (C, D)$ counts the number of intersections of C and D, which can be of several types:

(104)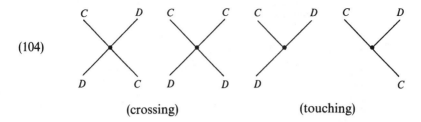

(crossing)   (touching)

By S_1 we denote the unit circle in the complex plane $\mathbb{C}$. A closed curve $D : S_1 \longrightarrow \mathbb{R}^2$ is called *doubly-odd* if it is the concatenation of two closed curves $D_1, D_2 : S_1 \longrightarrow \mathbb{R}^2$, with $D_1(1) = D_2(1) \notin G$, so that

$$(105) \qquad cr \ (G, D_1) + \sum_{i=1}^{k} kr \ (C_i, D_1) \ \text{is odd, and}$$

$$cr \ (G, D_2) + \sum_{i=1}^{k} kr \ (C_i, D_2) \ \text{is odd.}$$

Here $kr \ (C, D)$ denotes the number of crossings of C and D (cf. (104)).

An *obligatory point* is a point $p \in \mathbb{R}^2 \backslash (I_1 \cup \ldots \cup I_p)$ so that, for some $i = 1, \ldots, k$, each C_i' homotopic to C_i passes p.

Two closed curves $D, D' : S_1 \longrightarrow \mathbb{R}^2 \backslash (I_1 \cup \ldots \cup I_p)$ are called *homotopic* (or *freely homotopic*) denoted by $D \sim D'$, if there exists a continuous function $\Phi : S_1 \times [0, 1] \longrightarrow \mathbb{R}^2 \backslash (I_1 \cup \ldots \cup I_p)$ so that

$$(106) \qquad \Phi(z, 0) = D(z) \ \text{and} \ \Phi(z, 1) = D'(z)$$

for all $z \in S_1$ (so no base point is fixed). Again we denote:

$$(107) \qquad cr \ (G, D) := |\{z \in S_1 \mid D(z) \in G\}|,$$
$$cr \ (C, D) := |\{(y, z) \in [0, 1] \times S_1 \mid C(y) = D(z)\}|,$$
$$mincr \ (C, D) := min \ \{cr \ (\tilde{C}, \tilde{D}) \mid \tilde{C} \sim C, \tilde{D} \sim D\}.$$

Theorem 33 extends a theorem of Cole and Siegel [4] for grid graphs, and a theorem of Robertson and Seymour [33] for the case $p = 2$ (i.e., one proper hole). In these two cases we can delete condition (102) (iii).

To sketch the proof of Theorem 33, we note first that it is not difficult to see that the conditions (102) are necessary. To see sufficiency, observe that we may assume that $C_1', \ldots, C_k'$ in (102) (i) are in fact equal to $C_1, \ldots, C_k$, respectively, and that they are in the "blown up" graph H as above. Construct the system $Ax \leq b$ from this. Now each inequality

$$(108) \qquad x_u + x_{u'} \leq \beta_{u,u'} - 1$$

in (62) and (66) comes from a curve D in $\mathbb{R}^2 \backslash (I_1 \cup \ldots \cup I_p)$ connecting u and u' with $\varphi(D) = \beta_{u,u'}$. Similarly, the inequalities

$$(109) \qquad x_u \leq \beta_u$$

in (68) come from a curve D in $\mathbb{R}^2\backslash(I_1 \cup \ldots \cup I_p)$ connecting u and the boundary of some face $I_1, \ldots, I_p$ with $\varphi(D) = \beta_u$. The inequalities

$$(110) \qquad\qquad\qquad\qquad x_u + x_{u'} = 0$$

in (58) correspond to a line segment in $\mathscr{L}$ with end points u and u'. This implies that each link (94) in A corresponds to a curve D in $\mathbb{R}^2\backslash(I_1 \cup \ldots \cup I_p)$ connecting two points on $bd(I_1 \cup \ldots \cup I_p)$. Note that

(111) the number of inequalities in link (94) corresponding to a line segment in $\mathscr{L}$ is equal to $\frac{1}{2}(d-1)$.

Moreover, the length of the link is:

$$(112) \qquad\qquad\qquad \sum_{j=1}^{d} b_{e_j} = \varphi(D) - \frac{1}{2}(d-1).$$

It is not difficult to show further:

$$(113) \qquad\qquad\qquad \frac{1}{2}(d-1) = \sum_{i=1}^{k} mincr\ (C_i, D).$$

Hence condition (102) (ii) implies condition (101) (ii). It is also not difficult to see that condition (102) (ii) implies

$$(114) \qquad\qquad\qquad cr\ (G, D) \geq \sum_{i=1}^{k} mincr\ (C_i, D)$$

for each *closed* curve D in $\mathbb{R}^2\backslash(I_1 \cup \ldots \cup I_p)$: take any curve $E : [0,1] \longrightarrow \mathbb{R}^2\backslash(I_1 \cup \ldots \cup I_p)$ with $E(0) \in bd(I_1 \cup \ldots \cup I_p)$ and $E(1) = D(1)$. Then the curve

$$(115) \qquad\qquad\qquad E \cdot D^t \cdot E^{-1}$$

(for $t \in \mathbb{N}$) satisfies:

$$(116) \qquad\qquad cr\ (G, E \cdot D^t \cdot E^{-1}) = 2 \cdot cr\ (G, E) + t \cdot cr\ (G, D)$$

(assuming without loss of generality that $D(1) \notin G$). One can show that there exists a number S so that for each $t \geq 0$ and each $i = 1, \ldots, k$:

$$(117) \qquad\qquad mincr\ (C_i, E \cdot D^t \cdot E^{-1}) \geq t \cdot mincr\ (C_i, D) - S.$$

By (102) (ii) (applied to curve (115)) and (116), for each $t \geq 0$:

$$(118) \qquad\qquad t \cdot cr\ (G, D) \geq t \cdot \sum_{i=1}^{k} mincr\ (C_i, D) - kS - 2cr(G, E).$$

Hence (114) follows. In a similar way as above one can derive from (114) that (101) (i) holds. Moreover, condition (102) (iii) can be seen to imply condition (101) (iii).

Fixed Number of Holes

The following can be derived from Theorem 31:

Theorem 34. *For each fixed p there exists a polynomial-time algorithm for problem (3), whenever G is planar so that $r_1, s_1, \ldots, r_k, s_k$ can be covered by the boundaries of at most p faces.*

The idea of the proof is as follows. Let $r_1, s_1, \ldots, r_k, s_k$ be covered by the boundaries of faces $I_1, \ldots, I_p$ (including the unbounded face, without loss of generality). Consider $I_1, \ldots, I_p$ as holes. Now we can enumerate all "possibly feasible" homotopy classes of curves $C_1, \ldots, C_k$ (where C_i connects r_i and $s_i (i = 1, \ldots, k)$) in polynomial time.

Indeed, we only have to consider those curves which are pairwise disjoint and simple. Moreover, we can find curves $D_1, \ldots, D_{p-1}$, each connecting the boundaries of two of the faces $I_1, \ldots, I_p$, so that they form a "spanning tree" on $I_1, \ldots, I_p$. E.g.,

(119)

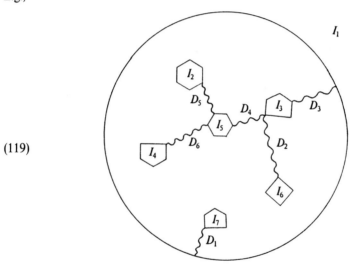

Note that the space obtained by deleting all holes $I_1, \ldots, I_p$ and the images of all curves $D_1, \ldots, D_{p-1}$ is simply connected.

We can take $D_1, \ldots, D_{p-1}$ so that $cr\ (G, D_j) \leq |V|$ for all $j = 1, \ldots, p-1$. Then we only have to consider those choices for the curves $C_1, \ldots, C_k$ for which

(120)
$$\sum_{i=1}^{k} mincr\ (C_i, D_j) \leq |V| \quad \text{for}\ \ j = 1, \ldots, p-1,$$

since other choices obviously are infeasible. It can be shown that there are at most $|V|^p$ such choices (up to homotopy). Hence we can restrict the enumeration to a polynomial number of choices.

Theorem 34 extends Robertson and Seymour's theorem (Theorem 1) for the case that G is planar: if k is fixed, we can cover $r_1, s_1, \ldots, r_k, s_k$ by a fixed number of faces, namely at most $2k$.

Surfaces

The following theorem from [45] can be proved in a way similar to the proof of Theorem 33 above.

Theorem 35. *Let* $G = (V, E)$ *be a graph, embedded on a compact surface S, and let* $C_1, \ldots, C_k$ *be closed curves on S, each not null-homotopic. Then there exist pairwise disjoint simple closed curves* $\tilde{C}_1, \ldots, \tilde{C}_k$ *in G so that $\tilde{C}_i$ is homotopic to C_i for* $i = 1, \ldots, k$, *if and only if:*

(121) (i) there exist pairwise disjoint simple closed curves $\tilde{C}_1, \ldots, \tilde{C}_k$ on
S so that $\tilde{C}_i$ is homotopic to C_i for $i = 1, \ldots, k$;
   (ii) for each closed curve $D : S_1 \longrightarrow S$:

$$cr\,(G, D) \geq \sum_{i=1}^{k} mincr\,(C_i, D);$$

   (iii) for each doubly-odd closed curve $D = D_1 \cdot D_2 : S_1 \longrightarrow S$
with $D_1(1) = D_2(1) \notin G$:

$$cr\,(G, D) > \sum_{i=1}^{k} mincr\,(C_i, D).$$

Here we use similar terminology as above. Thus a closed curve (on S) is a continuous function $C : S_1 \longrightarrow S$, where S_1 denotes the unit circle in the complex plane $\mathbb{C}$. It is *simple* if it is one-to-one. Two closed curves are *disjoint* if their images are disjoint.

Two closed curves C and $\tilde{C}$ are *(freely) homotopic (on S)*, in notation $C \sim \tilde{C}$, if there exists a continuous function $\Phi : S_1 \times [0, 1] \longrightarrow S$ so that $\Phi(z, 0) = C(z)$ and $\Phi(z, 1) = C(z)$ for all $z \in S_1$.

Again, we call a closed curve $D : S_1 \longrightarrow S$ *doubly-odd* (with respect to $G, C_1, \ldots, C_k$) if $D = D_1 \cdot D_2$ for some closed curves D_1, D_2 satisfying:

(122)
$$cr\,(G, D_1) \not\equiv \sum_{i=1}^{k} cr\,(C_i, D_1) \quad (mod\ 2),$$

$$cr\,(G, D_2) \not\equiv \sum_{i=1}^{k} cr\,(C_i, D_2) \quad (mod\ 2).$$

It is easy to see that the conditions (121) are necessary conditions. The essence of the theorem is sufficiency of (121).

Homotopic Trees

We can extend the polynomial-time algorithm for problem (49) to the following problem:

(123)   Given: — a planar graph G embedded in $\mathbb{R}^2$;
— faces $I_1, \ldots, I_p$ of G (including the unbounded face);
— pairwise disjoint sets $W_1, \ldots, W_k$ of vertices of G on the boundary of $I_1 \cup \ldots \cup I_p$;
— trees $T_1, \ldots, T_k$ embedded in $\mathbb{R}^2 \backslash (I_1 \cup \ldots \cup I_p)$, so that $W_i \subseteq V(T_i)$ for $i = 1, \ldots, k$;

find: — pairwise disjoint subtrees $\tilde{T}_1, \ldots, \tilde{T}_k$ of G so that for each $i = 1, \ldots, k : \tilde{T}_i$ is homotopic to T_i in $\mathbb{R}^2 \backslash (I_1 \cup \ldots \cup I_p)$ fixing W_i.

Here two trees T and $\tilde{T}$ embedded *in* $\mathbb{R}^2 \backslash (I_1 \cup \ldots \cup I_p)$ are called *homotopic* (in notation: $T \sim \tilde{T}$) in $\mathbb{R}^2 \backslash (I_1 \cup \ldots \cup I_p)$ *fixing* W if:

(124)   (i) W is a subset both of $V(T)$ and of $V(\tilde{T})$:
(ii)   for every pair of elements $w, w' \in W$, the unique simple curve in T connecting w and w' is homotopic to the unique simple curve in $\tilde{T}$ connecting w and w' (in $\mathbb{R}^2 \backslash (I_1 \cup \ldots \cup I_p)$).

In [44] we showed:

Theorem 36. *There exists a polynomial-time algorithm for problem* (123).

The idea of the algorithm is as follows. Again we blow up the graph G slightly, as in (54), to obtain H. We replace each tree T_i by $t_i := |W_i|$ paths, following the contours of T_i. E.g.,

(125)   becomes

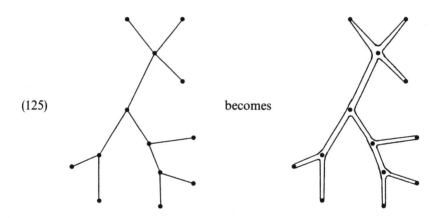

(assuming the end nodes are the elements of W_i). So T_i gives t_i paths $C_1, \ldots, C_{t_i}$, so that the concatenation

(126) $$K_i := C_1 \cdot C_2 \cdot \ldots \cdot C_{t_i}$$

is a simple closed curve, containing no face $I_1, \ldots, I_p$ in its interior. Let L_i denote this interior. Assuming the original $T_1, \ldots, T_k$ to be pairwise disjoint, the closed curves $K_1, \ldots, K_k$ are pairwise disjoint. We may assume that they are part of H.

Again we introduce line segments each time a curve K_i passes any disk D_v (cf. (57)). Let $\mathscr{L}$ be the set of these line segments, and let U be the set of end points of line segments in $\mathscr{L}$.

Now if u and u' are end points of one line segment in $\mathscr{L}$, crossing C_{ij} say, then one of the end points is in L_i and the other not. Call the first one the *inner* end point, and the other one the *outer* end point. Let U' be the set of inner end points, and let U'' be the set of outer end points.

Again we have $x_u + x_{u'} = 0$ for each two mates u, u'. Similarly, we have inequalities as in (62), (66) and (68) if u and u' are outer end points.

Moreover, we have for each pair of inner end points u, u' belonging to one and the same L_i :

$$(127) \qquad\qquad x_u + x_{u'} \le \beta_{u,u'}$$

where

$$(128) \qquad \beta_{u,u'} := min \ \{\varphi(D) \ | D \ \text{is a curve in} \ \mathbb{R}^2 \backslash (I_1 \cup \ldots \cup I_p)$$
$$\text{connecting} \ u \ \text{and} \ u', \ \text{homotopic to}$$
$$\text{some curve in} \ L_i\}.$$

Again, this gives us a system $Ax \le b$ of linear inequalities satisfying (69). Hence we can solve it in integers in polynomial time. The integer values are called the *shift numbers*. We shift each C_{ij} according to these shift numbers. After this shift we obtain curves $C'_{ij} (i = 1, \ldots, k; j = 1, \ldots, t_i)$ so that for each $i = 1, \ldots, k$ the closed curve

$$(129) \qquad\qquad K'_i := C'_1 \cdot C'_2 \cdot \ldots \cdot C'_{t_i}$$

does not enclose any $I_1, \ldots, I_p$, and so that no two different K'_i share the same disk D_v. Each K'_i gives in G a cycle K''_i so that two different K''_i are vertex-disjoint. Taking an arbitrary tree T'_i in K''_i spanning W_i gives a solution of problem (123).

Fixed Number of Holes

We can derive an extension of Theorem 34. Consider the problem:

(130)   Given: $-$ a graph $G = (V, E)$,
　　　　　　　　$-$ sets $W_1, \ldots, W_k$ of vertices of G,
　　　　　find: $-$ pairwise vertex-disjoint trees $T_1, \ldots, T_k$ in G
　　　　　　　　so that the vertex set of T_i contains
　　　　　　　　W_i (for $i = 1, \ldots, k$).

This problem clearly is NP-complete, as the case $|W_1| = \ldots = |W_k| = 2$ is just the disjoint paths problem. Problem (130) is important to solve in VLSI-layout - it means that we must connect several sets of pins by pairwise disjoint interconnections.

Now the following can be derived from Theorem 36:

Theorem 37. *For each fixed p, there exists a polynomial-time algorithm for problem* (130) *if G is planar and* $W_1 \cup \ldots \cup W_k$ *can be covered by the boundaries of at most p faces of G.*

The idea is again to enumerate all "possibly feasible" choices of homotopies of trees $T_1, \ldots, T_k$ covering $W_1, \ldots, W_k$, respectively, similar to that used in deriving Theorem 34 from Theorem 31.

6. Edge-Disjoint Homotopic Paths

We finally consider the problem:

(131) Given: – a planar graph $G = (V, E)$ embedded in $\mathbb{R}^2$,
   – faces $I_1, \ldots, I_p$ of G (including the unbounded face),
   – curves $C_1, \ldots, C_k$ with end points on bd $(I_1 \cup \ldots \cup I_p)$,
   find: – pairwise edge-disjoint paths $P_1, \ldots, P_k$ where P_i is
   – homotopic to C_i in $\mathbb{R}^2 \backslash (I_1 \cup \ldots \cup I_p)$ $(i = 1, \ldots, k)$.

(Here "pairwise edge-disjoint" is assumed to include that no path uses the same edge twice.) This problem is NP-complete, as was shown by Kaufmann and Maley [17]. A main difference with the vertex-disjoint case is that for the edge-disjont case the given curves $C_1, \ldots, C_k$ might necessarily cross, so that the natural ordering of the curves in the vertex-disjoint case does not occur.

Clearly, a necessary condition for the solvability of (131) is:

(132) (*Cut condition*) for each curve D in $\mathbb{R}^2 \backslash (I_1 \cup \ldots \cup I_p)$ with end points on bd $(I_1 \cup \ldots \cup I_p)$ and not intersecting V one has:

$$cr \ (G, D) \geq \sum_{i=1}^{k} mincr \ (C_i, D).$$

This condition is not sufficient, as is shown by a very simple example:

(133)

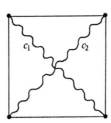

So this gives no hope for obtaining interesting special cases where the cut condition is sufficient. However, under a parity condition, the problem turns out better to handle:

(134) (*Local parity condition*) for each $v \in V$:
   $deg \ (v) + |\{i \in \{1, \ldots, k\}| \ C_i$ begins at $v\}| + |\{i \in \{1, \ldots, k\} \mid C_i$ ends at $v\}|$
   is even.

More general is the following condition:

(135) *(Global parity condition)* for each curve D in $\mathbb{R}^2 \setminus (I_1 \cup \ldots \cup I_p)$, with end points on $bd(I_1 \cup \ldots \cup I_p)$, not intersecting V and not touching edges, one has:

$$cr\,(G, D) \equiv \sum_{i=1}^{k} mincr\,(C_i, D) \quad (mod\ 2).$$

It is not difficult to derive (134) from (135). Kaufmann and Maley [17] showed that even under the local parity condition (134), problem (131) is NP-complete. It is not known whether this is also the case under the global parity condition (135). It turns out that the cut condition and one of the parity conditions are sufficient in some special cases.

Theorem 38. *If $p \le 2$ and the local parity condition is satisfied, then problem (131) has a solution if and only if the cut condition (132) is satisfied.*

For $p = 1$ this is just the Okamura-Seymour theorem (Theorem 18). For $p = 2$, this is shown by Van Hoesel and Schrijver [10]. They also gave a polynomial-time method.

Theorem 39. *If*

(136) *(i) G is part of the rectangular grid,*
   (ii)   each face of G of area larger than 1 belongs to $I_1, \ldots, I_p$,
   (iii)   each vertex of degree 4 incident to exactly one face in $I_1, \ldots, I_p$ is not an end point of any of the curves $C_1, \ldots, C_k$,
   (iv)   the global parity condition holds,

   then: problem (131) has a solution, if and only if the cut condition (132) holds.

This was shown by Kaufmann and Melhorn [18], who also gave a polynomial-time algorithm for the corresponding problem. For an extension to "straight-line" planar graphs, see [41].

Example (136) shows that we cannot extend Theorem 38 to the case $p = 3$ (even if we assume the global parity condition):

(137)

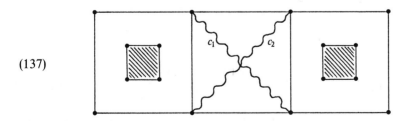

Example (138) shows that in Theorem 39 it is not sufficient to assume just the local parity condition instead of the global parity condition:

(138)

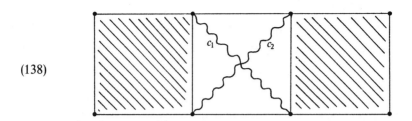

In fact, Kaufmann and Maley [17] showed that problem (131) is NP-complete if (136) holds with (iv) replaced by the local parity condition.

Although no solution exists in (137) and (138), there exists a "fractional" solution: we can find paths $P_1' \sim C_1, P_1'' \sim C_1, P_2' \sim C_2, P_2'' \sim C_2$ and scalars $\lambda_1' = \lambda_1'' = \lambda_2' = \lambda_2'' = \frac{1}{2}$ so that for each edge e:

$$(139) \qquad \lambda_1' \chi^{P_1'}(e) + \lambda_1'' \chi^{P_1''}(e) + \lambda_2' \chi^{P_2'}(e) + \lambda_2'' \chi^{P_2''}(e) \leq 1.$$

It turns out that, for any number of holes, the existence of such a fractional solution is equivalent to the cut condition, as was shown in [40]:

Theorem 40. *Let* $G = (V, E)$ *be a planar graph embedded in* $\mathbb{R}^2$. *Let* $I_1, \ldots, I_p$ *be some of the faces of* G, *including the unbounded face. Let* $P_1, \ldots, P_k$ *be paths in* G *with end points on the boundary of* $I_1 \cup \ldots \cup I_p$. *Then there exist paths* $P_{11}, \ldots, P_{1t_1}, P_{21}, \ldots, P_{2t_2}, \ldots, P_{k1}, \ldots, P_{kt_k}$ *in* G *and rationals* $\lambda_{11}, \ldots, \lambda_{1t_1}, \lambda_{21}, \ldots, \lambda_{2t_2}, \ldots, \lambda_{k1}, \ldots, \lambda_{kt_k} \geq 0$ *so that:*

$$(140) \qquad \text{(i)} \quad P_{ij} \sim P_i \text{ in } \mathbb{R}^2 \backslash (I_1 \cup \ldots \cup I_p) \quad (i = 1, \ldots, k; j = 1, \ldots, t_i),$$

$$\text{(ii)} \quad \sum_{j=1}^{t_i} \lambda_{ij} = 1 \qquad\qquad (i = 1, \ldots, k),$$

$$\text{(iii)} \quad \sum_{i=1}^{k} \sum_{j=1}^{t_i} \lambda_{ij} \chi^{P_{ij}}(e) \leq 1 \qquad (e \in E),$$

if and only if the cut condition (132) is satisfied.

Note that the λ_{ij} being integer would give a solution of (131).

Since the λ_{ij} can be found in polynomial time, with the help of the ellipsoid method (cf. [9]), we have as a consequence:

Theorem 41. *The cut condition (132) can be tested in polynomial time.*

We finally sketch the proof of Theorem 40. It is convenient to transform the space $\mathbb{R}^2 \backslash (I_1 \cup \ldots \cup I_p)$ into a compact orientable surface S : for each curve C_i, connecting I_j and $I_{j'}$ say, we add a "handle" between I_j and $I_{j'}$ and make C_i into a closed curve C_i' over this handle. Moreover, we extend the graph with an edge over the handle connecting the two end points of C_i. We do this for each C_i. In this way we obtain a compact orientable surface S. Then Theorem 40 follows from the following "homotopic circulation theorem":

Theorem 42. *Let* $G = (V, E)$ *be a graph embedded on a compact orientable surface* S. *Let* $C_1, \ldots, C_k$ *be closed curves on* S. *Then there exist cycles* $B_{11}, \ldots, B_{1t_1}, \ldots,$ $B_{k1}, \ldots, B_{kt_k}$ *in* G *and rationals* $\lambda_{11}, \ldots, \lambda_{1t_1}, \ldots, \lambda_{k1}, \ldots, \lambda_{kt_k} \geq 0$ *so that:*

(141)   (i) $\displaystyle\sum_{j=1}^{t_i} \lambda_{ij} = 1$ $(i = 1, \ldots, k),$

               (ii) $\displaystyle\sum_{i=1}^{k} \sum_{j=1}^{t_i} \lambda_{ij} \chi^{B_{ij}}(e) \leq 1$ $(e \in E),$

if and only if for each closed curve D *on* S *not intersecting* V *we have:*

$$(142) \qquad\qquad cr\,(G, D) \geq \sum_{i=1}^{k} mincr\,(C_i, D).$$

A *cycle* in G is a sequence

$$(143) \qquad\qquad (v_0, e_1, v_1, e_2, v_2, \ldots, e_\ell, v_\ell),$$

where $v_0, \ldots, v_\ell$ are vertices, with $v_0 = v_\ell$, and where e_i is an edge connecting v_{i-1} and $v_i (i = 1, \ldots, \ell)$. We identify in the obvious way such a cycle with a closed curve on S.

In fact, if S is the torus, we can take the λ_{ij} to be integers - see [8].

Basic in proving Theorem 42 is the following:

Theorem 43. *Let* G *be an eulerian graph embedded on a compact orientable surface* S. *Then the edges of* G *can be decomposed into cycles* $C_1, \ldots, C_t$ *in such a way that for each closed curve* D *on* S:

$$(144) \qquad\qquad mincr\,(G, D) = \sum_{i=1}^{t} mincr\,(C_i, D).$$

Decomposing the edges into cycles $C_1, \ldots, C_t$ *means that each edge occurs in exactly one of the* C_i, *while in each* C_i *all edges are different. Moreover,* $mincr\,(G, D) :=$ $min\,\{cr(G, \tilde{D}) \mid \tilde{D}, S_1 \longrightarrow S \backslash V(G); \tilde{D} \sim D\}.$

Our proof for this theorem is quite long, and uses some classical theorems in topology of Baer [2], Brouwer [3], von Kerékjártó [19] and Poincaré [32].

We do not know if Theorem 43 also holds for all compact nonorientable surfaces. In fact, it holds for the projective plane, in which case it is equivalent to Lins' theorem (Theorem 27 above).

In order to derive Theorem 42 from Theorem 43, we first derive the following from Theorem 43, using the duality of graphs on surface:

Theorem 44. *Let* $G = (V, E)$ *be a bipartite graph embedded on a compact orientable surface* S, *and let* $C_1, \ldots, C_k$ *be cycles in* G. *Then there exist closed curves* $D_1, \ldots, D_t : S_1 \longrightarrow S$ *so that (i) no* D_j *intersects* V, *(ii) each edge of* G *is intersected by exactly one* D_j *and by that* D_j *only once, (iii) for each* $i = 1, \ldots, k$:

(145) $$minlength_G \ (C_i) = \sum_{j=1}^{t} mincr \ (C_i, D_j).$$

Here we denote for any cycle C in G :

(146) $length_G \ (C) := \ell$, if $C = (v_0, e_1, v_1, \ldots, e_\ell, v_\ell)$,

$minlength_G \ (C) := min \ \{length_G(\tilde{C}) \mid \tilde{C} \sim C, \tilde{C} \text{ cycle in } G\}.$

(Cycles $\tilde{C}$ and C are allowed to pass one edge several times.)

Proof. We can extend (the embedded) G to a bipartite graph L embedded on S, containing G as a subgraph, so that each face of L (i.e., component of $S\backslash L$) is simply connected (i.e., homeomorphic to $\mathbb{R}^2$). Let
$d := max \ \{minlength_G \ (C_i) \mid i = 1,\ldots,k\}$. By inserting d new vertices on each edge of L not occuring in G, we obtain a bipartite graph H satisfying

(147) $$minlength_G \ (C_i) = minlength_H \ (C_i)$$

for $i = 1,\ldots,k$.

Consider a dual graph H^* of H on S. Since H is bipartite, H^* is eulerian. Hence by Theorem 43 the edges of H^* can be decomposed into cycles $D_1,\ldots,D_t$ so that for any closed curve C on S :

(148) $$mincr \ (H^*, C) = \sum_{j=1}^{t} mincr \ (D_j, C).$$

Now for each $i = 1,\ldots,k, mincr(H^*, C_i) = minlength_H(C_i) = minlength_G \ (C_i)$, and (145) follows. □

Using the polarity relation of convex cones in eulerian space we derive finally Theorem 42 from Theorem 44. Necessity of (142) being trivial, we only show sufficiency.

Suppose (142) is satisfied for each closed curve D not intersecting V. Let K be the convex cone in $\mathbb{R}^k \times \mathbb{R}^E$ generated by the vectors:

(149) $(\epsilon_i : \chi^\Gamma)$ $(i = 1,\ldots,k; \Gamma$ cycle in G with $\Gamma \sim C_i)$;

$(0;\epsilon_e)$ $(e \in E).$

Here ϵ_i denotes the i-th unit bases vector in $\mathbb{R}^k$. Similarly, ϵ_e denotes the e-th unit basis vector in $\mathbb{R}^E$. 0 denotes the origin in $\mathbb{R}^k$.

Although (149) gives infinitely many vectors, K is finitely generated. This can be seen as follows. For each fixed i, call a cycle $\Gamma \sim C_i$ *minimal* if there is no cycle $\Gamma' \sim C_i$ with $\chi^{\Gamma'}(e) \leq \chi^\Gamma (e)$ for each edge e, and with strict inequality for at least one edge e. So the set $\{\chi^\Gamma \mid \Gamma$ minimal cycle with $\Gamma \sim C_i\}$ forms an antichain in $\mathbb{Z}_+^E$ and is therefore finite. Since we can restrict, for each $i = 1,\ldots,k$, the χ^Γ in (149) to those with Γ minimal, K is finitely generated.

What we must show is that the vector $(1;1) = (1,\ldots,1;1,\ldots,1)$ belongs to K.

By Farkas' lemma, it suffices to show that for each vector $(p, b) \in \mathbb{Q}^k \times \mathbb{Q}^E$ with nonnegative inner product with each of the vectors (149), also the inner product with $(1; 1)$ is nonnegative. So let $(p; b)$ have nonnegative inner product with each of (149). This is equivalent to:

(150)   (i) $p_i + \sum_{e \in E} b(e) \chi^\Gamma (e) \geq 0$ $(i = 1, \ldots, k; \Gamma$ cycle in G with $\Gamma \sim C_i)$;
           (ii) $b(e) \geq 0$ $(e \in E)$.

Without loss of generality, each entry in $(p; b)$ is an even integer. Let G' be the graph arising from G by replacing each edge e by a path of length $b(e)$ (that is, $b(e) - 1$ new vertices are inserted on e, if $b(e) \geq 1$; e is contracted if $b(e) = 0$). Each cycle C_i in G directly gives a cycle C_i' in G'. Then by (150) (i):

(151) $-p_i \leq minlength_{G'}(C_i')$ for $i = 1, \ldots, k$.

Since G' is bipartite, by Theorem 44, there exist closed curves $D_1, \ldots, D_t$ on S so that (i) each D_j intersects G' only in edges of G', (ii) each edge of G' is intersected by exactly one D_j and only once by that $D_{j'}$ and (iii) for each $i = 1, \ldots, k$:

(152) $$minlength_{G'}(C_i') = \sum_{j=1}^{t} mincr(C_i', D_j).$$

Note that (ii) is equivalent to:

(153) $$b(e) = \sum_{j=1}^{t} \chi^{D_j}(e)$$

for each edge e of G. Therefore, using (142), (151), (152) and (153):

(154) $$\sum_{e \in E} b(e) = \sum_{j=1}^{t} \sum_{e \in E} \chi^{D_j}(e) = \sum_{j=1}^{t} cr(G, D_j) \geq$$

$$\sum_{j=1}^{t} \sum_{i=1}^{k} mincr(C_i, D_j) = \sum_{i=1}^{k} \sum_{j=1}^{t} mincr(C_i, D_j) =$$

$$\sum_{i=1}^{k} minlength_{G'}(C_i') \geq - \sum_{i=1}^{k} p_i.$$

So $(p; b) \cdot (1; 1)^T \geq 0.$ $\square$

References

[1] Adelson-Velskij, G.M., Dinits, E.A., Karzanov, A.V. (1975): Flow algorithms. Nauka, Moscow (In Russian)
[2] Baer, R. (1927): Kurventypen auf Flächen. J. Reine Angew. Math. **156**, 231–246
[3] Brouwer, L.E.J. (1910): Über eindeutige stetige Transformation von Flächen in sich. Math. Ann. **69**, 176–180

[4] Cole, R., Siegel, A. (1984): River routing every which way, but loose. Proc. 25th IEEE FOCS, pp. 65–73

[5] Ford, Jr., L.R., Fulkerson, D.R. (1956): Maximal flow through a network. Can. J. Math. **8**, 399–404

[6] Frank, A. (1990): Packing paths, circuits and cuts - a survey. This volume

[7] Frank, A., Schrijver, A.: Vertex-disjoint simple paths of given homotopy in a planar graph. Preprint

[8] Frank, A., Schrijver, A. (1988): Disjoint homotopic cycles in a graph on the torus. Preprint

[9] Grötschel, M., Lovász, L., Schrijver, A. (1988): Geometric algorithms and combinatorial optimization. Springer-Verlag, Berlin, Heidelberg (Algorithms Comb., Vol. 2)

[10] van Hoesel, C., Schrijver, A. (1990): Edge-disjoint homotopic paths in a planar graph with one hole. J. Comb. Theory, Ser. B **48**, 77–91

[11] Hu, T.C. (1963): Multicommodity network flows. Oper. Res. **11**, 344–360

[12] Hu, T.C. (1973): Two-commodity cut-packing problem. Discrete Math. **4**, 108–109

[13] Hurkens, C.A.J., Schrijver, A., Tardos, É. (1988): On fractional multicommodity flows and distance functions. Discrete Math. **73**, 99-109

[14] Karzanov, A.V. (1984): A generalized MFMC-property and multicommodity cut problems. In: Hajnal, A., Lovász, L., Sós, V. (eds.): Finite and infinite sets II. North-Holland, Amsterdam, pp. 443–486 (Colloq. Math. Soc. János Bolyai, Vol. 37)

[15] Karzanov, A.V. (1985): Metrics and undirected cuts. Math. Program. **32**, 183–198

[16] Karzanov, A.V. (1987): Half-integral five-terminus flows. Discrete Appl. Math. **18**, 263–278

[17] Kaufmann, M., Maley, F.M. (1988): Parity conditions in homotopic knock-knee routing. Preprint

[18] Kaufmann, M., Mehlhorn, K. (1986): On local routing of two-terminal nets. Technical Report 03/1986, FB 10, Universität des Saarlandes

[19] von Kerékjártó, B. (1923): Vorlesungen über Topologie I: Flächentopologie. Springer-Verlag, Berlin

[20] Leiserson, C.E., Maley, F.M. (1985): Algorithms for routing and testing routability of planar VLSI-layouts. Proc. 17th ACM STOC, pp. 69–78

[21] Lins, S. (1981): A minimax theorem on circuits in projective graphs. J. Comb. Theory, Ser. B **30**, 253–262

[22] Lomonosov, M.V. (1976): Solutions for two problems on flows in networks. (Submitted to Problemy Peredachi Informatsii)

[23] Lomonosov, M.V. (1979): Multiflow feasibility depending on cuts. Graph Theory Newsl. **9**, 4

[24] Lomonosov, M.V. (1985): Combinatorial approaches to multiflow problems. Discrete Appl. Math. **11**, 1–94

[25] Lovász, L. (1976): On some connectivity properties of Eulerian graphs. Acta Math. Acad. Sci. Hung. **28**, 129–138

[26] Lynch, J.F. (1975): The equivalence of theorem proving and the interconnection problem. ACM SIGDA Newslett. **5**, 31–65

[27] Menger, K. (1927): Zur allgemeinen Kurventheorie. Fundam. Math. **10**, 96–115

[28] Okamura, H. (1983): Multicommodity flows in graphs. Discrete Appl. Math. **6**, 55–62

[29] Okamura, H., Seymour, P.D. (1981): Multicommodity flows in planar graphs. J. Comb. Theory, Ser. B **31**, 75–81

[30] Papernov, B.A. (1976): Feasibility of multicommodity flows. In: Friedman, A.A. (ed.): Studies in discrete optimization. Nauka, Moscow, pp. 230–261 (In Russian)

[31] Pinter, R.Y. (1983): River routing: Methodology and analysis. In: Bryant, R. (ed.): 3rd CalTech Conf. on Very Large-Scale Integration. Springer-Verlag, Berlin, pp. 141–163

[32] Poincaré, H. (1904): 5e complément à l'Analysis Situs. Rend. Circ. Mat. Palermo **18**, 45–110

[33] Robertson, N., Seymour, P.D. (1986): Graph minors VI: Disjoint paths across a disc. J. Comb. Theory, Ser. B **41**, 115–138

[34] Robertson, N., Seymour, P.D. (1988): Graph minors VII: Disjoint paths on a surface. J. Comb. Theory, Ser. B **45**, 212–254

[35] Robertson, N., Seymour, P.D. (1986): Graph minors XIII: The disjoint paths problem. Preprint (J. Comb. Theory, Ser. B, to appear)

[36] Robertson, N., Seymour, P.D.: 1989. An outline of a disjoint path algorithm This volume

[37] Rothschild, B., Whinston, A. (1966): On two-commodity network flows. Oper. Res. **14**, 377–387

[38] Rothschild, B., Whinston, A. (1966): Feasibility of two-commodity network flows. Oper. Res. **14**, 1121–1129

[39] Schrijver, A. (1986): Theory of linear and integer programming. J. Wiley & Sons, Chichester

[40] Schrijver, A. (1987): Decomposition of graphs on surfaces and a homotopic circulation theorem. Report OS-R8719, Mathematical Centre, Amsterdam (J. Comb. Theory, Ser. B, to appear)

[41] Schrijver, A. (1987): Edge-disjoint homotopic paths in straight-line planar graphs. Report OS-R8718, Mathematical Centre, Amsterdam (SIAM J. Discrete Math., to appear)

[42] Schrijver, A. (1988): Disjoint homotopic paths and trees in a planar graph: Description of the method. Preprint

[43] Schrijver, A. (1988): Disjoint homotopic paths and trees in a planar graph II: Proof of the theorem. Preprint

[44] Schrijver, A. (1988): Disjoint homotopic paths and trees in a planar graph III: Disjoint homotopic trees. Preprint

[45] Schrijver, A. (1988): Disjoint circuits of prescribed homotopies in a graph on a compact surface. Report OS-R8812, Mathematical Centre, Amsterdam (J. Comb. Theory, Ser. B, to appear)

[46] Schrijver, A. (1988): The Klein bottle and multicommodity flows. Report OS-R8810, Mathematical Centre, Amsterdam (Combinatorica, to appear)

[47] Schrijver, A. (1988): Short proofs on multicommodity flows and cuts. Preprint

[48] Schrijver, A.: An $O(n^2 \log^2 n)$ algorithm for finding disjoint homotopic trees in a planar graph. (To appear)

[49] Schrijver, A. (1989): Distances and cuts in planar graphs. J. Comb. Theory, Ser. B **46**, 46-57

[50] Seymour, P.D. (1978): A two-commodity cut theorem. Discrete Math. **23**, 177–181

[51] Seymour, P.D. (1978): Sums of circuits. In: Bondy, J.A., Murty, U.S.R. (eds.): Graph theory and related topics. Academic Press, New York, NY, pp. 341–355

[52] Seymour, P.D. (1979): A short proof of the two-commodity flow theorem. J. Comb. Theory, Ser. B **26**, 370–371

[53] Seymour, P.D. (1980): Four-terminus flows. Networks **10**, 79–86

[54] Seymour, P.D. (1981): On odd cuts and planar multicommodity flows. Proc. Lond. Math. Soc. (3) **42**, 178–192

Author Index

Subject Index

Algorithms and Combinatorics

Editors: R. L. Graham, B. Korte, L. Lovász

Combinatorial mathematics has substantially influenced recent trends and developments in the theory of algorithms and its applications. Conversely, research on algorithms and their complexity has established new perspectives in discrete mathematics. This new series is devoted to the mathematics of these rapidly growing fields with special emphasis on their mutual interactions.

The series will cover areas in pure and applied mathematics as well as computer science, including: combinatorial and discrete optimization, polyhedral combinatorics, graph theory and its algorithmic aspects, network flows, matroids and their applications, algorithms in number theory, group theory etc., coding theory, algorithmic complexity of combinatorial problems, and combinatorial methods in computer science and related areas.

The main body of this series will be monographs ranging in level from first-year graduate up to advanced state-of-the-art research. The books will be conventionally type-set and bound in hard covers. In new and rapidly growing areas, collections of carefully edited mono-graphic articles are also appropriate for this series. A subseries with the subtitle "Study and Research Texts" will be published in softcover and camera ready form. This will be mainly on outlet for seminar and lecture notes, drafts of textbooks with essential novelty in their presentation, and preliminary drafts of monographs. Refereed proceedings of meetings devoted to special topics may also be considered for this subseries. The main goals of this subseries are very rapid publication and the wide dissemination of new ideas.

Prospective readers of the series ALGORITHMS AND COMBINATORICS include scientists and graduate students working in discrete mathematics, operations research and computer science.

Volume 1

K. H. Borgwardt

The Simplex Method

A Probabilistic Analysis

1987. XI, 268 pp. 42 figs. in 115 sep. illustrations.
Softcover DM 79,–
ISBN 3-540-17096-0

Springer-Verlag
Berlin
Heidelberg
New York
London
Paris
Tokyo
Hong Kong
Barcelona

Algorithms and Combinatorics

Editors: R. L. Graham, B. Korte, L. Lovász

Springer-Verlag
Berlin
Heidelberg
New York
London
Paris
Tokyo
Hong Kong
Barcelona

THE UNIVERSITY OF MICHIGAN